T0351106

Introducing Financial Mathematics

Introducing Financial Mathematics: Theory, Binomial Models, and Applications seeks to replace existing books with a rigorous stand-alone text that covers fewer examples in greater detail with more proofs. The book uses the fundamental theorem of asset pricing as an introduction to linear algebra and convex analysis. It also provides example computer programs, mainly Octave/MATLAB® functions but also spreadsheets and Macsyma scripts, with which students may experiment on real data. The text's unique coverage is in its contemporary combination of discrete and continuous models to compute implied volatility and fit models to market data. The goal is to bridge the large gaps among nonmathematical finance texts, purely theoretical economics texts, and specific software-focused engineering texts.

Mladen Victor Wickerhauser is professor of mathematics and statistics at Washington University, St. Louis. He holds a PhD from Yale University. Professor Wickerhauser's research interests include harmonic analysis, wavelets, and numerical algorithms for data compression. He has six US patents and 118 publications, one of which led to an algorithm used by the FBI to encode fingerprint images.

Chapman & Hall/CRC Financial Mathematics Series

Aims and scope:
The field of financial mathematics forms an ever-expanding slice of the financial sector. This series aims to capture new developments and summarize what is known over the whole spectrum of this field. It will include a broad range of textbooks, reference works and handbooks that are meant to appeal to both academics and practitioners. The inclusion of numerical code and concrete real-world examples is highly encouraged.

Series Editors

Rama Cont
Department of Mathematics
Imperial College, UK

Robert A. Jarrow
Lynch Professor of Investment Management
Johnson Graduate School of Management
Cornell University, USA

M.A.H. Dempster
Centre for Financial Research
Department of Pure Mathematics and
Statistics
University of Cambridge, UK

Dilip B. Madan
Robert H. Smith School of Business
University of Maryland, USA

Machine Learning for Factor Investing: R Version
Guillaume Coqueret, Tony Guida

Malliavin Calculus in Finance: Theory and Practice
Elisa Alos, David Garcia Lorite

Risk Measures and Insurance Solvency Benchmarks: Fixed-Probability Levels in Renewal Risk Models
Vsevolod K. Malinovskii

Financial Mathematics: A Comprehensive Treatment in Discrete Time, Second Edition
Giuseppe Campolieti, Roman N. Makarov

Pricing Models of Volatility Products and Exotic Variance Derivatives
Yue Kuen Kwok, Wendong Zheng

Quantitative Finance with Python: A Practical Guide to Investment Management, Trading, and Financial Engineering
Chris Kelliher

Introducing Financial Mathematics: Theory, Binomial Models, and Applications
Mladen Victor Wickerhauser

UPCOMING

Foundations of Quantitative Finance, Book I: Measure Spaces and Measurable Functions
Robert R. Reitano

Foundations of Quantitative Finance, Book II: Probability Spaces and Random Variables
Robert R. Reitano

For more information about this series please visit: https://www.crcpress.com/Chapman-and-HallCRC-Financial-Mathematics-Series/book series/CHFINANCMTH

Introducing Financial Mathematics

Theory, Binomial Models, and Applications

Mladen Victor Wickerhauser

CRC Press
Taylor & Francis Group
Boca Raton London New York

CRC Press is an imprint of the
Taylor & Francis Group, an **informa** business

A CHAPMAN & HALL BOOK

First edition published 2023
by CRC Press
6000 Broken Sound Parkway NW, Suite 300, Boca Raton, FL 33487-2742

and by CRC Press
4 Park Square, Milton Park, Abingdon, Oxon, OX14 4RN

CRC Press is an imprint of Taylor & Francis Group, LLC

Library of Congress Cataloging-in-Publication Data

Names: Wickerhauser, Mladen Victor, author.
Title: Introducing financial mathematics : theory, binomial models, and applications / Mladen Victor Wickerhauser.
Description: First edition. | Boca Raton : CRC Press, 2022. | Series: Chapman & Hall/CRC financial mathematics series | Includes bibliographical references and index.
Identifiers: LCCN 2022022648 (print) | LCCN 2022022649 (ebook) | ISBN 9781032359854 (hardback) | ISBN 9781032359908 (paperback) | ISBN 9781003329695 (ebook)
Subjects: LCSH: Finance--Mathematical models.
Classification: LCC HG106 .W53 2022 (print) | LCC HG106 (ebook) | DDC 332.6--dc23/eng/20220519
LC record available at https://lccn.loc.gov/2022022648
LC ebook record available at https://lccn.loc.gov/2022022649

ISBN: 978-1-032-35985-4 (hbk)
ISBN: 978-1-032-35990-8 (pbk)
ISBN: 978-1-003-32969-5 (ebk)

DOI: 10.1201/9781003329695

Typeset in CMR10
by KnowledgeWorks Global Ltd.

Publisher's note: This book has been prepared from camera-ready copy provided by the authors.

To Lauren Grace Plummer.

Contents

Preface

Financial mathematics comprises theorems, models, and algorithms, all the ingredients necessary for a rich course in applied mathematics. For many students, however, the subject needs an introduction, as it has its own peculiar terminology and goals which may be unfamiliar to newcomers.

This textbook is intended to provide such an introduction, with the precision needed to convert theorems and models into working computer programs. It gathers together material that has been used for the past several years at Washington University in St. Louis. It is designed for undergraduate mathematics or statistics majors who are comfortable with what is usually covered in basic required courses:

- differential and integral calculus,
- linear algebra and matrix computations,
- elementary probability and statistics,
- calculation with some programming, and
- techniques of proof including induction and proof by contradiction.

Students will gain strength using these tools to price assets, construct novel investment products, and find and test implied financial models. They should expect to learn plenty, but also to put prior knowledge to use in new ways, merging ideas from different branches of mathematics and gaining insight and proficiency through programming and computation. Toward this last goal, algorithms are presented as open source computer programs for the free, well supported, and widely available Octave,[1] Macsyma, and LibreOffice systems.

This book is not an encyclopedia of financial mathematics, which is a huge, rapidly growing, and evolving subject. Instead, it contains topics selected for relevance, accessibility, rigor, and computational usefulness. All of its chapters can be presented in a typical 15-week semester, including time for midterm examinations. But there are also features to make the book suitable for independent study or for mentored undergraduate thesis projects: suggested Further Reading after each chapter, model solutions to all the chapter exercises, and a list of projects that apply coursework to meaningful amounts of real data.

---

[1] The Octave programs are simplified to run without modification on the commercial software system MATLAB®, which is a registered trademark of The MathWorks, Inc. See https://www.mathworks.com for further trademark and product information.

1

Basics

1.1 Assets and Portfolios

The fundamental object of financial mathematics is the *asset*, something that can be bought or sold, for an agreed price in some money units, or *cash*.

An asset has a time-varying price $A(t)$. For simplicity, the present time is set to be $t = 0$, so $A(0)$ is the current market price, or *spot price*, of the asset.

Assets come in a great variety. Currencies of many nations, commodities such as metals, grains, or minerals, shares of stock in corporations, and government or corporate bonds are all assets whose prices are determined in markets around the world. Two large markets for stocks and bonds in the United States are NYSE and NASDAQ which currently list about 2400 and 3800 corporations, respectively.

A *risky asset* is one for which the price $A(t)$ for $t > 0$ is unknown, such as a share of common stock in a company.

A *riskless asset* is one for which the price $A(t)$ is known for at least some times $t > 0$, such as a government bond which may be exchanged for its face value at some future maturity date.

A *portfolio* is a linear combination of risky and riskless assets, possibly with time-varying proportions. The coefficient of an asset in a portfolio will be positive if the asset is held *long*, namely purchased for cash. It will be negative if the asset is sold *short*, namely borrowed from someone else and then sold. The negative sign indicates that there is a liability, or unmet obligation, reducing the price of the portfolio.

1.1.1 Stocks and Bonds

If B is a government bond and S is a share of common stock, then

$$h_0 B(t) + h_1 S(t) \tag{1.1}$$

is the price of a portfolio consisting of h_0 units of the riskless asset B and h_1 shares of the risky asset S.

For example, the portfolio $2B - S$ consists of two units of B plus one share of S sold short. Similarly, the portfolio $(-3)B + S$ consists of 3 units of B sold short, equivalent to borrowing the amount $3B$ from a bank, plus one share of S held long.

1.1.2 Foreign Exchange

A *foreign exchange portfolio* contains riskless assets B_d and B_f denominated in domestic (DOM) and foreign (FRN) currency, respectively. The risk comes from the *exchange rate* $X = X(t)$, which is the number of units of DOM equal to one unit of FRN at time t. It may change unpredictably with time. An investment of h_d DOM of domestic bond B_d and h_f FRN of foreign bond B_f is then worth

$$h_d B_d(t) + h_f X(t) B_f(t), \qquad (1.2)$$

denominated in DOM.

1.1.3 Derivatives

A *derivative*, or *contingent claim*, is also an asset. Its price is related to some other *underlying asset*. The relationship is often a contract that creates rights and obligations that depend upon the underlying asset prices. These may be almost anything, but there are some standard contract types to study:

- Forward contract: an agreement to buy an asset at a future time for a specified price;
- Option: the right, but not the obligation, to buy (*Call*) or sell (*Put*) an asset by or on a future *expiry time* for a stated *strike price*;
- Swap: an obligation to exchange the *cash flow*, or income, from one portfolio for that of another;
- Future: a forward contract with a required *margin account*, offering some protection against default.

A portfolio of assets may contain derivatives of these and other kinds, in any amounts, positive or negative.

1.1.4 Riskless Return

The time value of money must be included when computing prices. One way is to assume that there is a *riskless asset* which, over one unit of time, increases in price by a factor $R > 1$ regardless of the state of the world. This factor is called the *riskless return*.

It is assumed that cash deposited into a bank account earns the riskless return, and that money may be borrowed from the bank at an interest rate corresponding to the riskless return.

1.1.5 Interest Rates and Present Value

Riskless return is related to interest rates which may change unpredictably. One market-traded asset that sets interest rates is the *zero-coupon bond*,[1]

---

[1] *Zero-coupon* means that no money is paid back until maturity.

usually issued by a government or a large institution. It pays a fixed *face value* F at a fixed *maturity* $T > 0$. Denote by $Z(0, T)$ the spot price of such a zero-coupon bond with nominal face value $F = 1$. This is a market price, set by an auction for various $T > 0$. It is clear that $Z(0, 0) = 1$.

Normally $Z(0, T) < 1$, since a promise to pay 1 at a future time $T > 0$ is worth less than 1 right now. Thus $Z(0, T)$ is a *discount* on the face value. This discount should be deeper for longer T, so expect

$$T' > T \implies Z(0, T') < Z(0, T). \tag{1.3}$$

This holds when interest rates are positive, as will always[2] be assumed.

The price of a zero-coupon bond varies with time up to its maturity T. So, define $Z(t, T)$ to be the price, at time $0 \le t \le T$, of a zero-coupon bond with face value $F = 1$ that was purchased at time 0 and matures at time $T > 0$. Such a bond is a risky asset with spot price $Z(0, T)$ and known future payoff $Z(T, T) = 1$. It is a random variable[3] for $0 < t < T$, assumed to have an expected value $\bar{Z}(t, T)$ determined by spot discounts:

$$\bar{Z}(t, T) \stackrel{\text{def}}{=} \frac{Z(0, T)}{Z(0, t)}, \qquad 0 < t < T. \tag{1.4}$$

Riskless return may be computed from US Treasury zero-coupon bond discounts. Such a bond with face value F and maturity T, sold at a discount price of $FZ(0, T)$, yields a riskless return $R = F/(FZ(0, T)) = 1/Z(0, T)$ when it matures at time T and is redeemed for its face value.

The *annual percentage rate r*, or APR, implied by a return factor R over T years is

$$e^{rT} = R, \qquad \Longleftrightarrow \qquad r = \frac{\log R}{T}. \tag{1.5}$$

This assumes *continuous compounding*, the limit as $N \to \infty$ of the return R_N from collecting and then reinvesting the principal and interest at N equispaced times between 0 and T. Each reinvestment has a return factor $1 + rT/N$, so

$$R_N = \overbrace{(1 + \frac{rT}{N}) \cdots (1 + \frac{rT}{N})}^{N \text{ times}} = \left(1 + \frac{rT}{N}\right)^N \to e^{rT}, \quad \text{as } N \to \infty,$$

by a well-known result from Calculus.

Equation 1.5 implies that the *riskless interest rate* over time $0 \le t \le T$ satisfies $Z(0, T) = e^{-rT}$, since $R = 1/Z(0, T)$. More generally,

$$\bar{Z}(t, T) = Z(0, T)/Z(0, t) = e^{-r(T-t)}, \qquad 0 \le t \le T, \tag{1.6}$$

which values the endpoints $t = 0$ and $t = T$ correctly.

---

[2]Zero or negative interest rates have occured, but are considered extraordinary.
[3]See Section 1.3.1 for a definition of random variable.

Continuous compounding or reinvestment implies that zero-coupon bond discounts are multiplicative. For example, if the riskless return over one unit of time is R and will be constant forever, then

$$Z(0,T) = 1/R^T = Z(0,1)^T,$$

though this assumption is not likely to hold for long T.

If $Z(0,T)$ is known for a sequence of future times $0 < T_1 < \cdots < T_N$, then it determines a fair price $A(0)$, or *present value*, for an asset[4] that pays a *cash flow* of X_n at time T_n, $1 \le n \le N$:

$$A(0) = \sum_{n=1}^{N} X_n Z(0,T_n) = X_1 Z(0,T_1) + \cdots X_N Z(0,T_N). \qquad (1.7)$$

This follows from Axiom 1 below. The proof is left as an exercise.

1.2 Payoff and Profit Graphs

The spot price $X(0)$ of a derivative X depends on its future price $X(T)$ and thus on the possible future prices $S(T)$ of its underlying asset S. The plot of $X(T)$ as a function of $S(T)$ is called the *payoff graph*. The plot of net proceeds $X(T) - X(0)$ versus $S(T)$ is called the *profit graph*.

1.2.1 Payoff Graphs for Forward Contracts

A forward contract to buy a risky asset S for an agreed price K at a future expiry time $t = T$ confers both a right and an obligation. The strike price K is mutually agreeable to the buyer (the Long side) and the seller (the Short side) of S, and thus the cost of such a contract is zero.

The payoffs for the Long and Short sides of a forward contract are plotted in Figure 1.1 as a function of the market price $S(T)$ at expiry. Writing $F_{\text{long}}(T)$ and $F_{\text{short}}(T)$ for the payoffs of the Long and Short forward contracts, respectively, the formulas are

$$\begin{aligned} F_{\text{long}}(T) &= S(T) - K, \\ F_{\text{short}}(T) &= K - S(T) = -F_{\text{long}}(T). \end{aligned}$$

Because there is no initial cost, the payoff and profit are the same.

1.2.2 Payoff and Profit Graphs for Options

An option to buy ("Call") or sell ("Put") an asset has a time-dependent price that will be determined only in the future.

---

[4]An annuity is such an asset.

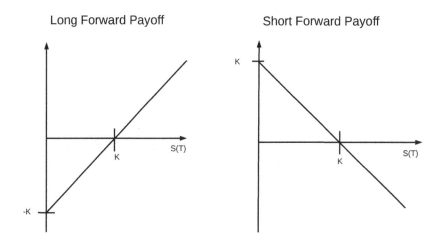

FIGURE 1.1
Payoff graphs for the Long and Short sides of a Forward contract to buy asset S for strike price K at time T.

Both Call and Put options have a *strike price* K and an *expiry date* T. Some common variations are *European style*, where the right may only be exercised at the expiry date, or *American style*, which may be exercised any time up to expiry. More exotic options include *Bermuda style*, with a restricted set of early exercise dates, or *Asian style*, where the strike price is an average of the asset prices before expiry.

At expiry $t = T$, a Call option will only be exercised if $S(T) > K$. This makes the value of a Call option

$$C(T) = \max\{0, S(T) - K\} \overset{\text{def}}{=} [S(T) - K]^{+}, \qquad (1.8)$$

which is read "the plus-part of $S(T) - K$." Similarly, a Put option will only be exercised if $S(T) < K$, and then its value is

$$P(T) = \max\{0, K - S(T)\} = [K - S(T)]^{+}. \qquad (1.9)$$

These functions giving the value of an option at expiry are nonnegative in all states of the world, and positive in some states. It will be shown below, by Theorem 1.1, that therefore an option must cost a positive *premium*, $C(0) > 0$ or $P(0) > 0$, at $t = 0$ when it is bought or sold.

The graph of $C(T)$ as a function of $S(T)$ is the *Call payoff graph*. If $C(0)$ is the premium paid for the Call option, then the *Call profit graph* is the graph of $C(T) - C(0)$ as a function of $S(T)$. Simlarly, plot $P(T)$ as a function of $S(T)$ to get the *Put payoff graph*. Plot $P(T) - P(0)$ to get the *Put profit graph*. These may be seen in Figure 1.2.

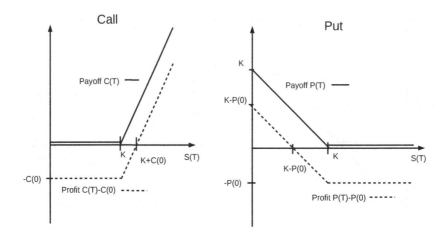

FIGURE 1.2
Payoff and profit for European-style Call and Put options on asset S, with
strike price K, expiry T, and respective premiums $C(0)$ and $P(0)$.

1.2.3   Payoff and Profit Graphs for Contingent Options

A *contingent premium option* costs nothing at $t = 0$ but incurs an obligation
to pay a premium α if it is "in the money," namely has a positive payoff,
at expiry or when exercised. Both Call and Put contingent premium options
have a *strike price K* and an *expiry date T*.

   The payoff and profit graphs for the European style of these options, which
may only be exercised at the expiry date, may be seen in Figure 1.3.

1.3   Arbitrage

Traditionally, an *arbitrage* is a trade that yields profit with no net investment.
An arbitrage arose when equivalent assets had different prices at the same
time. *Arbitrageurs* would buy at the lower price and simultaneously sell at
the higher price, thus making an instant riskless profit. Equivalently, they
could sell short at the higher price and buy to cover at the lower. Such trades
increased the lower-price demand and higher-price supply, causing equivalent
assets to converge in price across communicating markets.[5]

---

[5]Arbitrageurs were big users of (initially expensive) rapid electronic communication!

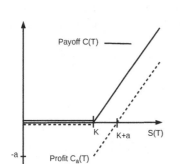

Contingent Premium Call

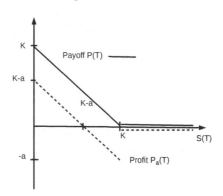

Contingent Premium Put

FIGURE 1.3
Payoff and profit graphs for European-style contingent premium Call and Put options on asset S, with strike price K, expiry T, and contingent premium a.

A more general definition of arbitrage is "an asset or portfolio with no initial investment that, over some time interval, yields profit without risk." If such portfolios include contingent claims and other derivatives, then exploitation by arbitrageurs will cause their market prices to converge. It is a goal of financial mathematics to compute the *no-arbitrage prices* that are the presumed limits of such market action.

Arbitrage portfolios of this more general type may have guaranteed profit with no risk of loss, or only the expectation of profit but no guarantee. More precise definitions, from which prices may be calculated, require a few notions from probability theory.

1.3.1 Random Variables and Stochastic Processes

The future at $t > 0$ is unknown, but it can be modeled by a *probability space*, a set Ω of possible futures with a *probability function*, Pr, defined on certain well-behaved *measurable subsets* of Ω and satisfying:

- $0 \leq \Pr(S) \leq 1$ for any measurable $S \subset \Omega$;

- $\emptyset$ and Ω are both measurable, with $\Pr(\emptyset) = 0$ and $\Pr(\Omega) = 1$;

- If S is measurable, then its complement $\Omega \setminus S \overset{\text{def}}{=} \{\omega \in \Omega : \omega \notin S\}$ is measurable, with
$$\Pr(\Omega \setminus S) = 1 - \Pr(S).$$

- If $S_1, S_2 \subset \Omega$ are both measurable, and $S_1 \subset S_2$, then $\Pr(S_1) \leq \Pr(S_2)$. This is called *monotonicity*.

- If $\{S_n : n = 1, 2, \dots \}$ is any countable collection of measurable sets, then both $\bigcup_n S_n$ and $\bigcap_n S_n$ are measurable. Furthermore, if $S_i \cap S_j = \emptyset$ for all $i \neq j$, then

$$\Pr\left(\bigcup_n S_n\right) = \sum_n \Pr(S_n).$$

This is called *countable additivity*. It implies monotonicity.

If Ω is a finite set, then every subset is measurable. Each one-element subset $\{\omega\} \subset \Omega$ may be assigned some probability $0 \leq \Pr(\{\omega\}) \leq 1$, with the only other constraint being

$$1 = \Pr(\Omega) = \sum_{\omega \in \Omega} \Pr(\{\omega\}).$$

Then any $S \subset \Omega$ will have probability

$$\Pr(S) \stackrel{\text{def}}{=} \sum_{\omega \in S} \Pr(\{\omega\}),$$

which is a finite sum. It is easy to check that Pr has the requisite properties.

Similary, if Ω is a countably infinite set, then any subset is countable and Pr may be similarly defined using one-point sets. The probability of any infinite subset will be given by summing an infinite series of nonnegative terms, with partial sums that increase but are bounded above (by $\Pr(\Omega) = 1$), making the series convergent.

Uncountably infinite Ω require ideas from measure theory to understand fully, something beyond the scope of a basic introduction. Fortunately, most calculations involving such models may be done with real-valued functions of real variables, not abstract Ω or Pr, and thus need only Calculus methods.

Random variables, distributions, and densities

A *random variable*, often abbreviated *r.v.*, is a function $X : \Omega \to \mathbf{R}$ such that for each $s \in \mathbf{R}$, the subset $\{\omega \in \Omega : X(\omega) \leq s\}$ is measurable,[6] so it is in the domain of Pr and may be used to define the *cumulative distribution function (c.d.f.)* of X at t:

$$F_X(s) \stackrel{\text{def}}{=} \Pr(\{\omega \in \Omega : X(\omega) \leq s\}), \qquad s \in \mathbf{R}.$$

By monotonicity of Pr, $F_X(s)$ is nondecreasing, lies between 0 and 1, tends to 0 as $s \to -\infty$, and tends to 1 as $s \to +\infty$.

---

[6]Such a function is also called measurable, so an r.v. is a *measurable function*.

If Ω is finite or countably infinite, then F_X for any X will have countably many jump discontinuities, with positive jumps that sum to 1. Those jumps will be at discrete values s_n where $X^{-1}(s_n) \stackrel{\text{def}}{=} \{\omega : X(\omega) = s_n\}$ is nonempty, and the size of each jump will be $\Pr(X^{-1}(s_n))$. Without loss of generality it may be assumed that X is one-to-one on $\Omega = \{\omega_1, \omega_2, \dots\}$, with $X(\omega_n) = s_n$ so that $X^{-1}(s_n) = \{\omega_n\}$ is a one-point set. Then $\Pr(\{\omega_n\})$ may be defined to be the jump of F_X at s_n. In this way, the model $\Omega, \Pr$ is completely determined by the cumulative distribution function of X. This suggests an equivalence relation: say that random variables X, Y have the *same distribution*, and write

$$X \sim Y, \qquad \Longleftrightarrow \quad (\forall s \in \mathbf{R}) \, F_X(s) = F_Y(s), \qquad (1.10)$$

namely iff their cumulative distribution functions are the same.

If F_X is continuous everywhere, then X is called a *continuous random variable*. In this case Ω must be uncountably infinite, but a standard one may be constructed for X as follows:

- Let $\Omega = \mathbf{R}$.

- Define the probability function $\Pr$ on intervals $(a, b] \subset \Omega$ using F_X:

$$\Pr((a, b]) \stackrel{\text{def}}{=} F_X(b) - F_X(a).$$

Observe that $\Pr((-\infty, b]) = F_X(b)$ and $\Pr(\Omega) = 1$. From this definition on intervals, $\Pr$ may be extended to a countably additive probability function on a collection of measurable subsets of $\mathbf{R}$ called the *Borel sets*.[7] If F_X is also continuously differentiable, then X has a bounded, continuous *probability density function (p.d.f.)*, defined on $\Omega = \mathbf{R}$ by

$$f_X(s) \stackrel{\text{def}}{=} \frac{d}{ds} F_X(s), \qquad s \in \Omega.$$

This p.d.f. provides another way to compute probabilities:

$$\Pr(S) = \int_S f_X(s) \, ds, \quad \text{for any Borel set } S \subset \mathbf{R}.$$

Hence, f_X completely determines the model $\Omega, \Pr$, just as F_X does.

The *median* of a random variable X is $F_X^{-1}(1/2)$, if this is a single number, or else it is defined to be the midpoint of the set $F_X^{-1}(1/2)$ which must be a bounded interval. Thus every r.v. has a finite median. However, the *mean*, or *expected value*, may not exist as it is given by an improper integral:[8]

$$\mathrm{E}(X) \stackrel{\text{def}}{=} \int_{-\infty}^{\infty} s \, dF_X(s). \qquad (1.11)$$

---

[7]See Bass, p.9, in Further Reading from Chapter 8.
[8]This form, with dF_X, is called a *Riemann-Stieltjes integral*.

Expected values of functions of X are likewise defined as improper integrals:

$$\mathrm{E}(g(X)) \stackrel{\mathrm{def}}{=} \int_{-\infty}^{\infty} g(s)\, dF_X(s). \tag{1.12}$$

It is a necessary condition for certain results that X have *finite expectation and finite variance*, which is true iff both $\mathrm{E}(X)$ and $\mathrm{E}(X^2)$ exist and are finite. Fortunately, that holds for many important examples of random variables, where explicit c.d.f., p.d.f., median, mean, and variance formulas are known.

Stochastic processes

The price of an asset A may be considered a function of two variables $A(t, \omega)$, where the price at time $t > 0$ depends on the state ω of the world. The price $A(0, \omega)$ at the present time $t = 0$ is assumed known and independent of ω. Write $A(0, \omega) = A(0)$ for this spot price, or *initial price*.

Modeling the future means choosing the set Ω of possible states ω and assigning them probabilities to define Pr. Then at each fixed $t > 0$, the price of A is a random variable $\omega \mapsto A(t, \omega)$ on Ω. It will always be assumed to have finite expectation and variance.

Such a function of t and ω is called a *stochastic process*.

As seen above, the distribution function of the r.v. completely determines a model Ω, Pr, and vice versa. Which to use is a matter of convenience. Discrete and finite models are simpler as Ω, Pr, while continuous models are more simply described using c.d.f.s.

The model-dependent *expected price* $\mathrm{E}(A(T))$ of an asset A at time $T > 0$ is a weighted average. When the probability space Ω is finite, it may be computed with a sum:

$$\mathrm{E}(A(T)) \stackrel{\mathrm{def}}{=} \sum_{\omega \in \Omega} \mathrm{Pr}(\{\omega\}) A(T, \omega)$$

The weights $\{\mathrm{Pr}(\{\omega\}) : \omega \in \Omega\}$ are probabilities of individual states, constrained only by $0 \leq \mathrm{Pr}(\{\omega\}) \leq 1$ and $\sum_{\omega \in \Omega} \mathrm{Pr}(\{\omega\}) = 1$.

Models with countably infinite Ω typically define nonnegative $\mathrm{Pr}(\{\omega_n\})$ using some formula involving n that gives convergent infinite sums, normalized so that $\mathrm{Pr}(\Omega) = 1$. Models with uncountable Ω are typically specified with a c.d.f. and use Eq.1.11 to compute expectations.

1.3.2   Deterministic Arbitrages

This is a somewhat misleading name for portfolios that are arbitrages for every probability function on the future states Ω. For ease of definition, assume that

- Ω is a finite set, and

- $\mathrm{Pr}(\omega) > 0$ for all $\omega \in \Omega$.

For such finite financial models with no zero-probability future states, there are two types of deterministic arbitrages:

Definition 1 (IA) *A type one arbitrage, or immediate arbitrage, is an asset or portfolio with price $A(t,\omega)$ such that*

IA1: $A(0) < 0$.

IA2: $A(t,\omega) \geq 0$ *for all $t > 0$ in all states ω.*

This kind of arbitrage leaves a surplus as it is assembled but costs nothing to settle its liabilities in any future state. Of all the generalizations, it is the one most similar to a traditional instant arbitrage.

A less certain investment, but still an arbitrage, is one that cannot lose but wins only in some future states:

Definition 2 (AO) *A type two arbitrage, or arbitrage opportunity, is an asset or portfolio with price $A(t,\omega)$ satisfying*

AO1: $A(0) \leq 0$.

AO2: $A(t,\omega) \geq 0$ *for all $t > 0$ in all states ω.*

AO3: *There exists some $t > 0$ and some state ω such that $A(t,\omega) > 0$.*

This costs nothing to assemble and cannot lose value, but has a positive payoff in some future state.

For infinite Ω with arbitrary Pr, condition AO3 must be strengthened to

AO3+ : There exists $t > 0$ such that $\Pr(\{\omega \in \Omega : A(t,\omega) > 0\}) > 0$.

(A single ω with $A(t,\omega) > 0$, from AO3, may not suffice to give $\Pr(\{\omega\}) > 0$.) At the same time, conditions AO2 and IA2 may be weakened to

AO2– and IA2– : For all $t > 0$, $\Pr(\{\omega \in \Omega : A(t,\omega) < 0\}) = 0$.

(A single ω with $A(t,\omega) < 0$ but $\Pr(\{\omega\}) = 0$ would pose no risk of loss.)

Any type one immediate arbitrage is also a type two arbitrage opportunity. Including the surplus cash[9] in the no-loss portfolio guarantees that there will be a positive payoff in every future state.

1.3.3 Arbitrage and Expected Value

Any particular probability function in the model of the future may expose another, weaker kind of arbitrage.

Definition 3 (AE) *An arbitrage expectation is an asset or portfolio with price $A(t,\omega)$ satisfying*

AE1: $A(0) = 0$

---

[9]Not all markets have cash. It is assumed, however, that there is some riskless asset, used to price all other assets, whose value is the same at all times and states. Such an asset plays the role of cash but is more generally called a *numeraire*.

AE2: $E(A(t)) \geq 0$ *for all* $t > 0$.

AE3: $E(A(T)) > 0$ *for some* $T > 0$.

A type one or two arbitrage is an arbitrage expectation, though an arbitrage expectation need not be an immediate arbitrage or even an arbitrage opportunity. That is because averaging over future states allows losses in some states so long as the expected value is positive. Thus, arbitrage expectations not only are model-dependent but are also sensitive to the probability function. The relationship may be stated as

$$\exists \text{ IA} \implies \exists \text{ AO} \implies \exists \text{ AE}. \tag{1.13}$$

1.4   No Arbitrage and Its Consequences

The universal desire for profit creates unlimited demand for arbitrages so it is assumed that if assets are freely traded, then prices will adjust instantly to consume any supply. This may be stated as an axiom:

Axiom 1   *There are no arbitrages.*

The chain of implications for no arbitrages is the reverse of Eq.1.13:

$$\nexists \text{ AE} \implies \nexists \text{ AO} \implies \nexists \text{ IA}. \tag{1.14}$$

(These reversed implications are the equivalent *contrapositive* of the implications in Eq.1.13.) Forbidding even the weakest kind of arbitrage, no **AE**, thus yields the strongest results, though they will in general depend on the model of the future by which expectations are calculated. It is better to employ so-called *model independent* arguments using no **AO**.

Theorems about prices are deduced from Axiom 1 using proof by contradiction. Begin by assuming that the claimed price relationship does not hold, enumerate the violating cases, then construct an arbitrage for each case.

A first example of this argument is to show that the profit graph at any time T for a portfolio with initial cost 0 cannot lie entirely on one side of the horizontal axis. That is because the points on that axis represent all future states, so either the long or the short position for such a portfolio would offer an arbitrage opportunity.

Theorem 1.1 *If X is an option with expiry $T > 0$, then $X(t) > 0$ for all times $0 \leq t < T$.*

Proof: Suppose not, namely suppose that $X(t) \leq 0$ at some $0 \leq t < T$.

- If $X(t) < 0$, then X is an immediate arbitrage since it leaves a surplus when assembled (at time t) but incurs no liabilities: $X(T,\omega) \geq 0$ in all states and future times $T > t$ since it need not be exercised.

- If $X(t) = 0$, then X is an arbitrage opportunity since $X(t,\omega) \geq 0$ for all states and all times $0 < t \leq T$, and it will pay off $X(T) > 0$ at expiry (be "in the money" at T) with nonzero probability in the mind of some hopeful speculator, so $\Pr(\{\omega : X(T,\omega) > 0\}) > 0$.

These cases contradict Axiom 1. Conclude that $X(t) > 0$ for all $t < T$. $\quad\square$

Since the payoff graph of $X(T)$ versus $S(T)$ for an option on S, either Call or Put, lies entirely on or above the $S(T)$ axis, this theorem implies that the profit graph must be shifted partially below the axis by a positive amount $X(t)$ for any purchase time $t < T$.

The same type of proof gives the *One Price Theorem:*

Theorem 1.2 (One Price) *If at some time $T > 0$, two assets A, B have the same price $A(T,\omega) = B(T,\omega)$ in all states ω, then $A(t) = B(t)$ for all times $0 \leq t < T$.*

Proof: Fix a time t with $0 \leq t < T$.

- If $A(t) > B(t)$, then assemble a portfolio $-A + B$ by selling A short and buying B, leaving the surplus $-A(t) + B(t) < 0$. At time T, sell B for $B(T,\omega)$ and buy A for $A(T,\omega)$ to cover the short. By hypothesis this costs $-A(T,\omega) + B(T,\omega) = 0$ in all states ω. The assembled portfolio would thus be an immediate arbitrage, contradicting Axiom 1.

- Similarly, if $A(t) < B(t)$, then the portfolio $A - B$, namely long one A and short one B, would be an immediate arbitrage.

Conclude that $A(t) = B(t)$ for all times. $\quad\square$

The proof of the following very similar result is left as an exercise:

Corollary 1.3 *If at some time $T > 0$, $A(T,\omega) > B(T,\omega)$ in all states ω, then $A(t) > B(t)$ for all times $0 \leq t < T$.* $\quad\square$

1.4.1 Hedging

Theorem 1.2 provides prices for derivatives through the technique of *hedging*, or *portfolio replication*. Begin with a contingent claim X on a set of risky assets $\{A_i : i = 1, 2, \ldots, M\}$. Suppose also that there is riskless asset $A_0 = B$. Then a *hedge* for X is a portfolio containing assets $\{A_i : i = 0, 1, 2, \ldots, M\}$ in proportions $\{h_i : i = 0, 1, \ldots, M\}$ chosen so that its price matches the price of X in every time and state:

$$X(t,\omega) = \sum_{i=0}^{M} h_i(t) A_i(t,\omega), \qquad 0 \leq t \leq T, \quad \omega \in \Omega.$$

Finding these proportions is not always possible. However, there are special cases where the *hedge ratios* $\{h_i\}$ are determined, at least in some model of the future. So suppose that X is a contingent claim on a single risky asset $A_1 = S$. In that case, the hedge portfolio satisfies

$$X = h_0 B + h_1 S.$$

- If X is an option with expiry $T > 0$ on a risky asset S, then a one-step, two-state (discrete) model of the future determines the constants h_0, h_1 uniquely in terms of $S(0)$, $B(T)$, and the two modeled values of $S(T)$. The One-Price Theorem then gives $X(0) = h_0 B(0) + h_1 S(0)$.

- If X is a European-style option on S priced by the Black-Scholes model, then $h_1 = \frac{\partial X}{\partial S}$, which is called the *Delta* of the option.

- If X is an option priced with a multistep binomial model of the future, then h_1 is a discrete approximation to $\frac{\partial X}{\partial S}$.

In the last two cases, where there are intermediate times between 0 and T, the process of adjusting h_0 and h_1 to maintain price equality is called *dynamic hedging*. Since these adjustments depend on the model of the future, there is *model risk* with both procedures.

1.4.2 Martingales and Fair Prices

If there are no arbitrage expectations, then any portfolio $P = P(t, \omega)$ with initial price $P(0) = 0$ must be a *martingale*, which is a stochastic process satisfying, for all $t > 0$, the conditional expectation condition

$$E(P(t) \,|\, P(0) = 0) = 0.$$

Namely, the expected value of $P(t, \omega)$, given that only the value $P(0) = 0$ at present is known and there is no certainty about the future, is just $P(0)$. In particular, no portfolio assembled for zero cost is guaranteed to gain or lose value on average. This may be called the *Fair Price Theorem:*

Theorem 1.4 *If S is an asset and R is the riskless return over the time $t \in [0, T]$ and there are no arbitrage expectations, then*

$$E(S(T)) = RS(0).$$

Proof: Suppose first that $E(S(T)) > RS(0)$. With no initial cash at time $t = 0$, borrow $S(0)$ from the bank and buy one S at $S(0)$. Then at time $t = T$, sell S and repay the bank $RS(0)$, the borrowed capital plus interest. The expected value is then

$$E(S(T)) - RS(0) > 0,$$

an arbitrage expectation violating Axiom 1.

Similarly, if $E(S(T)) < RS(0)$, then at $t = 0$ sell one S short and deposit the proceeds $S(0)$ into the bank. At $t = T$, withdraw $RS(0)$ from the bank, the initial deposit plus interest, and buy S to cover the short. The expected value is then

$$RS(0) - E(S(T)) > 0,$$

again violating Axiom 1. □

The spot price of an asset is determined in free markets by the collective action of many traders. This "fair price" may be interpreted as an instantaneous consensus about the future.

1.4.3 No-Arbitrage Price Equalities

Assuming no arbitrage leads to price parity formulas for stock and bond portfolios and their derivatives.

Forward contracts

An agreed price K for a forward contract to buy a risky asset S at future time T is negotiated so that there is no premium for either party. It is thus the expected[10] value of $S(T)$, which by the Fair Price Theorem above is

$$E(S(T)) = RS(0),$$

which depends on the current price $S(0)$ and the riskless return R over the period $t = 0$ to $t = T$. The fair strike price is then:

$$K = RS(0). \tag{1.15}$$

But this formula actually follows from the weaker Axiom 1, regardless of the model of the future. It may be proved by constructing two arbitrages.

Case 1: $K > RS(0)$. At $t = 0$, starting with no money, do the following:

- Borrow $S(0)$ cash from the bank at riskless rate R.
- Buy one share of S for $S(0)$ cash.
- Sign a (short) Forward contract to sell S for K at time T.

These transactions have zero net cost, though of course they incur liabilities and obligations. But at $t = T$, clear all these debts as follows:

- Execute the forward contract to obtain K in exchange for S.
- Repay the bank loan for $RS(0)$ cash, which includes interest.

That leaves $K - RS(0) > 0$ with no unfunded liabilities, an arbitrage profit prohibited by Axiom 1.

Case 2: $K < RS(0)$.

---

[10]The people doing the expecting are the traders of S.

At $t = 0$, starting with no money, do the following:

- Sell S short for $S(0)$ cash.
- Deposit $S(0)$ cash in the bank at rate R.
- Sign a (long) Forward contract to buy S for K at time T.

Again, these transactions incur obligations but have zero net cost. At $t = T$, clear all liabilities as follows:

- Withdraw $RS(0)$ cash from the bank, which includes interest.
- Execute the forward contract to buy S for K, which is less than the $RS(0)$ cash just obtained.

That leaves $RS(0) - K > 0$ cash with no unfunded liabilities, an arbitrage profit forbidden by Axiom 1. Conclude that Eq.1.15 holds.

Call-Put parity formula

Axiom 1 implies a relationship between premiums on same-expiry Call and Put options at same strike price K for an asset S. If R is the riskless return between now ($t = 0$) and expiry, then the premiums $C(0)$ and $P(0)$ satisfy

$$C(0) - P(0) = S(0) - K/R. \tag{1.16}$$

To prove equality, show that inequality in either direction results in an arbitrage opportunity.

Case 1: $C(0) - P(0) > S(0) - K/R$.

At $t = 0$, starting with no money, do the following:

- Short-sell C for $C(0)$ cash.
- Buy P for $P(0)$ cash.
- Borrow K/R cash from the bank at riskless rate R.
- Buy S for $S(0)$ cash.

That leaves $C(0) - P(0) - S(0) + K/R > 0$, a net profit to keep. Then at $t = T$, clear all debts as follows:

- If $S(T) > K$, then cash-settle the short-sold C for $S(T) - K$. Otherwise there is no liability as C expires worthless.
- If $K > S(T)$, then exercise P to earn $K - S(T) > 0$. Otherwise P expires worthless.
- Sell S for $S(T)$ cash.
- Repay the bank $RK/R = K$ cash, including interest.

That leaves $-[S(T) - K]^+ + [K - S(T)]^+ - K + S(T) = 0$ and no unfunded liabilities. The positive amount obtained at $t = 0$ is therefore an arbitrage profit prohibited by Axiom 1.

Case 2: $C(0) - P(0) < S(0) - K/R$.

It is left as an exercise to show how this leads to an arbitrage profit. Conclude that Eq.1.16 holds.

Forwards versus options

The plus-part function satisfies

$$[X]^+ - [-X]^+ = X, \tag{1.17}$$

for any number X. This may be proved by checking cases and is left as an exercise. Applying this identity to the payoff values of European-style Call and Put options for S at strike price K and expiry T yields another parity formula:

$$C(T) - P(T) = [S(T) - K]^+ - [K - S(T)]^+ = S(T) - K = F_{\text{long}}(T). \tag{1.18}$$

Thus a long Call and a short Put have the same payoff as a long Forward contract, if all strike prices and expiries are equal. Applying Theorem 1.2 gives another way to choose the agreed Forward price:

Corollary 1.5 *The agreed price for a Forward contract to buy S at time $T > 0$ is that K for which the Call and Put options on S with strike price K and expiry T have the same premium.*

Proof: By Eq.1.17, portfolio $C - P$ and asset F_{long} have the same price at time T in all states. Now let K be a strike price for which $F_{\text{long}}(0) = 0$. By Theorem 1.2, conclude that $C(0) - P(0) = F_{\text{long}}(0) = 0$. □

Forward exchange rate parity

The agreed rate K for a forward contract to exchange foreign currency is negotiated so that there is no premium for either party. For expiry T, the price is called the *T-forward exchange rate*, and it depends on the current exchange rate $X(0)$ and the riskless returns R_d and R_f of domestic and foreign bonds, respectively, over the period $t = 0$ to $t = T$. The T-forward exchange rate is given by the *(Covered) Interest Rate Parity Formula:*

$$K = \frac{R_d}{R_f} X(0). \tag{1.19}$$

It may be established by constructing arbitrages for the two inequalities.

Case 1: $K > \frac{R_d}{R_f} X(0)$.

At $t = 0$, starting with no money, perform the following trades:

- Borrow $X(0)/R_f$ DOM from the domestic bank.
- Spend $X(0)/R_f$ DOM to buy $1/R_f$ FRN and invest it in the foreign bank.
- Sign a (short) Forward contract to sell 1 FRN for K DOM at time T.

These transactions have zero net cost.

At $t = T$, clear all debts as follows:

- Withdraw 1 FRN from the foreign bank, including interest.
- Execute the forward contract to obtain K DOM.
- Return $X(0)R_d/R_f$ to the domestic bank, which includes interest.

That leaves $K - X(0)R_d/R_f > 0$ DOM with no unfunded liabilities, an arbitrage profit prohibited by Axiom 1.

Case 2: $K < \frac{R_d}{R_f} X(0)$.

At $t = 0$, starting with no money, perform the following trades:

- Borrow $1/R_f$ FRN from the foreign bank.
- Spend the $1/R_f$ FRN to buy $X(0)/R_f$ DOM and invest it in the domestic bank at riskless return R_d.
- Sign a (long) Forward contract to buy 1 FRN for K DOM at time T.

These transactions also have zero net cost.

At $t = T$, clear all debts as follows:

- Withdraw $X(0)R_d/R_f$ DOM from the domestic bank, including interest.
- Execute the forward contract to obtain 1 FRN in exchange for K DOM, which is less than the $X(0)R_d/R_f$ DOM in hand.
- Return 1 FRN to the foreign bank, which includes interest.

That leaves $X(0)R_d/R_f - K > 0$ DOM with no unfunded liabilities, an arbitrage profit forbidden by Axiom 1.

Conclude that Eq.1.19 holds.

Call-Put parity for foreign exchange options

The relationship between Call and Put premiums on foreign exchange options is similar to that on stock options:

$$C(0) - P(0) = \frac{X(0)}{R_f} - \frac{K}{R_d}, \tag{1.20}$$

where $C(0), P(0)$ are the respective premiums for Call and Put options for the exchange rate X at strike price K, expiring at time T, and R_d, R_f are the respective riskless returns for domestic currency DOM and foreign currency FRN. The proof of this formula is left as an exercise.

1.4.4 No-Arbitrage Inequalities

Suppose that $C_E(t)$ and $C_A(t)$ are the prices of European-style and American-style Call options, respectively, at time t. Suppose that $0 \leq t \leq T$, where the expiry time is $T > 0$ for both, and that the options have the same strike price on the same underlying asset S.

Theorem 1.6 *For all* $0 \leq t \leq T$, $C_A(t) \geq C_E(t)$.

Proof: Suppose not, namely that $C_A(t) < C_E(t)$ at some time t before expiry. Then the portfolio $C_A - C_E$, namely one European Call sold short to buy one American Call, satisfies $C_A(t) - C_E(t) < 0$, leaving a surplus when it is assembled at time t. But $C_A(T) - C_E(T) = 0$, since at expiry both are worth $[S(T) - K]^+$. Hence this portfolio is an immediate arbitrage, contradicting the no-arbitrage axiom. □

For the next two results, let $Z(t, T)$ be the discount factor, at time t, of a zero-coupon bond maturing at time T. Assume that riskless interest rates are positive so that $Z(t, T) < 1$.

Theorem 1.7 *For all* $0 \leq t \leq T$, $C_E(t) \geq [S(t) - Z(t, T)K]^+$.

Proof: Suppose not, namely that $C_E(t) < [S(t) - Z(t, T)K]^+$, at some time t before expiry. There are two cases to consider:

- If $S(t) \leq Z(t, T)K$, then $C_E(t) < [S(t) - Z(t, T)K]^+ = 0$. But at expiry, $C_E(T) = [S(T) - K]^+ \geq 0$.

- If $S(t) > Z(t, T)K$, then $[S(t) - Z(t, T)K]^+ = S(t) - Z(t, T)K$, so $C_E(t) < S(t) - Z(t, T)K$. The portfolio $C_E - S - ZK$, namely one Call C_E, one share of S sold short, and ZK borrowed from the bank and invested in a riskless bond, satisfies $C_E(t) - S(t) + Z(t, T)K < 0$, leaving a surplus when it is assembled at time t. But at expiry, the bond matures to yield K, exercising the Call buys S for K, and delivering S covers the short and satisfies all liabilities.

In either case, there is an immediate arbitrage. □

Theorem 1.8 *For all* $0 \leq t \leq T$, $C_A(t) > [S(t) - K]^+$.

Proof: Consider the two cases:

- If $S(t) > K$, then, since interest rates are assumed positive,

$$S(t) - Z(t, T)K > S(t) - K > 0,$$

so by Theorems 1.6 and 1.7,

$$
\begin{aligned}
C_A(t) &\geq C_E(t) \\
&\geq [S(t) - Z(t, T)K]^+ \\
&= S(t) - Z(t, T)K \\
&> S(t) - K = [S(t) - K]^+.
\end{aligned}
$$

- If $S(t) \leq K$, then $[S(t) - K]^+ = 0$. But $C_A(t) > 0$ for all $t < T$ by Theorem 1.1, so $C_A(t) > [S(t) - K]^+$.

In either case, the claimed inequality holds.   □

1.5   Exercises

1. Suppose that a risky asset S has spot price $S(0) = 100$ and that the riskless return to $T = 1$ year is $R = 1.0112$. Assuming there are no arbitrages, compute the following:

   (a) current zero-coupon bond discount $Z(0, T)$,

   (b) Forward price for one share of S at expiry T,

   (c) riskless annual interest rate given continuous compounding.

2. With $S(0)$, R, and T as in Exercise 1, suppose that $S(T)$ is modeled by the following table:

$S(T)$	90	95	98	100	102	105	110
$\Pr(S(T))$	0.01	0.04	0.15	0.30	0.30	0.18	0.02

   (a) Use this finite probability space model to estimate premiums $C(0)$ and $P(0)$ for European-style Call and Put options, respectively, with strike price $K = 101$ and expiry T.

   (b) Does Call-Put Parity hold in this model? What might cause it to be inaccurate?

3. Use the no-arbitrage Axiom 1 to prove that Eq.1.7 holds.

4. Prove Corollary 1.3. (Hint: review the proof of Theorem 1.2.)

5. Suppose that $C(0) - P(0) < S(0) - K/R$, in contradiction with Eq.1.16. Construct an arbitrage.

6. Prove Eq.1.20, the Call-Put Parity Formula for foreign exchange options:

$$C(0) - P(0) = \frac{X(0)}{R_f} - \frac{K}{R_d}.$$

Use the no-arbitrage axiom.

7. (a) Prove that the plus-part function satisfies Eq.1.17:

$$[X]^+ - [-X]^+ = X,$$

for any number X.

(b) Apply the identity in part (a) to the payoff values of European-style Call and Put options for S at strike price K and expiry T to show Eq.1.18:

$$C(T) - P(T) = S(T) - K.$$

8. Plot the payoff and profit graphs for the following colorfully named option portfolios as a function of the price $S(T)$ at expiry time T:

(a) *Long straddle:* buy one Call and one Put on S with the same expiry T and at-the-money strike price $K \approx S(0)$. For what values of $S(T)$ will this be profitable?

(b) *Long strangle:* buy one Call at K_c and one Put at K_p with the same expiry T but with out-of-the-money strike prices $K_p < S(0) < K_c$. How does its profitability compare with that of a long straddle?

9. A *butterfly spread* is a portfolio of European-style Call options purchased at time $t = 0$ with the same expiry $t = T$ but at three strike prices $L < M < H$, where $M = \frac{1}{2}(L + H)$:

- buy one Call C_L at strike price L for $C_L(0)$,
- buy one Call C_H at strike price H for $C_H(0)$,
- sell two Calls C_M short at strike price M for $2C_M(0)$.

(a) Plot the payoff graph for the butterfly spread at expiry when its price is $C_L(T) + C_H(T) - 2C_M(T)$. Mark the three strike prices on the $S(T)$ axis.

(b) Conclude that $C_M(0) < \frac{1}{2}[C_L(0) + C_H(0)]$. (Hint: apply the no-arbitrage Axiom 1 to the graph plotted in part (a).)

10. An *iron condor* is a portfolio $C_1 - C_2 - P_3 + P_4$ of four European-style options. To construct it, simultaneously buy one Call at K_1, sell one Call at K_2, sell one Put at K_3, and buy one Put at K_4, all with the same expiry T but with $K_1 < K_2 < K_3 < K_4$.

(a) Describe the payoff graph for an iron condor portfolio at expiry.

(b) Assuming no arbitrage, prove that the portfolio must have a positive net premium.

(c) Assuming no arbitrage, find inequalities bounding the maximum profit and the maximum loss of an iron condor portfolio at expiry.

1.6   Further Reading

- Freddy Delbaen and Walter Schachermayer. "What is... a Free Lunch?" *Notices of the AMS* 51:5 (2004), pp.526–528.

- Robert James Elliott and Peter Ekkehard Kopp. *Mathematics of Financial Markets (2nd Edition)*. Springer-Verlag, New York (2004).

- William Feller. *An Introduction to Probability Theory and Its Applications, Volume I (3rd Edition)*. John Wiley & Sons, Inc., New York (1968).

- John C. Hull. *Options, Futures and Other Derivatives (5th Edition)*. Prentice Hall, London (2003).

- Stanley R. Pliska. *Introduction to Mathematical Finance: Discrete Time Models*. Blackwell, Oxford, U.K., and Malden, MA (1997).

2

Continuous Models

Asset prices are functions of time. In some financial models, prices and times lie in a *continuum*, an interval of the real number line **R**.

In a continuous time model, changes in price are described by differential equations. For example, a riskless asset $B = B(t)$ with return rate r satisfies

$$\frac{dB}{B} = r\, dt, \tag{2.1}$$

since the relative or percentage change in price, per unit of time, is just the return rate r. This is the continuous version of "interest equals principal times the rate times time:"

$$\Delta B = Br\Delta t \tag{2.2}$$

Its initial value problem for constant rate r and initial principal $B_0 = B(0)$ is solved by

$$B(t) = B_0 \exp(rt), \qquad t \geq 0. \tag{2.3}$$

This describes the return from continuous compounding.

Now let $S = S(t, \omega)$ be the price of a risky asset at time t in state ω of the world. The relative change of its price has two parts: a riskless rate r, plus random fluctuations modeled by a *stochastic process* $W(t, \omega)$. Without loss of generality, it may be assumed that $W(0, \omega) = 0$ in all states and that $\mathrm{Var}(W(1, \cdot)) = 1$ so that it may be scaled by volatility v. The stochastic term adds risk to the return rate and gives a generalization of Eq.2.2:

$$\Delta S = S\left[r\Delta t + v\Delta W\right]. \tag{2.4}$$

The continuous time version of this model is the *stochastic differential equation*

$$\frac{dS}{S} = r\, dt + v\, dW, \tag{2.5}$$

which, if solved, gives one solution for every state ω of the world. This simplifies to Eq.2.1 for B in the riskless world where $v = 0$.

2.1 Some Facts from Probability Theory

Asset prices in a bustling free market behave like the average of very many independent random variables, each having bounded variance around the same mean. Such an average is itself a random variable, with an approximately normal distribution as a result of the Central Limit Theorem.[1]

A useful fact about any random variable X with finite expectation $E(X)$ and variance $\text{Var}(X)$ is that

$$E(a + bX) = a + bE(X); \qquad \text{Var}(a + bX) = b^2\text{Var}(X). \qquad (2.6)$$

Applied to normal distributions, in the notation of Eq.1.10, this gives

$$\mathcal{N}(\mu, \sigma^2) \sim \mu + \mathcal{N}(0, \sigma^2) \sim \mu + \sigma\mathcal{N}(0, 1),$$

showing how $\mathcal{N}(\mu, \sigma^2)$, the *normally distributed* random variable with mean μ and variance σ^2, is easily obtained from $\mathcal{N}(0, 1)$, the *standard normal* random variable with mean 0 and variance 1. Such a formula reduces the amount of software that needs to be implemented for simulation and calculation.

The expected value $E(X)$ of a contingent claim X depending on $S(T, \omega)$ is computed by averaging over the states ω weighted by $\Pr(\omega)$. Equivalently, if the cumulative distribution function Q of $S(T)$ is known,

$$Q(s) \overset{\text{def}}{=} \Pr\{\omega : S(T, \omega) \leq s\}, \qquad -\infty < s < \infty,$$

then

$$E(X) = \int_{-\infty}^{\infty} X(s)\, dQ(s),$$

which is a Riemann-Stieltjes integral. If the c.d.f. Q is differentiable with probability density function $q(s) \overset{\text{def}}{=} Q'(s)$, then the integral may be rewritten with the p.d.f. as a weight:

$$E(X) = \int_{-\infty}^{\infty} X(s)q(s)\, ds$$

It remains to find this c.d.f. or p.d.f. and then compute the integral, if possible.

2.2   Understanding Brownian Motion

Unfortunately, Eq.2.5 may not be solvable for a given W. Many continuous stochastic processes are not differentiable with respect to time, so dW/dt may not exist for most or all states ω. The naïve integration method will fail for

---

[1]See Feller, Volume II, p.258, in Further Reading after Chapter 3.

those states:

$$\frac{dS}{S} = r\,dt + v\,dW$$

$$\implies \quad \frac{d}{dt}\log S = r + v\frac{dW}{dt}$$

$$\implies \quad \log S(t,\omega) = \log S(0,\omega) + \int_0^t [r + v\frac{dW(t,\omega)}{dt}]\,dt$$

$$\implies \quad S(t,\omega) = S(0,\omega)\exp\int_0^t [r + v\frac{dW(t,\omega)}{dt}]\,dt.$$

In the case of constant r and v this gives the invalid solution

$$S(t,\omega) = S(0,\omega)\exp\left(rt + vW(t,\omega)\right),$$

which is wrong because an intermediate step requires a nonexistent quantity.

Fortunately, there is a natural choice of continuous stochastic process that satisfies sufficient conditions for the existence of solutions. It is called *Brownian motion*, after botanist Robert Brown (1773-1858) who described the random motion of pollen grains viewed under a microscope, but is also called the *Wiener process*, after mathematician Norbert Wiener (1894-1964), and is most often denoted by $W(t,\omega)$. It may be understood using *random walks*.

Let n be a fixed positive integer and let Ω be the (finite) probability space of equally probable n-bit positive integers $\omega = \omega_{n-1}\cdots\omega_1\omega_0$ (base 2), namely

$$\omega = \omega_0 + 2\omega + \cdots + 2^{n-1}\omega_{n-1} = \sum_{i=0}^{n-1}\omega_i 2^i,$$

with $\omega_i \in \{0,1\}$ for all i. The number of states $\#\Omega$ in this space is 2^n.

Let $X(t,\omega)$ be the continuous piecewise linear function on $0 \le t \le n$ defined recursively by

$$X(0,\omega) = 0, \quad \text{all } \omega \in \Omega;$$

$$X(i+1,\omega) = \begin{cases} X(i,\omega) - 1, & \omega_i = 0, \\ X(i,\omega) + 1, & \omega_i = 1, \end{cases} \quad i = 0,1,\ldots,n-1;$$

$$X(t,\omega) = (t-i)X(i+1,\omega) + (i+1-t)X(i,\omega), \quad i < t < i+1.$$

This is called a random walk. If every $\omega \in \Omega$ is equally probable, then for each i, $\Pr(\omega_i = 0) = \Pr(\omega_i = 1) = \frac{1}{2}$ so at each integer point $t = i$ there is an equal probability of going up by 1 or going down by 1. The terminal points of the random walk are $X(t,\omega)$ with $t = n$ and they lie in the range $[-n,n]$, with the extreme points reached uniquely by $X(n,00\cdots0) = -n$ and $X(n,11\cdots1) = n$.

The complete set of reachable points (there are $n+1$ of them) is

$$\{-n, -n+2, \ldots, -n+2k, \ldots, -n+2(n-1), n\}.$$

The terminal point $X(n, \omega) = -n + 2k$ is reached by just those paths where $\#\{i : \omega_i = 0\} = n - k$ and $\#\{i : \omega_i = 1\} = k$. There are $\binom{n}{k}$ ways to have exactly k 1-bits in a set of n bits, so

$$
\begin{aligned}
\Pr(X(n, \omega) = -n + 2k) &= \frac{\#\{\omega \in \Omega : \#\{i : \omega_i = 1\} = k\}\}}{\#\Omega} \\
&= \frac{1}{2^n}\binom{n}{k} = \binom{n}{k}\left(\frac{1}{2}\right)^k\left(\frac{1}{2}\right)^{n-k}.
\end{aligned}
$$

This is evidently the binomial probability of k successes in n independent trials, each with success probability $\frac{1}{2}$. Hence the $n+1$ terminal values have a binomial distribution. The mean and variance of this distribution are known to be

$$
\mathrm{E}(X(n, \cdot)) = 0; \qquad \mathrm{Var}(X(n, \cdot)) = \frac{n}{4}. \tag{2.7}
$$

It may be normalized to have variance 1 by multiplying by $1/\sqrt{n/4} = 2/\sqrt{n}$, using Eq.2.6, rescaling X. This is exactly the same as using jump size $\pm 2/\sqrt{n}$ instead of ± 1.

Likewise, the $[0, n]$ time interval may be normalized to $[0, 1]$ by changing the time step to $1/n$ from 1. It is left as an exercise to rewrite the recursive definition of X with these normalizations.

Now let $n \to \infty$ for the standardized X and consider $X(1, \omega)$ for some very large value of n. By the Central Limit Theorem, $X(1, \cdot)$ are distributed approximately like $\mathcal{N}(0, 1)$. Individual paths $t \mapsto X(t, \omega)$ will be piecewise linear with slopes

$$
\frac{\Delta X}{\Delta t} = \frac{X(\frac{i+1}{n}) - X(\frac{i}{n})}{\frac{i+1}{n} - \frac{i}{n}} = \frac{\pm 2/\sqrt{n}}{1/n} = \pm 2\sqrt{n},
$$

which tends to $\pm\infty$ as $n \to \infty$. Hence the limit paths are not differentiable.

However, the *quadratic variation* of this standardized random walk is bounded for all n:

$$
\frac{[\Delta X]^2}{\Delta t} = \frac{[X(\frac{i+1}{n}) - X(\frac{i}{n})]^2}{\frac{i+1}{n} - \frac{i}{n}} = \frac{4/n}{1/n} = 4,
$$

Now fix a time s with $0 < s < 1$ and consider $X(s, \omega)$. For very large n, $X(s, \cdot)$ is distributed very much like $\mathcal{N}(0, s)$, with equality in the limit as $n \to \infty$. Also, the position $X(s, \omega)$ was reached by a path determined by the first $\lfloor sn \rfloor$ of the bits, namely $\omega_0, \omega_1, \ldots, \omega_{\lfloor sn \rfloor}$. Where X goes after that is determined by the remaining bits $\omega_{\lfloor sn \rfloor + 1}, \ldots, \omega_n$, which are independent in the sense that knowing the first bits of unknown $\omega \in \Omega$ gives no information about the last bits. Hence the random variable $X(1, \omega) - X(s, \omega)$ will be independent of $X(s, \omega) = X(s, \omega) - X(0, \omega)$. More generally, for $0 \le s < t < u \le 1$, $X(u, \omega) - X(t, \omega)$ is independent of $X(t, \omega) - X(s, \omega)$. This is called the *independent increments property*.

Finally, for $s < t$, given the position $x = X(s,\omega)$ without knowing ω, it is not possible to predict $X(t,\omega)$. The set $\Omega' \stackrel{\text{def}}{=} \{\omega \in \Omega : X(s,\omega) = x\}$ indexes paths that lead to positions $X(t,\omega)$ which are distributed approximately like $\mathcal{N}(x, t - s)$. In particular, this gives the conditional expectation

$$\mathrm{E}(X(t,\omega)|X(s,\omega) = x) = x.$$

More generally, the conditional expectation is the last known value, irrespective of prior positions:

$$\mathrm{E}\left(X(t,\omega)|\{X(u,\omega) : u \leq s < t\}\right) = X(s,\omega).$$

This is called the *martingale property*. It means that knowledge of the current and past positions does not provide certainty about future movements. It is a standard assumption that excludes arbitrage expectations.

The limit of the standardized random walk X as $n \to \infty$ is the Wiener process, or standardized Brownian motion, and it shares the normal distribution, variance t, bounded quadratic variation, independent increments, and martingale properties. This makes it suitable for modeling financial markets.

2.3   The Black-Scholes Formula

If W is standard Brownian motion, then even though dW/dt fails to exist there is an alternative integral, *Itô's stochastic integral*, which takes advantage of bounded quadratic variation and independence of increments to solve for S. Assuming r and v to be independent of t and ω, the initial value problem for Eq.2.5 has the solution[2]

$$S(t,\omega) = S_0 \exp\left((r - \frac{v^2}{2})t + vW(t,\omega)\right), \tag{2.8}$$

where $S_0 = S(0,\omega)$ is the spot price of S, assumed to be the same for all ω. This is an explicit function of W, so the distribution function for S may be computed from that of Brownian motion. This is a good practical reason to model risky price fluctuations with Brownian motion.

S and S_0 will have the same sign, since $\exp()$ takes only positive values, so without loss of generality both sides may be assumed positive and the logarithm function applied to Eq.2.8 to give

$$\log S(t,\omega) = \log S_0 + (r - \frac{v^2}{2})t + vW(t,\omega). \tag{2.9}$$

---

[2]See Example 1, p.213 of Lawler, in Further Reading below.

Thus, for each fixed time $t \geq 0$, $\log S(t, \cdot)$ is a random variable with the same distribution[3] as $W(t, \cdot)$ scaled by the constant volatility v and then shifted by the constant mean $\log S_0 + (r - \frac{v^2}{2})t$.

Now, $W(t, \cdot)$, being the result of Brownian motion starting at 0 and running for time t, is normally distributed with mean 0 and variance t:

$$W(t, \cdot) \sim \mathcal{N}(0, t), \quad \implies \quad vW(t, \cdot) \sim \mathcal{N}(0, v^2 t), \tag{2.10}$$

in the notation of Eq.1.10, since multiplying a random variable by a constant v multiplies its variance by v^2. Likewise, adding a constant m to a random variable adds m to its mean. Thus,

$$\begin{aligned} \log S(t, \cdot) \quad &\sim \quad \log S_0 + (r - \frac{v^2}{2})t + \mathcal{N}(0, v^2 t) \\ &\sim \quad \mathcal{N}\left(\log S_0 + (r - \frac{v^2}{2})t, \ v^2 t\right), \end{aligned}$$

so $\log S$ in this model, at each fixed time $t \geq 0$, is a normal random variable with mean and variance

$$\mu \overset{\text{def}}{=} \log S_0 + (r - \frac{v^2}{2})t; \qquad \sigma^2 \overset{\text{def}}{=} v^2 t. \tag{2.11}$$

Since $\log S$ is normally distributed, S itself has a *lognormal distribution*:

$$S(t, \cdot) \sim \text{Lognormal}\left(\mu, \ \sigma^2\right), \tag{2.12}$$

where, by convention, the parameters are the mean and variance, respectively, of the normal random variable $\log S$. That is what must be used to compute expected values of functions of S such as options payoffs.

It is known[4] that Lognormal(μ, σ^2) has a probability density function

$$q(s) = \frac{1}{s\sigma\sqrt{2\pi}} \exp\left(-\frac{[\log s - \mu]^2}{2\sigma^2}\right), \tag{2.13}$$

for $s \in (0, \infty)$. As a p.d.f., it satisfies the usual requirements:

$$(\forall s)\, q(s) \geq 0; \qquad \int_0^\infty q(s)\, ds = 1. \tag{2.14}$$

Its mean $\text{E}(S)$ is not equal to μ but rather

$$\text{E}(S) = \int_0^\infty s q(s)\, ds = \exp\left(\mu + \frac{\sigma^2}{2}\right). \tag{2.15}$$

Its cumulative distribution function is

$$Q(s) \overset{\text{def}}{=} \int_0^s q(x)\, dx = \Phi\left(\frac{\log s - \mu}{\sigma}\right), \tag{2.16}$$

---

[3] The state variable ω is replaced by a dot since only distributions are being compared.
[4] See Forbes, *et al.*, in Further Reading below.

where Φ is the c.d.f. of $\mathcal{N}(0,1)$, a special function that is implemented in many software packages:

$$\Phi(z) \stackrel{\text{def}}{=} \frac{1}{\sqrt{2\pi}} \int_{-\infty}^{z} \exp(-\frac{x^2}{2}) \, dx. \tag{2.17}$$

The integrand in Eq.2.17 for Φ is the standard normal p.d.f.:

$$\phi(x) \stackrel{\text{def}}{=} \frac{1}{\sqrt{2\pi}} \exp(-\frac{x^2}{2}). \tag{2.18}$$

Because $\phi(-x) = \phi(x)$, it is easily shown that

$$\Phi(-z) = 1 - \Phi(z). \tag{2.19}$$

Also, the normal random variable $X \sim \mathcal{N}(a, b^2)$, with mean a and variance b^2, has respective p.d.f. and c.d.f.

$$\frac{1}{b} \phi\left(\frac{x-a}{b}\right) \quad \text{and} \quad \Phi\left(\frac{x-a}{b}\right),$$

which may be derived by substitution within the integral defining Φ.

2.3.1 Option Pricing

Now let $C = C(t, \omega)$ be the price of a European Call option with strike price K and expiry time T. This price is known at expiry, at least in terms of the underlying price $S(T, \omega)$, to be

$$C(T, \omega) = [S(T, \omega) - K]^+.$$

The Call premium, under the assumptions of no expected arbitrage and constant riskless rate r, is thus

$$C(0) = \exp(-rT)\mathrm{E}(C(T)), \tag{2.20}$$

where ω has been removed by averaging over all states. The averaging is weighted by the lognormal distribution of S, and the technique is to integrate $[s - K]^+ q(s)$ over all positive prices s that S may take:

$$\begin{aligned}
C(0) &= \exp(-rT) \int_0^\infty [s-K]^+ q(s) \, ds \\
&= \exp(-rT) \int_K^\infty (s-K) q(s) \, ds \\
&= \exp(-rT) \left[\int_K^\infty s q(s) \, ds - K \int_K^\infty q(s) \, ds \right].
\end{aligned}$$

Apply Eqs.2.14 and 2.16 to compute the rightmost integral:

$$\int_K^\infty q(s) \, ds = 1 - Q(K) = 1 - \Phi\left(\frac{\log K - \mu}{\sigma}\right) = \Phi\left(\frac{\mu - \log K}{\sigma}\right) \tag{2.21}$$

The other integral may be evaluated by substituting $s \leftarrow e^t$ in Eq.2.13, then completing squares and factoring out constant terms:

$$
\begin{aligned}
\int_K^\infty sq(s)\,ds &= \frac{1}{\sigma\sqrt{2\pi}} \int_K^\infty \exp\left(-\frac{[\log s - \mu]^2}{2\sigma^2}\right) ds \\
&= \frac{1}{\sigma\sqrt{2\pi}} \int_{\log K}^\infty \exp\left(-\frac{[t-\mu]^2}{2\sigma^2}\right) \exp(t)\,dt \\
&= \exp\left(\mu + \frac{\sigma^2}{2}\right) \frac{1}{\sigma\sqrt{2\pi}} \int_{\log K}^\infty \exp\left(-\frac{[t-(\mu+\sigma^2)]^2}{2\sigma^2}\right) dt \\
&= \exp\left(\mu + \frac{\sigma^2}{2}\right) \Phi\left(\frac{(\mu+\sigma^2)-\log K}{\sigma}\right),
\end{aligned}
$$

after recognizing the c.d.f. of $\mathcal{N}((\mu+\sigma^2),\sigma^2)$.

Combining the two parts gives

$$
C(0) = \exp(-rT)\left[\exp\left(\mu + \frac{\sigma^2}{2}\right)\Phi\left(\frac{(\mu+\sigma^2)-\log K}{\sigma}\right) - K\Phi\left(\frac{\mu-\log K}{\sigma}\right)\right].
$$

Substituting from Eq.2.11 with $t = T$ gives

$$
\begin{aligned}
\mu + \sigma^2 &= \log S_0 + \left(r - \frac{v^2}{2}\right)T + v^2 T = \log S_0 + \left(r + \frac{v^2}{2}\right)T, \\
\mu + \frac{\sigma^2}{2} &= \log S_0 + \left(r - \frac{v^2}{2}\right)T + \frac{v^2 T}{2} = \log S_0 + rT, \\
\exp\left(\mu + \frac{\sigma^2}{2}\right) &= S_0 \exp(rT), \tag{2.22}
\end{aligned}
$$

so the Call premium should be $C(0) =$

$$
S_0 \Phi\left(\frac{\log \frac{S_0}{K} + \left(r + \frac{v^2}{2}\right)T}{v\sqrt{T}}\right) - e^{-rT} K\Phi\left(\frac{\log \frac{S_0}{K} + \left(r - \frac{v^2}{2}\right)T}{v\sqrt{T}}\right). \tag{2.23}
$$

Eq.2.23 is the *Black-Scholes formula* for European Call options.

The spot price S_0 and the present value $e^{-rT}K$ of the strike price are combined in this formula with the positive weights $\Phi(d_1)$ and $\Phi(d_2)$, respectively, where

$$
d_1 \stackrel{\text{def}}{=} \frac{\log \frac{S_0}{K} + \left(r + \frac{v^2}{2}\right)T}{v\sqrt{T}}, \qquad d_2 \frac{\log \frac{S_0}{K} + \left(r - \frac{v^2}{2}\right)T}{v\sqrt{T}}. \tag{2.24}
$$

These weights are called *risk-adjusted probabilities*, and they take values between 0 and 1. Much of the complexity of the Black-Scholes formula is eliminated by writing it as

$$
C(0) = S_0 \Phi(d_1) - e^{-rT} K\Phi(d_2). \tag{2.25}
$$

Computations with the formula are facilitated by noting that $d_1 - d_2 = v\sqrt{T}$.

The similar Black-Scholes formula for European Put options, where the payoff is $[K - S(T)]^+$, is

$$P(0) = e^{-rT}K\Phi(-d_2) - S_0\Phi(-d_1). \tag{2.26}$$

Its derivation is left as an exercise.

Taking the difference between these formulas yields the *Call-Put parity formula*:

$$
\begin{aligned}
C(0) - P(0) &= S_0[\Phi(d_1) + \Phi(-d_1)] - e^{-rT}K[\Phi(d_2) + \Phi(-d_2)] \\
&= S_0 - e^{-rT}K,
\end{aligned}
$$

since $\Phi(z) + \Phi(-z) = 1$ for every z.

2.3.2 Historical Note

There is an alternative derivation of the Black-Scholes formula from a partial differential equation that avoids mentioning Brownian motion. Rather than compute C directly, it replicates C with a portfolio consisting of S and B in ratios chosen to match the price of C at all times in all states. Denote this portfolio by

$$V(t, S) \overset{\text{def}}{=} h_0(t)B(t) + h_1(t)S(t),$$

where h_0, h_1 are hedge ratios for the two-asset portfolio. The dependence on ω is unwritten, though it is still there. If this is to be a self-financing hedge portfolio, then it must satisfy the *Black-Scholes Equation*:

$$\dot{V}(t, x) + \frac{1}{2}x^2 v^2 V''(t, x) + rxV'(t, x) - rV(t, x) = 0, \tag{2.27}$$

where $\dot{V}(t, x) = \frac{\partial}{\partial t}V(t, x)$, $V'(t, x) = \frac{\partial}{\partial x}V(t, x)$, and $V''(t, x) = \frac{\partial^2}{\partial x^2}V(t, x)$.

Deriving this equation or its solution is beyond the scope of this book, but for the terminal condition $V(T, x) = [x - K]^+$, namely for a European Call option with strike price K, it is

$$
\begin{aligned}
V(T - t, x) &= x\Phi\left(\frac{\log\left(\frac{x}{K}\right) + (r + \frac{v^2}{2})t}{v\sqrt{t}}\right) \\
&\quad -Ke^{-rt}\Phi\left(\frac{\log\left(\frac{x}{K}\right) + (r - \frac{v^2}{2})t}{v\sqrt{t}}\right),
\end{aligned}
$$

where Φ is the standard normal distribution function as before. In particular, replacing $t \leftarrow T$ and $x \leftarrow S_0$ gives the premium $C(0)$ for a European Call option on S with expiry T and strike price K:

$$
\begin{aligned}
C(0) = V(0, S_0) \;&=\; S_0 \Phi\left(\frac{\log\left(\frac{S_0}{K}\right) + (r + \frac{v^2}{2})T}{v\sqrt{T}} \right) \\
&\quad - Ke^{-rT} \Phi\left(\frac{\log\left(\frac{S_0}{K}\right) + (r - \frac{v^2}{2})T}{v\sqrt{T}} \right) \\
&=\; S_0 \Phi(d_1) - Ke^{-rT} \Phi(d_2),
\end{aligned}
$$

where $\Phi(d_1)$ and $\Phi(d_2)$ are the same risk-adjusted probabilities as before.

2.3.3   Black-Scholes Greeks

"Greeks" is a colloquial expression for partial derivatives of options premiums because they are traditionally denoted by Greek letters.

The Delta of an option

Delta (Δ) denotes the partial derivative of the option premium with respect to S_0, the spot price of the underlying asset. It is left as an exercise to compute

$$
\Delta_C \stackrel{\text{def}}{=} \frac{\partial C(0)}{\partial S_0} \;=\; \Phi(d_1); \tag{2.28}
$$

$$
\Delta_P \stackrel{\text{def}}{=} \frac{\partial P(0)}{\partial S_0} \;=\; \Phi(d_1) - 1. \tag{2.29}
$$

Note that $\Delta_C - \Delta_P = 1$, which may also be derived directly from the Call-Put Parity Formula.

The Gamma of an option

Gamma (Γ) denotes the second derivative of the option premium with respect to S_0, or equivalently the partial derivative of Δ with respect to S_0. It is again left as an exercise to show that

$$
\Gamma_C \stackrel{\text{def}}{=} \frac{\partial^2 C(0)}{\partial S_0^2} = \frac{\partial \Delta_C}{\partial S_0} \;=\; \frac{\phi(d_1)}{v S_0 \sqrt{T}}; \tag{2.30}
$$

$$
\Gamma_P \stackrel{\text{def}}{=} \frac{\partial^2 P(0)}{\partial S_0^2} = \frac{\partial \Delta_P}{\partial S_0} \;=\; \frac{\phi(d_1)}{v S_0 \sqrt{T}}; \tag{2.31}
$$

Here $\phi(z) = \Phi'(z)$ is the standard normal p.d.f. The equality $\Gamma_C = \Gamma_P$ is also evident from $\Delta_C - \Delta_P = 1$.

The Theta of an option

Theta (Θ) denotes the rate of change of the option premium with respect to expiry T, traditionally computed with a negative sign:

$$\Theta_C \overset{\text{def}}{=} -\frac{\partial C(0)}{\partial T} = -\frac{vS_0\phi(d_1)}{2\sqrt{T}} - re^{-rT}K\Phi(d_2); \qquad (2.32)$$

$$\Theta_P \overset{\text{def}}{=} -\frac{\partial P(0)}{\partial T} = -\frac{vS_0\phi(d_1)}{2\sqrt{T}} + re^{-rT}K\Phi(-d_2). \qquad (2.33)$$

Then relation

$$\Theta_C - \Theta_P = -re^{-rT}K[\Phi(d_2) + \Phi(-d_2)] = -re^{-rT}K$$

may be computed from the Call-Put Parity Formula. A less obvious relation comes from the Black-Scholes Equation:

$$\Theta_C + rS_0\Delta_C + \frac{1}{2}v^2S_0^2\Gamma_C - rC(0) = 0. \qquad (2.34)$$

It is left as an exercise to verify this and also to find the equivalent relation for Puts.

The Vega of an option

"Vega" is not a Greek letter, it is just the colloquial term for this quantity usually denoted by kappa (κ). It is the partial derivative of the option premium with respect to volatility v:

$$\kappa_C \overset{\text{def}}{=} \frac{\partial C(0)}{\partial v} = S_0\phi(d_1)\sqrt{T}; \qquad (2.35)$$

$$\kappa_P \overset{\text{def}}{=} \frac{\partial P(0)}{\partial v} = S_0\phi(d_1)\sqrt{T}. \qquad (2.36)$$

Thus $\kappa_C = \kappa_P$. Equality may also be deduced from the observation that $C(0) - P(0) - S_0 - e^{-rT}K$ is independent of volatility v.

The Rho of an option

Rho (ρ) denotes the partial derivative of the option premium with respect to riskless rate r:

$$\rho_C \overset{\text{def}}{=} \frac{\partial C(0)}{\partial r} = Te^{-rT}K\Phi(d_2); \qquad (2.37)$$

$$\rho_P \overset{\text{def}}{=} \frac{\partial P(0)}{\partial r} = -Te^{-rT}K\Phi(-d_2). \qquad (2.38)$$

Their difference may be simplified as

$$\rho_C - \rho_P = Te^{-rT}K[\Phi(d_2) + \Phi(-d_2)] = Te^{-rT}K,$$

a relationship that may also be deduced immediately from the Call-Put Parity Formula. These derivations are all left as exercises in differential Calculus.

2.4   Implementation

The Black-Scholes formulas for Calls and Puts break up nicely into stepwise computations in Octave (or MATLAB) using Eqs.2.24, 2.25, and 2.26:

```
 1  function [C0,P0] = BS(T,S0,K,r,v)
 2  % Octave/MATLAB function to evaluate
 3  % the no−dividend European−style
 4  % Call and Put option price using the
 5  % Black−Scholes formulas.
 6  % INPUTS:                     (Example)
 7  %    T  = time to expiry      (1 year)
 8  %    S0 = asset spot price    ($90)
 9  %    K  = strike price        ($95)
10  %    r  = riskless APR        (0.02)
11  %    v  = volatility          (0.15)
12  % OUTPUTS:
13  %    C0 = European−style Call premium C(0)
14  %    P0 = European style Put premium P(0)
15  % EXAMPLE:
16  %    [C,P] = BS(1,90,95,0.02,0.15)
17  %
18     normcdf = @(x) 0.5*(1.0+erf(x/sqrt(2.0)));
19     d1 = (log(S0/K)+T*(r+v^2/2)) / (v*sqrt(T));
20     d2 = (log(S0/K)+T*(r−v^2/2)) / (v*sqrt(T));
21     % Alternatively, d2=d1−v*sqrt(T);
22     C0 = S0*normcdf(d1)−K*exp(−r*T)*normcdf(d2);
23     P0 = K*exp(−r*T)*normcdf(−d2)−S0*normcdf(−d1);
24     return
25  end
```

Macsyma (or wxMaxima) performs the same calculations symbolically, so the last step to get numerical values is to substitute parameters and evaluate in floating-point arithmetic:

```
 1  /* BLACK−SCHOLES FORMULA */
 2  /* Define Phi() using built−in function erf(): */
 3  normcdf(z):= (1+erf(z/sqrt(2)))/2;
 4  /* Intermediate quantities: */
 5  d1: ( log(S0/K)+T*(r+v^2/2) ) / ( v*sqrt(T) );
 6  d2: ( log(S0/K)+T*(r−v^2/2) ) / ( v*sqrt(T) );
 7  /* Rather complicated expanded formulas: */
 8  BSCall: S0*normcdf(d1) − K*exp(−r*T)*normcdf(d2);
 9  BSPut: K*exp(−r*T)*normcdf(−d2) − S0*normcdf(−d1);
10  /* Numerical values: */
11  float(ev(BSCall,K=100, S0=100, r=0.01, T=1, v=0.15));
12  float(ev(BSPut, K=105, S0=100, r=0.01, T=1, v=0.15));
```

Additional Macsyma expressions give the Black-Scholes Greeks:

```
1  /* BLACK–SCHOLES "GREEKS" */
2  /* Define phi() using built−in function exp(): */
3  normpdf(z):= exp(−z^2/2)/sqrt(2*%pi); /* N(0,1) pdf */
4  DeltaC: normcdf(d1); DeltaP: normcdf(d1)−1;
5  GammaC: normpdf(d1)/(v*S0*sqrt(T)); GammaP: GammaC;
6  boTheta: −v*S0*normpdf(d1)/(2*sqrt(T)); /* common */
7  ThetaC: boTheta − r*exp(−r*T)*K*normcdf(d2);
8  ThetaP: boTheta + r*exp(−r*T)*K*normcdf(−d2);
9  VegaC: S0*normpdf(d1)*sqrt(T);   VegaP: VegaC;
10 RhoC: T*exp(−r*T)*K*normcdf(d2);
11 RhoP: −T*exp(−r*T)*K*normcdf(−d2);
```

2.4.1 Numerical Differentiation

To compute the Greeks for contingent claims, it often suffices to use numerical approximations. These are computed from evaluation of the price function, and thus generalize beyond the Black-Scholes formula to cases where the price is modeled numerically.

A differentiable function $f = f(x)$ of one variable has a derivative defined by

$$f'(x) \stackrel{\text{def}}{=} \lim_{h \to 0} \frac{f(x+h) - f(x)}{h}. \tag{2.39}$$

This formula involves a limit as $h \to 0$, so it may be approximated numerically with fixed small h. The difference between the limit and the approximation, which is called *discretization error*, decreases as $h \to 0$. But if h is too small and the calculations themselves are truncated to a certain number of digits, then there is also *truncation error* that grows as $h \to 0$. The total error in practice is the sum of these errors, so h must be a compromise value, not too small and not too large.

If f has continuous derivatives of all orders, and thus is well approximated by its Taylor polynomial of degree 2 or higher, then the *centered difference approximation*,

$$f'(x) = \frac{f(x+h) - f(x-h)}{2h} + O(h^2), \tag{2.40}$$

has smaller discretization error as $h \to 0$ than the formula in Eq.2.39. This is easily proved:

$$f(x+h) = f(x) + hf'(x) + \frac{h^2}{2}f''(x) + O(h^3),$$

$$f(x-h) = f(x) - hf'(x) + \frac{h^2}{2}f''(x) + O(h^3),$$

so by subtracting and dividing one gets

$$\frac{f(x+h) - f(x-h)}{2h} = f'(x) + O(h^3)/h = f'(x) + O(h^2),$$

whereas

$$\frac{f(x+h)-f(x)}{h} = f'(x) + O(h).$$

The *big-Oh* notation $O(h^k)$ means "smaller than a fixed multiple of $|h|^k$ for all sufficiently small $|h|$." For example, $h^2 = O(h)$ for all sufficiently small $|h|$. The reverse is not true: $O(h^2)$ error will decrease faster and become strictly smaller than $O(h)$ error as $h \to 0$.

If $g = g(x_1, x_2, \dots)$ is a function of two or more variables, then it is said to be differentiable if it has continuous partial derivatives

$$D_1 g \overset{\text{def}}{=} \lim_{h \to 0} \frac{g(x_1 + h, x_2, \dots) - g(x_1, x_2, \dots)}{h},$$

and likewise for D_2, D_3, and so on. It is said to be *smooth* if it has continuous higher order derivatives, and then the centered difference formula gives the more accurate approximation

$$D_1 g(x_1, x_2, \dots) = \frac{g(x_1 + h, x_2, \dots) - g(x_1 - h, x_2, \dots)}{2h} + O(h^2), \quad (2.41)$$

and likewise for D_2, D_3, and so on.

A similar argument with Taylor expansions out to $O(h^4)$ gives the centered difference formula for the second derivative:

$$f''(x) = \frac{f(x+h) - 2f(x) - f(x-h)}{h^2} + O(h^2). \quad (2.42)$$

Error estimation for $O(h^2)$ approximations

If $f(x)$ varies smoothly with x, then the error in centered difference formulas approximating f' or f'' also varies smoothly with h. This may be used to estimate the error.

Let $A(x, h)$ be an approximation, using small parameter h, to a quantity that is exactly equal to $A(x, 0) = \lim_{h \to 0} A(x, h)$. If $A(x, h) = A(x, 0) + O(h^2)$, and the dependence on h is smooth, then for small enough h the error will be very close to ch^2 for some fixed c. But then

$$
\begin{aligned}
A(x, h) &= A(x, 0) + ch^2, \\
A(x, 2h) &= A(x, 0) + c(2h)^2 = A(x, 0) + 4ch^2,
\end{aligned}
$$

so the error may itself be approximated by

$$\frac{|A(x, 2h) - A(x, h)|}{3} = |ch^2| \approx |A(x, h) - A(x, 0)|.$$

Namely, the absolute difference between two approximations, using h and $2h$, is roughly three times the error between the h approximation and the limiting (exact) value. There is no need to find c, since only the order of approximation, the exponent "2" in $O(h^2)$, plays a role.

Delta by formula versus centered differences

As an experiment, the Delta of the European Call option may be computed by centered differences with $h = 0.1$ and the result compared with the exact value given by Eq.2.28:

```
T=1; S0=100; K=100; r=0.02; v=0.10;
h=0.1; Axh=(BS(T,S0+h,K,r,v)-BS(T,S0-h,K,r,v))/(2*h)
Ax2h=(BS(T,S0+2*h,K,r,v)-BS(T,S0-2*h,K,r,v))/(2*2*h)
err = abs(Ax2h-Axh)/3
d1=(log(S0/K)+T*(r+v^2/2))/(v*sqrt(T)); Ax0=normcdf(d1)
```

The approximate value is Axh=0.59870, with an estimated err=0.000002. The exact value from the formula is Ax0=0.59871, to five decimal places. Simlarly, other first-derivative Greeks may be approximated by varying other inputs to function BS.

Gamma by formula versus centered differences

The second derivative Greek $\Gamma_C = \partial^2 C / \partial S_0^2$ may likewise be computed by centered differences with $h = 0.1$ and the result compared with the exact value given by Eq.2.30:

```
T=1; S0=100; K=100; r=0.02; v=0.10; h=0.1;
Gx=@(h)(BS(T,S0+h,K,r,v)+BS(T,S0-h,K,r,v)-2*BS(T,S0,K,r,v))/h^2;
Gxh=Gx(h),   Gx2h=Gx(2*h),   err=abs(Gx2h-Gxh)/3
d1=(log(S0/K)+T*(r+v^2/2))/(v*sqrt(T));
Gx0=normpdf(d1)/(v*S0*sqrt(T))
```

The approximate value is Gxh=0.038667, with an estimated err=0.0000003. The exact value from the formula is Gx0=0.038667, to six decimal places.

2.4.2   Interpolation

The values of a function $f(x)$ may only be known at certain values of x. Under various assumptions on f, its values elsewhere may be computed, or at least approximated, by *interpolation*. This is useful for functions such as asset prices in discrete models, or to estimate interest rates from specific published maturity times.

Polynomial interpolation

If a polynomial p of degree n or less has known values $y_0, y_1, \ldots, y_n$ at $n + 1$ distinct points $\{x_0, x_1, \ldots, x_n\}$, then its values are determined at every x. This is a generalization of the statement that "two points determine a line,"

$$p(x) = y_0 + \frac{y_1 - y_0}{x_1 - x_0}(x - x_0),$$

which is the point-slope equation of the line passing through (x_0, y_0) and (x_1, y_1). This also has a more symmetrical form:

$$p(x) = \frac{(x_1 - x)y_0 + (x - x_0)y_1}{x_1 - x_0}.$$

In both cases, the polynomial is $p(x) = p_1 x + p_0$ with p_1, p_0 determined by the interpolation points $(x_0, y_0), (x_1, y_1)$.

A degree n polynomial has the formula

$$p(x) = p_n x^n + p_{n-1} x^{n-1} + \cdots + p_1 x + p_0,$$

and the conditions $p(x_i) = y_i$ for $i = 0, 1, \ldots, n$ imply that the coefficients $\{p_n, p_{n-1}, \ldots, p_1, p_0\}$ satisfy the linear system of equations

$$\mathbf{Vp} = \mathbf{y}; \qquad \begin{pmatrix} x_0^n & x_0^{n-1} & \cdots & x_0 & 1 \\ x_1^n & x_1^{n-1} & \cdots & x_1 & 1 \\ \vdots & \vdots & \ddots & \vdots & \vdots \\ x_n^n & x_n^{n-1} & \cdots & x_n & 1 \end{pmatrix} \begin{pmatrix} p_n \\ p_{n-1} \\ \vdots \\ p_1 \\ p_0 \end{pmatrix} = \begin{pmatrix} y_0 \\ y_1 \\ \vdots \\ y_n \end{pmatrix}.$$

If the x_i's are distinct then $\mathbf{V}$, which is called the *Vandermonde* matrix of $\mathbf{x}$, is invertible. The system may be solved with a few Octave/MATLAB commands[5] such as in this example:

```
x=[0;1;2;3;4]; y=[9;-1;4;7;11]; p=vander(x)\y
% p =  0.83333  -7.83333   25.16667  -28.16667  9.00000
```

(The output p is actually a column vector, as are x,y.) But the system may also be solved with the Octave/MATLAB function polyfit():

```
deg=length(x)-1; p=polyfit(x,y,deg)  % output p is a row vector
```

The argument deg=length(x)-1 specifies the degree of the polynomial to use. The output for this method is a row vector $[p_n, p_{n-1}, \ldots, p_1, p_0]$:

```
p =   0.83333  -7.83333  25.16667  -28.16667   9.00000
```

Each of these outputs corresponds to the same polynomial

$$p(x) = 0.83333x^4 - 7.83333x^3 + 25.16667x^2 - 28.16667x + 9.00000$$

Having either row or column p, compute $p(z)$ at any single point $\mathbf{z}$ or list of points $\mathbf{z} = [z_1, z_2, \ldots, z_m]$ with the Octave/MATLAB function polyval(p,z). In particular, polyval(p,x) will return the vector y used to find p. Use this to visualize the graph of the interpolating polynomial:

```
z=0:0.1:4; yy=polyval(p,z); plot(z,yy)
```

The results may be seen as part of Figure 2.1.

---

[5]The usual notations $\mathbf{v}^T$ and A^T for the transposes of vectors and matrices are written as v' and A' in MATLAB and Octave.

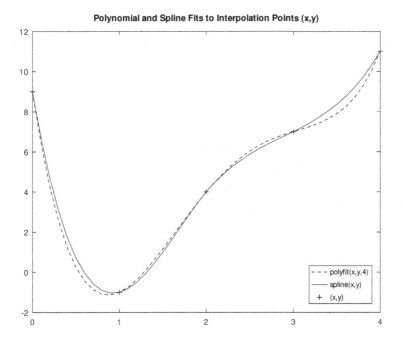

FIGURE 2.1
Comparison of two interpolation methods.

Spline interpolation

For large n, the polynomial interpolating $\{(x_0, y_0), \ldots, (x_n, y_n)\}$ may have undesirably high degree. This may be avoided by using piecewise cubic polynomials called *splines* between adjacent x-values. For $x_0 < x_1 < \cdots < x_n$, there will be a different cubic polynomial c_i for every interval $[x_{i-1}, x_i]$, $i = 1, \ldots, n$. Hence the interpolating function is

$$s(x) = c_i(x) \overset{\text{def}}{=} c_{i3}x^3 + c_{i2}x^2 + c_{i1}x + c_{i0}, \qquad \text{if } x_{i-1} \le x \le x_i,$$

and there are $4n$ coefficients $\{c_{ij} : i = 1, \ldots, n, \ j = 0, 1, 2, 3\}$ to determine.

The algorithm to compute s is well known.[6] It imposes $4n - 2$ continuity conditions plus two endpoint conditions for a total of $4n$ linear equations in the $4n$ unknowns. There are various endpoint conditions with the simplest being $s''(x_0) = 0$ and $s''(x_n) = 0$, the so-called *natural spline* conditions.

The Octave/MATLAB function that finds a cubic spline s through $\{(x_i, y_i)\}$ is pp=spline(x,y). It returns a data structure pp that includes

---

[6]See Mathews and Fink, p.284, in Further Reading below.

a matrix `coefs=(`c_{ij}`)`. A subsequent function call `ppval(pp,z)` returns the values of $s(z)$ at a point z or vector $z = [z_1, \ldots, z_k]$.

For example, the spline fitted to the above example is computed by

```
x=[0;1;2;3;4]; y=[9;-1;4;7;11]; pp=spline(x,y)
```

It may be visualized by plotting its values at many points z between the first and last x points:

```
z=0:0.1:4; s=ppval(pp,z); plot(z,s);
```

The polynomial and spline interpolation graphs may be compared by plotting them on the same axes, as seen in Figure 2.1:

```
plot(z,yy,"b--", z,s,"r-", x,y,"k+");
legend("polyfit(x,y,4)","spline(x,y)","(x,y)",...
       "location","southeast");
title("Polynomial and Spline Fits to Interpolation Points (x,y)");
```

2.4.3 Regression

The relationship between an independent variable x and another variable y may not be known, but a hypothesis like $y = f(x)$ can be tested by measurements $\{(x_i, y_i) : i = 1, \ldots, n\}$, with f adjusted to minimize the errors $|y_i - f(x_i)|$. This procedure is called *regression*, and the minimizing f is called the *best fit* to the data. Confidence in the hypothesized relationship grows if the errors remain small as the number n of measurements increases.

Linear and quadratic regression

Simple linear regression is fitting the line $f(x) = p_1 x + p_0$ to a data set $\{(x_i, y_i) : i = 1, \ldots, n\}$, where $n \geq 2$ and there are at least two distinct x values. The equation to be solved for $\mathbf{p} = (p_1, p_0)$ is

$$\mathbf{Xp} = \mathbf{y}; \qquad \begin{pmatrix} x_1 & 1 \\ x_2 & 1 \\ \vdots & \vdots \\ x_n & 1 \end{pmatrix} \begin{pmatrix} p_1 \\ p_0 \end{pmatrix} = \begin{pmatrix} y_1 \\ y_2 \\ \vdots \\ y_n \end{pmatrix}. \qquad (2.43)$$

When $n > 2$, this system is overdetermined and may not have a solution. However, if there are at least two distinct x values, then the matrix

$$\mathbf{X}^T\mathbf{X} = \begin{pmatrix} x_1 & x_2 & \cdots & x_n \\ 1 & 1 & \cdots & 1 \end{pmatrix} \begin{pmatrix} x_1 & 1 \\ x_2 & 1 \\ \vdots & \vdots \\ x_n & 1 \end{pmatrix} = \begin{pmatrix} \sum x_i^2 & \sum x_i \\ \sum x_i & \sum 1 \end{pmatrix} \qquad (2.44)$$

will be invertible, and then there will be a unique *least squares* solution to the *normal equations*

$$\mathbf{X}^T\mathbf{Xp} = \mathbf{X}^T\mathbf{y}. \qquad (2.45)$$

The system may be solved with a few Octave/MATLAB commands such as in the earlier example:

```
x=[0;1;2;3;4]; y=[9;-1;4;7;11]; X=[x ones(size(x))]; (X'*X)\(X'*y)
```

This same solution is returned by the function `polyfit()` with the degree parameter set to 1:

```
x=[0;1;2;3;4]; y=[9;-1;4;7;11]; deg=1; polyfit(x,y,deg)
```

The normal equations for the overdetermined system to fit a quadratic polynomial $f(x) = p_2 x^2 + p_1 x + p_0$ are similar:

$$\mathbf{Q}^T \mathbf{Q} \mathbf{p} = \mathbf{Q}^T \mathbf{y}, \qquad \text{for} \quad \mathbf{Q} = \begin{pmatrix} x_1^2 & x_1 & 1 \\ x_2^2 & x_2 & 1 \\ \vdots & \vdots & \vdots \\ x_n^2 & x_n & 1 \end{pmatrix}. \tag{2.46}$$

In this case, matrix $\mathbf{Q}^T \mathbf{Q}$ will be invertible if $n \geq 3$ and there are at least three distinct x values. This computation is performed by `polyfit()` with the degree parameter set to 2:

```
x=[0;1;2;3;4]; y=[9;-1;4;7;11]; deg=2; polyfit(x,y,deg)
```

The linear and quadratic regression graphs may be compared by plotting them on the same axes as in Figure 2.2:

```
x=[0;1;2;3;4]; y=[9;-1;4;7;11]; z=0:0.1:4;
poly1=polyfit(x,y,1); poly2=polyfit(x,y,2); poly4=polyfit(x,y,4);
r1=polyval(poly1,z); r2=polyval(poly2,z); r4=polyval(poly4,z);
plot(z,r1,"b--", z,r2,"r.", z,r4, "k-",x,y, "k+");
legend("polyfit(x,y,1)","polyfit(x,y,2)","polyfit(x,y,4)",...
        "location","southeast");
title("Linear and Quadratic Regression vs. Interpolation");
```

General least squares regression

More general than a polynomial is the linear combination

$$f(x) = p_1 f_1(x) + p_2 f_2(x) + \cdots + p_m f_m(x),$$

where $f_1, \ldots, f_m$ is a fixed set of functions, and $\mathbf{p} = (p_1, \ldots, p_m)$ is a vector of expansion coefficients that is chosen to minimize the squared error

$$E(\mathbf{p}) \stackrel{\text{def}}{=} \sum_{i=1}^{n} (f(x_i) - y_i)^2 = \sum_{i=1}^{n} \left(\sum_{j=1}^{m} p_j f_j(x_i) - y_i \right)^2.$$

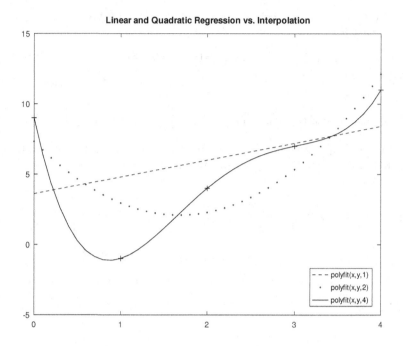

FIGURE 2.2
Comparison of regression and interpolation.

The minimum will be at the unique critical point $\nabla E(\mathbf{p}) = \mathbf{0}$, giving a system of m linear equations for the expansion coefficients $\mathbf{p}$ that is very similar to the normal Eq.2.45:

$$\mathbf{F}^T\mathbf{F}\mathbf{p} = \mathbf{F}^T\mathbf{y}, \qquad \text{for} \quad \mathbf{F} = \begin{pmatrix} f_1(x_1) & f_2(x_1) & \cdots & f_m(x_1) \\ f_1(x_2) & f_2(x_2) & \cdots & f_m(x_2) \\ \vdots & \vdots & \ddots & \vdots \\ f_1(x_n) & f_2(x_n) & \cdots & f_m(x_n) \end{pmatrix}. \qquad (2.47)$$

The matrix $\mathbf{F}^T\mathbf{F}$ is invertible whenever the rows are linearly independent vectors. This requires that $f_1, \ldots, f_m$ be linearly independent functions and that there be at least m distinct values in $\{x_1, \ldots, x_n\}$.

Special cases of this formula give linear and quadratic regression:

- Put $m = 2$ and use $f_1(x) = x$, $f_2(x) = 1$ to get $\mathbf{F} = \mathbf{X}$ as in Eq.2.43.

- Put $m = 3$ and use $f_1(x) = x^2$, $f_2(x) = x$, and $f_3(x) = 1$ to get $\mathbf{F} = \mathbf{Q}$ as in Eq.2.46.

For a more general example, put $m = 3$ and use $f_1(x) = 1$, $f_2(x) = \exp(x)$, and $f_3(x) = \cos(x)$ to fit the data from before. The least squares **p** may be found directly from Eq.2.47:

```
x=[0;1;2;3;4]; y=[9;-1;4;7;11]; % must be column vectors
f1=@(v) ones(size(v)); f2=@(v) exp(v); f3=@(v) cos(v); % functions
F=@(v) [f1(v),f2(v),f3(v)]; % regression matrix function
pF=(F(x)'*F(x))\(F(x)'*y); % least squares best fit coeffs
```

As an alternative, the Octave function `LinearRegression()` returns **p** directly from inputs $\mathbf{F}(\mathbf{x})$ and **y**. It is in the package `optim` that may be downloaded from a public repository (`https://octave.sourceforge.io/`, in this case) and installed via:

```
pkg install -forge optim % from Octave Forge
pkg load optim % contains LinearRegression()
```

(A particular Octave setup may require installing additional packages as well.) Its result agrees with the previous pF except for truncation error differences:

```
pFa=LinearRegression(F(x),y); norm(pF-pFa); % negligible diff
```

The graph of $f(z) = p_1 f_1(z) + p_2 f_2(z) + p_3 f_3(z)$ may be plotted by evaluating it at many intermediate points z. Since f is neither a polynomial nor a spline, neither `polyval()` nor `ppval()` can be used. Instead, each coordinate z_i in the column vector **z** gives a row $[f_1(z), f_2(z), f_3(z)]$ of the matrix $\mathbf{F}(\mathbf{z})$, so $\mathbf{F}(\mathbf{z})\mathbf{p}$ produces a column vector of values of f:

```
z=linspace(0,4,41)'; % must be a column vector to use with F
yF=F(z)*pF; % column vector of the best-fit function at z
plot(z,yF); % graph of the best-fit function at all z
```

Weighted least squares regression

Some of the y values may have less uncertainty than others and should be fitted more closely. This is the case, for example, in an options chain at strike prices $K_1 < \cdots < K_n$. The market-determined premium y_i at a strike price $x_i = K_i$ with large open interest is the consensus of many trades, making it less uncertain[7] than the premium at strike prices with small open interest.

To obtain the best unbiased[8] fit, the squared error to be minimized when fitting the data should have equal variance at each measurement. Thus, if the variance of measurement y_i is σ_i^2, then its contribution to the squared error should be weighted by $w_i = 1/\sigma_i^2 > 0$, which in the options chain example is just the open interest at x_i. The weighted squared error is therefore

$$E_w(\mathbf{p}) \overset{\text{def}}{=} \sum_{i=1}^{n} \frac{(f(x_i) - y_i)^2}{\sigma_i^2} = \sum_{i=1}^{n} w_i \left(\sum_{j=1}^{m} p_j f_j(x_i) - y_i \right)^2 .$$

---

[7] This conclusion is based on the *Law of Large Numbers*.
[8] The *Gauss-Markov theorem* gives a precise definition of "best" and "unbiased."

Setting $\nabla E_w(\mathbf{p}) = \mathbf{0}$ gives another system of linear equations for the unique critical point $\mathbf{p}$, generalizing the normal equations a bit further:

$$\mathbf{F}^T\mathbf{W}\mathbf{F}\mathbf{p} = \mathbf{F}^T\mathbf{W}\mathbf{y}; \qquad \mathbf{W} = \begin{pmatrix} w_1 & 0 & \cdots & 0 \\ 0 & w_2 & & \vdots \\ \vdots & & \ddots & 0 \\ 0 & \cdots & 0 & w_n \end{pmatrix}. \qquad (2.48)$$

Using the previous example with weights $\mathbf{w} = (1, 5, 8, 5, 1)$, the best unbiased fit is computed with the following Octave/MATLAB commands:

```
x=[0;1;2;3;4]; y=[9;-1;4;7;11]; z=linspace(0,4,41)'; % columns
f1=@(v) ones(size(v)); f2=@(v) exp(v); f3=@(v) cos(v); % functions
F=@(v) [f1(v),f2(v),f3(v)]; % regression matrix function
w=[1;5;8;5;1]; W=diag(w); % weights and diagonal weight matrix
pW=(F(x)'*W*F(x))\(F(x)'*W*y); % weighted least squares coeffs
```

Alternatively, the previously mentioned Octave function LinearRegression() allows a weight vector to be specified. However, it assumes that the weights are proportional to $1/\sigma_i$ rather than $1/\sigma_i^2$, so for the open interest example the values $\sqrt{w_i}$ must be used:

```
pWa=LinearRegression(F(x),y,sqrt(w)); norm(pW-pWa) % negligible
```

The special case with $\mathbf{F} = \mathbf{X}$ merits additional simplification:

$$\mathbf{X}^T\mathbf{W}\mathbf{X} = \begin{pmatrix} \sum w_i x_i^2 & \sum w_i x_i \\ \sum w_i x_i & \sum w_i \end{pmatrix}; \qquad \mathbf{X}^T\mathbf{W}\mathbf{y} = \begin{pmatrix} \sum w_i x_i y_i \\ \sum w_i y_i \end{pmatrix}.$$

Then

$$\begin{pmatrix} p_1 \\ p_2 \end{pmatrix} = \mathbf{p} = (\mathbf{X}^T\mathbf{W}\mathbf{X})^{-1}\mathbf{X}^T\mathbf{W}\mathbf{y}$$

give the slope and intercept of the best fit line $f(x) = p_1 x + p_2$:

$$\text{Slope } p_1 = \frac{\sum w_i x_i y_i - (\sum w_i x_i)(\sum w_i y_i)}{\sum w_i x_i^2 - (\sum w_i x_i)^2}$$

$$\text{Intercept } p_2 = \sum w_i y_i - m \sum w_i x_i.$$

Using the previous example with weights $\mathbf{w} = (1, 5, 8, 5, 1)$, the slope and intercept are computed in Octave/MATLAB as follows:

```
x=[0;1;2;3;4]; y=[9;-1;4;7;11]; w=[1;5;8;5;1]; W=diag(w);
X=[x ones(size(x))]; p=(X'*W*X)\(X'*W*y); slope=p(1),inter=p(2)
```

This evaluates slope= 2.4444 and inter= -0.78889.

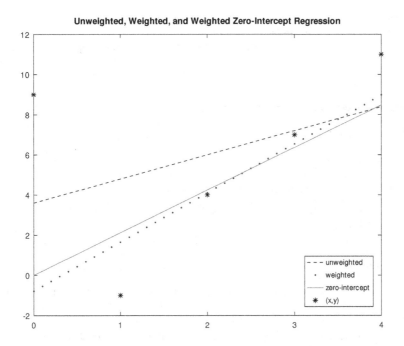

FIGURE 2.3
Comparison of unweighted, weighted, and weighted zero-intercept linear regression.

Another special case is *zero-intercept* weighted linear regression, where $m = 1$ and $f_1(x) = x$ and the fitted function is just $f(x) = p_1 x$ with

$$\text{Slope } p_1 = \frac{\sum w_i x_i y_i}{\sum w_i x_i^2}.$$

The Octave/MATLAB commands for this example are very simple:

```
x=[0;1;2;3;4]; y=[9;-1;4;7;11]; w=[1;5;8;5;1];
slope=sum(w.*x.*y)/sum(w.*x.*x)
```

which returns the somewhat different `slope= 2.1224`.

Unweighted, weighted, and weighted zero-intercept linear regression graphs may be compared by plotting them on the same axes as in Figure 2.3:

```
x=[0;1;2;3;4]; y=[9;-1;4;7;11];  w=[1;5;8;5;1]; W=diag(w);
z=0:0.1:4; punw=polyfit(x,y,1); yunw=polyval(punw,z);
X=[x ones(size(x))]; pw=(X'*W*X)\(X'*W*y); yw=polyval(pw,z);
pzi=sum(w.*x.*y)/sum(w.*x.*x); yzi=pzi*z;
plot(z,yunw,"b--", z,yw,"r.", z,yzi,"g-", x,y, "k*");
```

```
legend("unweighted","weighted","zero-intercept","(x,y)",
   "location","southeast");
title("Unweighted, Weighted, and Weighted Zero-Intercept Regression");
```

2.5   Exercises

1. Let
$$\Phi(x) = \frac{1}{\sqrt{2\pi}} \int_{-\infty}^{x} e^{-t^2/2}\, dt$$

   be the cumulative distribution function of the standard normal random variable. Prove that
   $$1 - \Phi(x) = \Phi(-x)$$

   for every x.

2. Rewrite the recursive definition of the random walk X in Section 2.2 with these normalizations:

   • Multiply by $1/\sqrt{n/4} = 2/\sqrt{n}$ to have variance 1.

   • Change the time step to $1/n$ so that the time interval is $[0,1]$.

3. Derive the Black-Scholes formula for European-style Put options, Eq.2.26:
   $$P(0) = e^{-rT} K \Phi(-d_2) - S_0 \Phi(-d_1).$$

   (Hint: follow the steps in Section 2.3.1, but use the Put payoff $[K - S(T)]^+$ at expiry.)

4. Let d_1, d_2 be defined as in Eq.2.24, and let ϕ be the standard normal p.d.f. defined in Eq.2.18. Show that
   $$S_0 \phi(d_1) - K e^{-rT} \phi(d_2) = 0.$$

   (Hint: Notice that $d_1 - d_2 = v\sqrt{T}$ and
   $$d_1 + d_2 = 2(\log \frac{S_0}{K} + rT)/(v\sqrt{T}),$$

   and thus $(d_1^2 - d_2^2)/2 = (d_1 - d_2)(d_1 + d_2)/2 = \log(S_0/K) + rT$.)

5. Derive Δ_C and Δ_P in Section 2.3.3 by differentiating the Black-Scholes formulas.

6. Derive Γ_C and Γ_P in Section 2.3.3 from Black-Scholes.

7. Derive Θ_C and Θ_P in Section 2.3.3 from Black-Scholes.

8. Derive κ_C and κ_P in Section 2.3.3 from Black-Scholes.

9. Derive ρ_C and ρ_P in Section 2.3.3 from Black-Scholes.

10. Implement the computation of all the Black-Scholes Greeks in Octave or MATLAB and add this functionality to the program BS() in Section 2.4. Apply it to compute the premiums and all Greeks for European-style Call and Put options on a risky asset with the following parameters: spot price \$90, strike price \$95, expiry in 1 year, annual riskless rate 2%, volatility 15%.

11. (a) Verify Eq.2.34 relating Θ_C, Δ_C, and Γ_C.

    (b) Find the equivalent relation for Puts and verify it.

12. (a) Find the coefficients p_1, p_2, p_3 that give the least-squares best fit

$$f(x) = p_1 + p_2 e^x + p_3 e^{-x}$$

to the data $\{(x, y)\} = \{(-2, 4), (-1, 1), (0, 0), (1, 1), (2, 4)\}$.

    (b) Plot f at 81 equispaced points on a graph showing the data.

13. At one instant, two days before expiry, near-the-money Call options for ABC common stock had the following prices:

Strike price	Premium	Open interest
44.00	3.80	3,260
45.00	2.77	4,499
46.00	1.77	3,862
47.00	0.78	6,271
48.00	0.18	10,156
49.00	0.03	10,619
50.00	0.01	14,219

The spot price for ABC is \$47.58. Estimate the premium for the at-the-money Call option in the following ways:

(a) Unweighted quadratic regression.

(b) Weighted quadratic regression.

(c) Polynomial interpolation.

(d) Spline interpolation.

2.6 Further Reading

- Catherine Forbes, Merran Evans, Nicholas Hastings, and Brian Peacock. *Statistical Distributions (4th Edition)*. John Wiley & Sons, New York (2010).

- Gregory F. Lawler. *Introduction to Stochastic Processes (2nd Edition)*. Chapman & Hall/CRC Press, Boca Raton (2006).

- John H. Mathews and Kurtis K. Fink. *Numerical Methods using MATLAB (4th Edition)*. Pearson, Boca Raton (2004).

- Lars Tyge Nielsen. "Understanding $N(d_1)$ and $N(d_2)$: Risk-Adjusted Probabilities in the Black-Scholes Model." *Revue Finance (Journal of the French Finance Association)* 14 (1993), pp.95–106.

- Mladen Victor Wickerhauser. *Mathematics for Multimedia*, Birkhäuser, Boston (2010).

3

Discrete Models

A probability space is called *discrete* if its set of possible futures can be indexed by integers:

$$\Omega = \{\omega_1, \ldots, \omega_N\}.$$

This name also applies to *countably infinite* sets Ω where N is allowed to tend to infinity. In both cases, it is possible to define a probability function[1] $\Pr(\omega)$ on individual states $\omega \in \Omega$. The requirement

$$1 = \sum_{\omega \in \Omega} \Pr(\omega)$$

makes sense for countably infinite Ω using Calculus to evaluate infinite sums.

A stochastic process is called discrete on the interval $[0, T] \subset \mathbf{R}$ if its time variable takes values in a finite set:

$$t \in \{t_i : i = 0, 1, \ldots, M\}; \qquad 0 = t_0 < t_1 < \cdots < t_M = T.$$

This name also applies to countably infinite sequences of times as are needed for $T = \infty$.

A model of future asset prices $A(t, \omega)$ is called discrete if both the time variable and state variable are discrete. It is reasonable to suppose that such *discrete models* can give realistic simulations of actual markets since spot prices are decided after finitely many actions and are posted at discrete times.

3.1 One-Step, Two-State Models

The simplest example $\Omega = \{\uparrow, \downarrow\}$ is a set of only two possible future states, denoted by the *up state* $\uparrow$ and the *down state* $\downarrow$ in order to boost intuition.

Likewise, the simplest discrete stochastic process is the *one-step model*. In it there are only two times, $t = 0$ and $t = T > 0$, and so only one time step. The state at $t = 0$ is known but the state at time $t = T$ is modeled using Ω.

For a riskless asset $B(t, \omega)$ where the riskless return over time $0 \le t \le T$ is R, the one-step two-state model assigns

$$B(T, \downarrow) = B(T, \uparrow) = RB(0).$$

---

[1] In this notation, ω means the *singleton*, or one-element set $\{\omega\}$.

For a risky asset $S(t,\omega)$, the model assigns different prices[2] at time $t = T$. For definiteness, and without loss of generality, it may be assumed that

$$S(T,\downarrow) < S(T,\uparrow).$$

This can always be arranged by relabeling the up and down states, if necessary. The *no-arbitrage* assumption imposes a relation with $S(0)$:

Theorem 3.1 $S(T,\downarrow) < RS(0) < S(T,\uparrow).$

Proof: This is left as Exercise 1, at the end of the chapter.   □

3.1.1   Risk Neutral Probabilities

Write $p = \Pr(\uparrow)$ and $1 - p = \Pr(\downarrow)$ for the probabilities of the two states in $\Omega = \{\uparrow,\downarrow\}$. Let R be the riskless return over $0 \leq t \leq T$. Then Theorem 1.4 implies

$$RS(0) = \mathrm{E}(S(T)) = pS(1,\uparrow) + (1-p)S(1,\downarrow).$$

Consider applying this model to decide whether to purchase a risky asset $S(t,\omega)$ with

- $S(0) = 100$,

- $S(T,\uparrow) = 120$,

- $S(T,\downarrow) = 90$,

- $R = 1.05$.

In the absence of any other information, the two states may be assigned equal probabilities, namely $\Pr(\uparrow) = \Pr(\downarrow) = 0.5$. Then the *expected price* of the asset at time $t = T$ is just the weighted average

$$
\begin{aligned}
\mathrm{E}(S(T)) \;&\overset{\text{def}}{=}\; \sum_{\omega \in \Omega} \Pr(\omega)S(T,\omega) \\
&=\; \Pr(\uparrow)S(T,\uparrow) + \Pr(\downarrow)S(T,\downarrow) \\
&=\; (0.5)(120) + (0.5)(90) \;=\; 105.
\end{aligned}
$$

In this model, an investment of 100 at $t = 0$ is expected, though not guaranteed, to produce a profit of 5, or 5% of the invested 100 when the asset is liquidated at $t = T$. That is a neutral outcome: the riskless return on the same capital is $RS(0) = 105$.

But another perspective is to estimate the probabilities $\Pr(\uparrow)$ and $\Pr(\downarrow)$ from the market price $S(0)$, the riskless return R over time T, and the free parameters $S(T,\uparrow)$ and $S(T,\downarrow)$. The main assumption will be Axiom 1: there

---

[2]Otherwise, there is no risk!

are no arbitrages. Its consequence is that market actions have moved $S(0)$ to the no-arbitrage price, averaging all opinions about $S(T)$. Solving for p in Theorem 1.4 then gives

$$\Pr(\uparrow) = p = \frac{RS(0) - S(T,\downarrow)}{S(T,\uparrow) - S(T,\downarrow)}; \qquad \Pr(\downarrow) = 1 - p = \frac{S(T,\uparrow) - RS(0)}{S(T,\uparrow) - S(T,\downarrow)}.$$

These are called *risk neutral probabilities*. Theorem 3.1 implies

$$S(T,\uparrow) - S(T,\downarrow) > RS(0) - S(T,\downarrow) > 0,$$

so $0 < p < 1$. Likewise

$$S(T,\uparrow) - S(T,\downarrow) > S(T,\uparrow) - RS(0) > 0,$$

so $0 < 1 - p < 1$. Clearly, $p + (1 - p) = 1$, so all the properties of a probability function on $\Omega = \{\uparrow, \downarrow\}$ are satisfied.

In the example above with $R = 1.05$, $S(0) = 100$, $S(T,\uparrow) = 120$, and $S(T,\downarrow) = 90$, these probabilities are $\Pr(\uparrow) = 0.5$ and $\Pr(\downarrow) = 0.5$.

Remark. Risk neutral probabilities are denoted by π in much of the financial mathematics literature.

3.1.2 Pricing Derivatives by Hedging

Suppose that $X(t)$ is a derivative, or contingent claim, on risky asset S, with expiry T, so that $X(T)$ is a known function of $S(T)$. One method to find the no-arbitrage *premium*, or price $X(0)$, is to *hedge*, or replicate, asset X with an equal-price portfolio containing S and a riskless asset B:

$$X(t) = h_0 B(t) + h_1 S(t).$$

The quantities h_0 and h_1 are called *hedge ratios*. In the one-step two-state model they are determined algebraically from the system of two simultaneous linear equations

$$
\begin{aligned}
X(T,\uparrow) &= h_0 B(T,\uparrow) + h_1 S(T,\uparrow) = h_0 R + h_1 S(T,\uparrow), \\
X(T,\downarrow) &= h_0 B(T,\downarrow) + h_1 S(T,\downarrow) = h_0 R + h_1 S(T,\downarrow),
\end{aligned}
$$

where there is one equation for each state at expiry T. Such a system, with two equations and two unknowns h_0, h_1, is solvable if $S(T,\uparrow) \neq S(T,\downarrow)$ and $B(T,\uparrow) = B(T,\downarrow) \overset{\text{def}}{=} R > 0$. Solving it, which is left as an exercise, gives

$$h_0 = \frac{S(T,\uparrow)X(T,\downarrow) - S(T,\downarrow)X(T,\uparrow)}{(S(T,\uparrow) - S(T,\downarrow))R}; \qquad h_1 = \frac{X(T,\uparrow) - X(T,\downarrow)}{S(T,\uparrow) - S(T,\downarrow)}. \qquad (3.1)$$

Applying Theorem 1.2, the one-price theorem, then gives

$$X(0) = h_0 B(0) + h_1 S(0), \qquad (3.2)$$

since $X(T) = h_0 B(T) + h_1 S(T)$ in all states.

The hedge ratios depend upon the model prices $S(T,\uparrow)$ and $S(T,\downarrow)$, which determine $X(T,\uparrow)$ and $X(T,\downarrow)$, and the riskless return $R = B(T)$. But it is also useful to link $X(0)$ directly to its own payoffs. Substituting the h_0, h_1 formulas into Eq.3.2 and rearranging in terms of $X(T,\uparrow)$ and $X(T,\downarrow)$ gives

$$
\begin{aligned}
X(0) &= h_0 R + h_1 S(0) \\
&= \frac{S(T,\uparrow)X(T,\downarrow) - S(T,\downarrow)X(T,\uparrow)}{(S(T,\uparrow) - S(T,\downarrow))R} R + \frac{X(T,\uparrow) - X(T,\downarrow)}{S(T,\uparrow) - S(T,\downarrow)} S(0) \\
&= \frac{S(0) - S(T,\downarrow)/R}{S(T,\uparrow) - S(T,\downarrow)} X(T,\uparrow) + \frac{S(T,\uparrow)/R - S(0)}{S(T,\uparrow) - S(T,\downarrow)} X(T,\downarrow) \\
&\stackrel{\text{def}}{=} \frac{pX(T,\uparrow) + (1-p)X(T,\downarrow)}{R},
\end{aligned}
$$

where p is evidently the risk neutral probability for the S model:

$$
p \stackrel{\text{def}}{=} \text{Pr}(\uparrow) = \frac{RS(0) - S(T,\downarrow)}{S(T,\uparrow) - S(T,\downarrow)}. \tag{3.3}
$$

Likewise,

$$
1 - p \stackrel{\text{def}}{=} \text{Pr}(\downarrow) = \frac{S(T,\uparrow) - RS(0)}{S(T,\uparrow) - S(T,\downarrow)}.
$$

Then the equation

$$
X(0) = \frac{pX(T,\uparrow) + (1-p)X(T,\downarrow)}{R}, \tag{3.4}
$$

together with Eq.3.3 for p, is called the *backward pricing formula* for derivatives. It is important to note that it depends on the price model for the underlying asset S and the riskless return R.

Example: Call-Put parity formula

Eq.3.4 applies to European-style Call and Put options on the same underlying risky asset S. It implies a relation between the premiums $C(0)$ and $P(0)$ for strike price K and spot price $S(0)$.

Let p and $1 - p$ be the risk neutral probabilities for S, and let R denote the riskless return to expiry T. Then

$$
\begin{aligned}
RC(0) &= pC(T,\uparrow) + (1-p)C(T,\downarrow) \\
&= p\left[S(T,\uparrow) - K\right]^+ + (1-p)\left[S(T,\downarrow) - K\right]^+.
\end{aligned}
$$

Similarly,

$$
\begin{aligned}
RP(0) &= pP(T,\uparrow) + (1-p)P(T,\downarrow) \\
&= p\left[K - S(T,\uparrow)\right]^+ + (1-p)\left[K - S(T,\downarrow)\right]^+.
\end{aligned}
$$

Now combine these to eliminate the explicit dependence on p:

$$
\begin{aligned}
RC(0) - RP(0) &= p\left([S(T,\uparrow) - K]^+ - [K - S(T,\uparrow)]^+\right) \\
&\quad + (1 - p)\left([S(T,\downarrow) - K]^+ - [K - S(T,\downarrow)]^+\right) \\
&= p\left(S(T,\uparrow) - K\right) + (1 - p)\left(S(T,\downarrow) - K\right) \\
&= pS(T,\uparrow) + (1 - p)S(T,\downarrow) - (p + 1 - p)K \\
&= RS(0) - K.
\end{aligned}
$$

The second equality follows from the identity in Eq.1.17 which implies that $[S - K]^+ - [K - S]^+ = S - K$. The last equality follows from Eq.3.4 applied to S itself. Divide by R to deduce the Call-Put parity formula of Eq.1.16:

$$
C(0) - P(0) = S(0) - K/R.
$$

Note that all mention of the model has vanished. This equation may be viewed as a prediction of the two-state model. Its agreement with the formula deduced from the no-arbitrage Axiom 1 is a successful test of the model.

Example: Pricing contingent premium options

Recall that a contingent premium option on a risky asset S costs nothing at $t = 0$ but includes the obligation to pay a premium $\alpha > 0$ if it is exercised or expires "in the money," namely with a positive payoff. This is a derivative which may be priced using the backward pricing formula, giving an equation that may be solved for α.

Consider first the European-style contingent premium Call option C on S with strike price K and expiry T. Its payoff is actually the profit graph depicted in Figure 1.3 on p.7, since the premium is charged at expiry. The payoff to be hedged is thus

$$
C_\alpha(T) = \begin{cases} 0, & S(T) \leq K, \\ S(T) - K - \alpha, & S(T) > K. \end{cases}
$$

Since $C_\alpha(0) = 0$, Eq.3.4 gives

$$
0 = C_\alpha(0) = \frac{pC_\alpha(T,\uparrow) + (1 - p)C_\alpha(T,\downarrow)}{R},
$$

with risk neutral probability p determined from the model for S. There are three cases to consider in the two-state model with $S(T,\uparrow) > S(T,\downarrow)$, but two of them may be excluded by the no-arbitrage axiom:

Case 1: If $K > S(T,\uparrow) > S(T,\downarrow)$, then the option expires worthless in all states yielding the uninformative identity

$$
0 = \frac{0p + 0(1 - p)}{R} = 0.
$$

But then short-selling the "vanilla" European Call option gives an arbitrage.
Case 2: If $S(T,\uparrow) > S(T,\downarrow) > K$, then

$$
\begin{aligned}
0 &= \frac{p[S(T,\uparrow) - K - \alpha] + (1-p)[S(T,\downarrow) - K - \alpha]}{R} \\
&= \frac{pS(T,\uparrow) + (1-p)S(T,\downarrow)}{R} - \frac{(p+1-p)(K+\alpha)}{R} \\
&= S(0) - (K+\alpha)/R,
\end{aligned}
$$

using Eq.3.4 on S for the last equality. Solving for α gives

$$
\alpha = RS(0) - K = E(S(T)) - K,
$$

using Theorem 1.4. The contingent premium option expires "in the money"
in all states, so the premium should just cancel the expected payoff. But in
this case, short-selling the "vanilla" European Put yields an arbitrage.

There remains one interesting, no-arbitrage case:
Case 3: If $S(T,\uparrow) > K > S(T,\downarrow)$, then

$$
0 = \frac{p[S(T,\uparrow) - K - \alpha] + (1-p)[0]}{R} = \frac{p[S(T,\uparrow) - K]^+}{R} - \frac{p\alpha}{R} = C(0) - \frac{p\alpha}{R},
$$

where $C(0)$ is the "vanilla" European Call premium. Solving for α gives

$$
\alpha = RC(0)/p.
$$

Unlike the previous example, this is a model-dependent price since p depends
on the model for $S(T)$.

Example: Unit portfolios in complete markets

The one-step, two-state model with riskless B and risky S gives unique hedge
ratios for any derivative X, precisely because its two-asset *payoff matrix* A is
nonsingular:

$$
A \;\overset{\text{def}}{=}\; \begin{pmatrix} B(T,\downarrow) & B(T,\uparrow) \\ S(T,\downarrow) & S(T,\uparrow) \end{pmatrix} = \begin{pmatrix} R & R \\ S(T,\downarrow) & S(T,\uparrow) \end{pmatrix};
$$

$$
\det A \;=\; RS(T,\uparrow) - S(T,\downarrow)R \;=\; R[S(T,\uparrow) - S(T,\downarrow)] \neq 0.
$$

The hedge ratios h_0, h_1 of derivative X solve the linear system

$$
(h_0\ h_1)A = (X(T,\downarrow)\ \ X(T,\uparrow)), \quad \Longrightarrow \quad (h_0\ h_1) = (X(T,\downarrow)\ \ X(T,\uparrow))\,A^{-1}.
$$

Its solution is given by Eq.3.1 above.

A collection of assets which can hedge any derivative is called a *complete
market*. It remains complete if more assets are added, though then the hedge
ratios may not be uniquely determined.

Since the one-step, two-state model with S, B is complete, there is a basis of elementary or unit portfolios[3] that can hedge any derivative. There is one unit portfolio for each state $\omega \in \Omega$ that pays off 1 in its state and 0 in all other states. Thus for $\Omega = \{\uparrow, \downarrow\}$ there are two of them, $W_\uparrow$ and $W_\downarrow$, with their payoffs given by the standard basis (row) vectors (1 0) and (0 1) for two-dimensional space:

$$W_\uparrow(T, \omega) = \begin{cases} 1, & \omega = \uparrow, \\ 0, & \omega = \downarrow; \end{cases} \qquad W_\downarrow(T, \omega) = \begin{cases} 0, & \omega = \uparrow, \\ 1, & \omega = \downarrow. \end{cases} \qquad (3.5)$$

The spot price for each unit portfolio may be computed by substituting W for X in Eq.3.4 and then using Eq.3.5:

$$W_\uparrow(0) = \frac{pW_\uparrow(T, \uparrow) + (1 - p)W_\uparrow(T, \downarrow)}{R} = \frac{p}{R};$$

$$W_\downarrow(0) = \frac{pW_\downarrow(T, \uparrow) + (1 - p)W_\downarrow(T, \downarrow)}{R} = \frac{1 - p}{R}.$$

Note that these are model-dependent prices, as p depends on the underlying assets and their modeled future prices.

Any derivative in a complete market may be hedged with unit portfolios, and thus priced at time $t = 0$ in terms of its payoffs at time $t = T$. The general form of this *Arrow-Debreu expansion*, for a derivative X, is

$$X(0) = \sum_{\omega \in \Omega} W_\omega(0) X(T, \omega),$$

where W_ω is the unit portfolio that pays 1 in state ω and 0 in all other states. The special case $\Omega = \{\uparrow, \downarrow\}$ of this expansion recovers Eq.3.4:

$$X(0) = W_\uparrow(0)X(T, \uparrow) + W_\downarrow(0)X(T, \downarrow) = \frac{p}{R}X(T, \uparrow) + \frac{1 - p}{R}X(T, \downarrow).$$

3.1.3   Pricing Foreign Exchange Derivatives by Hedging

Suppose now that V is a derivative whose underlying asset is a foreign exchange portfolio, whose payoff therefore depends on the future exchange rate.

Recall Eq.1.2, which defines a foreign currency portfolio containing riskless assets B_d in domestic currency DOM and B_f in foreign currency FRN, with a risky exchange rate X giving the number of units of DOM needed to purchase one unit of FRN. That portfolio may be used to hedge, or replicate, the foreign exchange derivative:

$$V(t) = h_d B_d(t) + h_f X(t) B_f(t).$$

---

[3]These are called *Arrow-Debreu securities* and will be defined more generally below.

The two-state model $\Omega = \{\uparrow, \downarrow\}$ gives two linear equations for h_d, h_f at $t = T$:

$$
\begin{aligned}
V(T,\uparrow) &= h_d B_d(T,\uparrow) + h_f X(T,\uparrow) B_f(T,\uparrow) = h_d R_d + h_f X(T,\uparrow) R_f, \\
V(T,\downarrow) &= h_d B_d(T,\downarrow) + h_f X(T,\downarrow) B_f(T,\downarrow) = h_d R_d + h_f X(T,\downarrow) R_f,
\end{aligned}
$$

where $R_d = B_d(T,\uparrow) = B_d(T,\downarrow)$ and $R_f = B_f(T,\uparrow) = B_f(T,\downarrow)$ are the riskless returns over time T in the two currencies. If $X(T,\uparrow) \neq X(T,\downarrow)$ and both R_d and R_f are nonzero, then this system has a unique solution. It may be found by hand, or computed symbolically using the following Macsyma code:

```
eq1: vTu=Hd*Rd+Hf*xTu*Rf;
eq2: vTd=Hd*Rd+Hf*xTd*Rf;
solve([eq1,eq2],[Hd,Hf]);
```

The result is

$$
h_d = \frac{V(T,\downarrow)X(T,\uparrow) - V(T,\uparrow)X(T,\downarrow)}{R_d(X(T,\uparrow) - X(T,\downarrow))}; \quad h_f = \frac{V(T,\uparrow) - V(T,\downarrow)}{R_f(X(T,\uparrow) - X(T,\downarrow))}.
$$

Applying Theorem 1.2, the one-price theorem, then gives

$$
V(0) = h_d B_d(0) + h_f X(0) B_f(0) = h_d + h_f X(0), \tag{3.6}
$$

since $V(T) = h_d B_d(T) + h_f X(T) B_f(T)$ in all states. Note that $B_d(0) = 1$ DOM and $B_f(0) = 1$ FRN, and that V is priced in DOM.

But Eq.3.6 may be rearranged to relate $V(0)$ to its payoffs. Substituting the formulas for proportions h_f and h_d yields:

$$
\begin{aligned}
V(0) &= h_d + h_f X(0) \\
&= \frac{V(T,\downarrow)X(T,\uparrow) - V(T,\uparrow)X(T,\downarrow)}{R_d(X(T,\uparrow) - X(T,\downarrow))} + \frac{(V(T,\uparrow) - V(T,\downarrow))X(0)}{R_f(X(T,\uparrow) - X(T,\downarrow))} \\
&= \frac{X(0)/R_f - X(T,\downarrow)/R_d}{X(T,\uparrow) - X(T,\downarrow)} V(T,\uparrow) + \frac{X(T,\uparrow)/R_d - X(0)/R_f}{X(T,\uparrow) - X(T,\downarrow)} V(T,\downarrow).
\end{aligned}
$$

This gives the same backward pricing formula as before:

$$
V(0) = \frac{pV(T,\uparrow) + (1-p)V(T,\downarrow)}{R_d}, \tag{3.7}
$$

except that the risk neutral probabilities in this case are

$$
p \stackrel{\text{def}}{=} \text{Pr}(\omega = \uparrow) = \frac{\frac{R_d}{R_f}X(0) - X(T,\downarrow)}{X(T,\uparrow) - X(T,\downarrow)}, \tag{3.8}
$$

and therefore

$$
1 - p \stackrel{\text{def}}{=} \text{Pr}(\omega = \downarrow) = \frac{X(T,\uparrow) - \frac{R_d}{R_f}X(0)}{X(T,\uparrow) - X(T,\downarrow)}.
$$

Example: Risk neutral probabilities for multiple currencies

Suppose now that two foreign currencies, FRX and FRY, have respective exchange rates X and Y. Let R_d, R_x and R_y be the riskless returns available for deposits in DOM, FRX, and FRY, respectively. The spot exchange rates $X(0)$ and $Y(0)$ are known at time 0, but at some future time T they are modeled by $X(T,\uparrow), X(T,\downarrow)$ and $Y(T,\uparrow), Y(T,\downarrow)$. Then there seem to be two risk neutral probabilities p_X and p_Y for the portfolios of foreign exchange assets FRX and FRY:

$$p_X = \frac{\frac{R_d}{R_x}X(0) - X(T,\downarrow)}{X(T,\uparrow) - X(T,\downarrow)}, \qquad p_Y = \frac{\frac{R_d}{R_y}Y(0) - Y(T,\downarrow)}{Y(T,\uparrow) - Y(T,\downarrow)}.$$

However, if there are no arbitrages, then these probabilities must in fact be the same:

Theorem 3.2 *In a one-step binomial model, the risk neutral probabilities for foreign exchange portfolios are independent of the currency.*

Proof: Let $W_\uparrow^X$ and $W_\uparrow^Y$ be the unit portfolios containing currencies FRX and FRY, respectively, that pay 1 DOM at time T in the $\uparrow$ state and 0 in all other states. Then the no-arbitrage backward pricing formula in the one-step, two-state model gives

$$W_\uparrow^X(0) = \frac{p_X W_\uparrow^X(T,\uparrow) + (1 - p_X)W_\uparrow^X(T,\downarrow)}{R_d} = \frac{p_X}{R_d} \text{ DOM},$$

$$W_\uparrow^Y(0) = \frac{p_Y W_\uparrow^Y(T,\uparrow) + (1 - p_Y)W_\uparrow^Y(T,\downarrow)}{R_d} = \frac{p_Y}{R_d} \text{ DOM}.$$

However, these two portfolios have the same payoff at time T in all states:

$$W_\uparrow^X(T,\uparrow) = W_\uparrow^Y(T,\uparrow) = 1 \text{ DOM}; \qquad W_\uparrow^X(T,\downarrow) = W_\uparrow^Y(T,\downarrow) = 0 \text{ DOM}.$$

Therefore, by the One Price Theorem 1.2, they must have the same price now:

$$W_\uparrow^X(0) = W_\uparrow^X(0), \qquad \Longrightarrow \quad \frac{p_X}{R_d} = \frac{p_Y}{R_d}.$$

Conclude that $p_X = p_Y$. □

3.1.4 Zero-Coupon Bonds of Different Maturity

Recall the notation used in Section 1.1.5: $Z(t,T)$ is the market price, at time $t \in [0,T]$, of a zero-coupon bond that pays 1 at maturity $t = T$. This is a risky asset with a known spot price $Z(0,T)$.

However, $Z(t,1)$ is a riskless asset in any discrete model where there are no intermediate times $0 < t < 1$. It provides a riskless return R over the short time $0 \le t \le 1$, which is related to zero-coupon bond discounts by

$$R = \frac{1}{Z(0,1)}. \tag{3.9}$$

For fixed longer maturities $T > 1$, consider a one-step binomial model for $Z(1,T)$. This assumes two possible future prices, denoted $Z(1,T,\uparrow)$ and $Z(1,T,\downarrow)$. The risk neutral probabilities of these states are

$$p = \Pr(\uparrow) = \frac{Z(0,T)R - Z(1,T,\downarrow)}{Z(1,T,\uparrow) - Z(1,T,\downarrow)},$$

$$1 - p = \Pr(\downarrow) = \frac{Z(1,T,\uparrow) - Z(0,T)R}{Z(1,T,\uparrow) - Z(1,T,\downarrow)}.$$

Note that $Z(1,T,\downarrow) < Z(0,T)R < Z(1,T,\uparrow)$ or else there is an arbitrage similar to the one constructed in the proof of Theorem 3.1.

Now let $T' > 1$ be another maturity date. Then $Z(t,T')$ is an interest rate contingent claim with a one-step binomial model no-arbitrage pricing formula

$$Z(0,T') = \frac{pZ(1,T',\uparrow) + (1-p)Z(1,T',\downarrow)}{R}.$$

The risk neutral probabilities p and $1 - p$ for all such claims are computed using $Z(1,T,\uparrow)$ and $Z(1,T,\downarrow)$. After some algebra, solve for p in terms of $Z(1,T',\uparrow)$ and $Z(1,T',\downarrow)$ to get

$$p = \frac{Z(0,T')R - Z(1,T',\downarrow)}{Z(1,T',\uparrow) - Z(1,T',\downarrow)}.$$

This is the same formula as the one for maturity T. It would be the formula used to find risk neutral probabilities for contingent claims with underlying asset $Z(t,T')$, and it gives the same result as for $Z(t,T)$. This proves:

Theorem 3.3 *In a one-step binomial model, the risk neutral probabilities for zero-coupon bonds are independent of the maturity date.*   □

Now suppose that $R(1,\uparrow)$ and $R(1,\downarrow)$ are the two possible riskless returns in the one-step binomial model of the future. They are related to zero-coupon bond discounts by Eq.3.9 translated one time step into the future:

$$R(1,\uparrow) = \frac{1}{Z(1,2,\uparrow)}, \qquad R(1,\downarrow) = \frac{1}{Z(1,2,\downarrow)}.$$

Since the risk neutral probabilities for $Z(0,T)$ are independent of T, they may be used to find the no-arbitrage price $Z(0,2)$:

$$Z(0,2) = \frac{pZ(1,2,\uparrow) + (1-p)Z(1,2,\downarrow)}{R} = \frac{1}{R}\left[\frac{p}{R(1,\uparrow)} + \frac{1-p}{R(1,\downarrow)}\right].$$

Substituting for R from Eq.3.9 gives a constraint:

Corollary 3.4 *In a one-step binomial model, the two future riskless returns $R(1,\uparrow)$ and $R(1,\downarrow)$ must satisfy*

$$\frac{p}{R(1,\uparrow)} + \frac{1-p}{R(1,\downarrow)} = \frac{Z(0,2)}{Z(0,1)},$$

where $p = \Pr(\uparrow)$ is the risk neutral probability, and $Z(0,T)$ is the spot price of a zero-coupon bond with maturity T and face value 1. □

Since the two future riskless returns have one constraint equation, they are determined by one parameter. Two such parametrizations are popular:

Ho and Lee: Fix a ratio $k \stackrel{\text{def}}{=} R(1,\uparrow)/R(1,\downarrow)$ between the future up and down returns. Then Corollary 3.4 determines $R(1,\uparrow)$ uniquely from k:

$$\frac{Z(0,2)}{Z(0,1)} = \frac{p}{R(1,\uparrow)} + \frac{1-p}{R(1,\downarrow)} = \frac{p}{R(1,\uparrow)} + \frac{(1-p)k}{R(1,\uparrow)} = \frac{[p + (1-p)k]}{R(1,\uparrow)},$$

so

$$R(1,\uparrow) = \frac{Z(0,1)}{Z(0,2)}\left[p + (1-p)k\right]; \qquad R(1,\downarrow) = R(1,\uparrow)/k.$$

Black, Derman and Toy: Write $r = \log(R)$, so $R = e^r$, and fix a ratio $\sigma \stackrel{\text{def}}{=} r(1,\uparrow)/r(1,\downarrow)$ between the future up and down rates. Then Corollary 3.4 determines $r(1,\downarrow)$ uniquely from σ:

$$\frac{Z(0,2)}{Z(0,1)} = \frac{p}{R(1,\uparrow)} + \frac{1-p}{R(1,\downarrow)} = \frac{p}{e^{\sigma r(1,\downarrow)}} + \frac{1-p}{e^{r(1,\downarrow)}} = pe^{-\sigma r(1,\downarrow)} + (1-p)e^{-r(1,\downarrow)},$$

which is rather complicated but may be solved numerically. However, using the degree-one Taylor approximation $e^{-x} \approx 1 - x$ gives a linear equation:

$$\frac{Z(0,2)}{Z(0,1)} = p(1 - \sigma r(1,\downarrow)) + (1-p)(1 - r(1,\downarrow)),$$

so

$$r(1,\downarrow) = \frac{1 - Z(0,2)/Z(0,1)}{1 + p(\sigma - 1)}, \qquad r(1,\uparrow) = \sigma r(1,\downarrow).$$

For $\sigma \approx 1$ and small values of $r(1,\downarrow)$, this is as accurate and faster to compute than the approximate numerical solution.

3.2 One-Step, Multistate Models

The simplest multistate models have $\Omega = \{\omega_1, \ldots, \omega_n\}$ while retaining the assumption that there are only two times, $t = 0$ and $t = T > 0$.

Payoffs at time T of m assets in this model may be arranged into a matrix:

$$A \stackrel{\text{def}}{=} \begin{pmatrix} R & \cdots & R \\ a_1(T,\omega_1) & \cdots & a_1(T,\omega_n) \\ \vdots & \ddots & \vdots \\ a_m(T,\omega_1) & \cdots & a_m(T,\omega_n) \end{pmatrix}.$$

Note that the top row of this matrix has the riskless asset a_0, whose payoff is R in every state.

Similarly, the asset spot prices may be arranged into a column vector:

$$\mathbf{q} \overset{\text{def}}{=} \begin{pmatrix} 1 \\ a_1(0) \\ \vdots \\ a_m(0) \end{pmatrix}.$$

Prices may be represented as matrix-vector products in this notation. A portfolio $x_0 a_0 + x_1 a_1 + \cdots + x_m a_m$ assembled from the assets $\{a_0, a_1, \ldots, a_m\}$ corresponds to a column vector

$$\mathbf{x} = \begin{pmatrix} x_0 \\ x_1 \\ \vdots \\ x_m \end{pmatrix},$$

and it has spot price

$$\mathbf{x}^T \mathbf{q} = \sum_{i=0}^{m} x_i a_i(0),$$

and payoffs

$$\mathbf{x}^T A = \left(\sum_{i=0}^{m} x_i a_i(T, \omega_1), \quad \ldots \quad , \sum_{i=0}^{m} x_i a_i(T, \omega_n) \right),$$

forming a row vector indexed by states in Ω.

The market is immediate-arbitrage-free (IA-free, using Definition 1) if every portfolio with a nonnegative payoff vector has a nonnegative spot price:

$$(\forall j) \; \mathbf{x}^T A(j) \geq 0 \quad \Longrightarrow \quad \mathbf{x}^T \mathbf{q} \geq 0.$$

It will be shown in Chapter 8 that in any arbitrage-free market, there is a nonnegative vector $\mathbf{k}$ giving the spot prices in terms of the payoffs:[4]

$$\mathbf{q} = A\mathbf{k}, \qquad (\forall j) k_j \geq 0.$$

The first equation in this system is

$$1 = A\mathbf{k}(0) = \sum_{j=1}^{n} R k_j = R \sum_{j=1}^{n} k_j,$$

---

[4]This is called the Fundamental Theorem of Asset Pricing.

so **k** may be normalized to give a risk neutral probability **p** $\stackrel{\text{def}}{=}$ R**k**, written as the column vector

$$\mathbf{p} = \begin{pmatrix} p_1 \\ \vdots \\ p_n \end{pmatrix} = \begin{pmatrix} \Pr(\omega_1) \\ \vdots \\ \Pr(\omega_n) \end{pmatrix}.$$

The market is *complete* if every payoff vector can be hedged, which requires that the row space of A is all of $\mathbf{R}^n$ and thus has the same dimension as the number n of states. This requires that the number of assets $m+1$ is at least n. In such a complete market, there exists a basis of unit portfolios $W_1, \ldots, W_n$ satisfying

$$W_i(T, \omega_j) = \begin{cases} 1, & i = j, \\ 0, & i \neq j. \end{cases}$$

Then any derivative X with payoffs $X(T, \omega_1), \ldots, X(T, \omega_n)$ has a spot price given by the Arrow-Debreu expansion

$$X(0) = \sum_{i=1}^{n} W_i(0) \, X(T, \omega_i).$$

3.3 Multistep Binomial Models

The $(N+1)$-state model $\Omega = \{\omega_0, \omega_1, \ldots, \omega_N\}$ arises naturally from the one-step, two-state model when the time interval $0 \leq t \leq T$ is divided into $N > 1$ subintervals and N independent, random, unseen changes occur. This has much in common with the random walk of Section 2.2.

The state labeled ω_k is reached at time $t_N = T$ after $k \uparrow$ steps and $N - k$ $\downarrow$ steps in any order. Thus there are

$$\binom{N}{k} \stackrel{\text{def}}{=} \frac{N!}{k!(N-k)!}$$

equivalent paths to state ω_k. The symbol $\binom{N}{k}$ is read as "N choose k" and its formula counts the number of ways to take exactly $k \uparrow$ steps in a total of N. If each $\uparrow$ step has probability p and each $\downarrow$ step has probability $1 - p$, and the steps are independent, then ω_k will have probability

$$\Pr(\omega_k) = b(k \mid N, p) \stackrel{\text{def}}{=} \binom{N}{k} p^k (1 - p)^{N-k}, \qquad k = 0, 1, \ldots, N. \quad (3.10)$$

This is the *binomial probability function* on Ω. A discrete random variable X on $\Omega = \{0, 1, \ldots, N\}$ with this probability function $\Pr(X = k) = b(k \mid N, p)$

is said to have the *binomial distribution*, on N independent trials, each with success probability p. This may be stated in the notation of Eq.1.10 as

$$X \sim \text{Binomial}(N, p).$$

If, in addition, each $\uparrow$ step increases the price of asset A by a factor $u > 1$, while each $\downarrow$ step decreases the price by a factor $d < 1$, then all future prices $A(T)$ are determined from the spot price $A(0)$:

$$A(T, \omega_k) = A(0)u^k d^{N-k},$$

The fair price formula then gives

$$RA(0) = \text{E}(A(T)) = \sum_{k=0}^{N} \text{Pr}(\omega_k)A(T, \omega_k) = A(0)\sum_{k=0}^{N} \binom{N}{k}(pu)^k([1-p]d)^{N-k},$$

after combining terms. Divide by $A(0)$ and apply the binomial theorem to get:

$$R = \sum_{k=0}^{N} \binom{N}{k}(pu)^k([1-p]d)^{N-k} = (pu + [1-p]d)^N,$$

which can only hold if $pu + [1-p]d = R^{1/N}$. Solving for p determines

$$p = \frac{R^{1/N} - d}{u - d}; \qquad 1 - p = \frac{u - R^{1/N}}{u - d}.$$

For these to be probabilities it is necessary that $0 \leq p \leq 1$, but that follows from

$$0 < d < R^{1/N} < u,$$

which itself may be deduced from the no-arbitrage Axiom 1.

Both one-step and multistep binomial models have many applications. The one-step case provides basic formulas that may be tested against the no arbitrage axiom. Multistep models may be used to price contingent claims from values at expiry. These can be compared to market prices in order to find best-fit parameters. They may also be used in simulations with varying parameters in order to estimate model risk.

Multistep binomial models converge to continuum models as the number of states increases. One of these limits is the *Black-Scholes formulas*, Eqs.2.25 and 2.26 of Chapter 2. A proof of convergence is provided below in Section 3.4.2. The limits may also be used in simulations and to fit model parameters to market data, applications that will be explored in Chapter 7.

3.3.1   Recombining Models

In a *recombining (binomial) model*, the intermediate time steps are visible. Time points $\{t_n\}$ are labeled by the discrete index n, where $0 \leq n \leq N$. It may be assumed that $0 = t_0 < t_1 < \cdots < t_N = T$. If the subintervals are of equal width, then $t_n = 0 + nh$, where $h = T/N$ is the *time step*.

The number of accessible states of the world grows linearly with the time index n. These states may be indexed by j with $0 \leq j \leq n$ at time t_n, so that $\omega(t_n) = j$ describes the world at time index $0 \leq n \leq N$ in state $0 \leq j \leq n$. The price of asset S at this time and state may be denoted $S(t_n, j)$ or, more simply, $S(n, j)$. The complete table of prices in such a model fits into the lower triangular part of an $N + 1$ by $N + 1$ matrix:

$$
\begin{array}{c}
j \quad \rightarrow \\
n \\
\downarrow
\end{array}
\begin{pmatrix}
S(0,0) & & & \\
S(1,0) & S(1,1) & & \\
\vdots & S(n,j) & \ddots & \\
S(N,0) & S(N,1) & \cdots & S(N,N)
\end{pmatrix}. \tag{3.11}
$$

There are $(N + 1)(N + 2)/2$ nonzero parameters to specify. The moderate complexity of this model allows $N \approx 10\,000$ on contemporary computers.

The riskless return $R(n, j)$ over time step $t_n \to t_{n+1}$ from state j is specified by a similar table of parameters:

$$
\begin{array}{c}
j \quad \rightarrow \\
n \\
\downarrow
\end{array}
\begin{pmatrix}
R(0,0) & & & \\
R(1,0) & R(1,1) & & \\
\vdots & R(n,j) & \ddots & \\
R(N,0) & R(N,1) & \cdots & R(N,N)
\end{pmatrix}. \tag{3.12}
$$

The no-arbitrage assumption imposes constraints on these models and thus on the tables. In particular, it implies the inequalities

$$
S(n+1, j) < R(n,j)S(n,j) < S(n+1, j+1), \tag{3.13}
$$

for all time steps $0 \leq n \leq N - 1$ and all states $0 \leq j \leq n$.

3.3.2 Generalized Backward Induction Pricing

Suppose now that risky asset prices $\{S(n,j)\}$ and riskless returns $\{R(n,j)\}$ are modeled in two recombining binomial trees of depth N. These tables of parameters determine the up factors, down factors, and risk neutral probabilities at each n, j with $0 \leq n < N$ and $0 \leq j \leq n$:

$$u(n,j) \overset{\text{def}}{=} S(n+1, j+1)/S(n,j); \tag{3.14}$$

$$d(n,j) \overset{\text{def}}{=} S(n+1, j)/S(n,j); \tag{3.15}$$

$$p(n,j) \overset{\text{def}}{=} [R(n,j) - d(n,j)]/[u(n,j) - d(n,j)]; \tag{3.16}$$

$$1 - p(n,j) \overset{\text{def}}{=} [u(n,j) - R(n,j)]/[u(n,j) - d(n,j)]. \tag{3.17}$$

Eq.3.13 guarantees that $0 < p(n,j) < 1$, so both $p(n,j)$ and $1 - p(n,j)$ are nonzero probabilities at all times in all states.

Octave code to compute risk neutral up probabilities

Eq.3.16 leads to a straightforward implementation:

```
 1  function pu = RiskNeut(S, R, N)
 2  % Octave/MATLAB function to compute risk neutral up
 3  % probabilities from recombining binomial trees S,R
 4  % of underlying asset prices and riskless returns.
 5  % INPUTS:                                          (Example)
 6  %    S = asset price matrix                        (N+1 x N+1)
 7  %    R = riskless returns matrix                   (N x N)
 8  %    N = binomial tree height, must be >1          (3)
 9  % OUTPUT:
10  %    pu = N row matrix of risk neutral up probs.
11  % EXAMPLE:
12  %   S=[100,0,0,0;87,115,0,0;75,100,133,0;65,87,115,152]
13  %   R=[1.025,0,0;1.03,1.02,0;1.035,1.025,1.01]
14  %   pu=RiskNeut(S,R,3)
15  %
16     pu = zeros(N,N); % allocate the output matrix
17     for(n=0:N−1)
18       for(j=0:n)
19         up = S(n+2,j+2)/S(n+1,j+1); % up factor
20         dn = S(n+2,j+1)/S(n+1,j+1); % down factor
21         pu(n+1,j+1) = (R(n+1,j+1)−dn)/(up−dn);
22       end
23     end
24     return
25  end
```

Note that:

- Value $S(n,j)$, for $0 \leq n \leq N$ and $0 \leq j \leq n$, is stored as S(n+1,j+1), and so on, since Octave and MATLAB do not allow 0 as an index.

- Unless $0 \leq p(n,j) \leq 1$ for all n, j, the S, R model is not arbitrage-free.

Recombining means that the price $S(n, j)$ does not depend on the path by which state j was reached. That imposes conditions on $\{u(n, j)\}$ and $\{d(n, j)\}$ such as

$$u(0, 0)d(1, 1) = \frac{S(2, 1)}{S(0, 0)} = d(0, 0)u(1, 0),$$

and similar identities relating longer sequences of up and down steps leading to the same price. One way to satisfy all these identities simultaneously is to model the future $\{S(n, j)\}$ assuming that the up factor u and down factor d are constant for all (n, j), and that

$$S(n, j) = S(0, 0)\, u^j d^{n-j}, \qquad 0 \le n \le N,\ 0 \le j \le n.$$

Thus $S(n+1, j+1) = uS(n, j)$ is the up state price and $S(n+1, j) = dS(n, j)$ is the down state price at time t_{n+1} given price $S(n, j)$ at time t_n. Then, to satisfy Eq.3.13, model the future $\{R(n, j)\}$ with a constant riskless return R satisfying $d < R < u$. Such *stationary* or *time-invariant* models have the additional advantage of not requiring storage for the S and R trees.

Backward pricing of derivatives in recombining models

Suppose that asset W is a contingent claim or other derivative of S. This relationship gives it a recombining model as well, with the same number of rows, or time steps, and the same number of columns, or states, in each row. The risk neutral probabilities for S and R apply[5] to derivative W of S as well, so they may be used to find the no-arbitrage spot price $W(0, 0)$ from known future prices by backward induction.

For example, if W is a European-style Call option C on S with strike price K and expiry T corresponding to time $n = N$, the bottom-row prices for W are determined by the parameters in the bottom row of the S tree and the Call payoff formula:

$$W(N, j) = C(N, j) \overset{\text{def}}{=} [S(N, j) - K]^+, \qquad j = 0, 1, \ldots, N.$$

Likewise, if W is a European-style Put option P, the bottom-row prices are determined by the parameters in the bottom row of the S tree and the Put payoff formula:

$$W(N, j) = P(N, j) \overset{\text{def}}{=} [K - S(N, j)]^+, \qquad j = 0, 1, \ldots, N.$$

After populating the bottom row of the W tree with these values, the risk neutral price $W(0, 0)$ is computed by backward induction:

$$W(n, j) = \frac{p(n, j)W(n+1, j+1) + [1 - p(n, j)]W(n+1, j)}{R(n, j)}, \qquad (3.18)$$

where $n = N - 1, \ldots, 2, 1, 0$ and $j = 0, 1, \ldots, n$.

Note that values from the R tree only down to depth $N - 1$ are needed.

---

[5]See the remark on complete markets just before Section 8.2.3, p.199.

3.3.3 Arrow-Debreu Securities

In a discrete model of the future with finitely many states, some computations may be simplified by using unit portfolios, assets constructed to have value 1 in a single state and 0 in all other states at some fixed future time $T > 0$. These were introduced in 1954 by Kenneth Joseph Arrow and Gérard Debreu, so they are called *Arrow-Debreu securities*. Other names for them are *pure securities*, *primitive securities*, and *state-price securities*.

There will be N such unit portfolios in the case $\Omega = \{\omega_1, \ldots, \omega_N\}$. They may be indexed as W_j, $j = 1, \ldots, N$. They have known future values in each state:

$$W_j(T, \omega) = \begin{cases} 1, & \omega = \omega_j, \\ 0, & \omega \neq \omega_j, \end{cases} \qquad j = 1, 2, \ldots, N. \qquad (3.19)$$

The price of an arbitrary portfolio W at any time t may be expanded in terms of Arrow-Debreu securities $\{W_j\}$:

$$W(t, \omega) = \sum_{j=1}^{N} W(T, \omega_j) W_j(t, \omega).$$

Notice that the Arrow-Debreu security W_j is the "vector" here while the future price $W(T, \omega_j)$ is a coefficient. Equality for all t follows from the *One Price Theorem*, 1.2 above: if two assets have the same price in all states at some future time, then they must have the same price between now and then.

In particular, the Arrow-Debreu decomposition gives a spot price for an asset from knowledge of its price in all modeled future states:

$$W(0) = \sum_{j=1}^{N} W(T, \omega_j) W_j(0). \qquad (3.20)$$

Of course, this requires computing the spot prices $\{W_j(0) : j = 1, \ldots, N\}$ for all the Arrow-Debreu securities. That is equivalent to finding the probabilities for all the states in Ω, since Fair Price Theorem 1.4 asserts

$$RW(0) = \mathrm{E}(W(T)) = \sum_{j=1}^{N} W(T, \omega_j) \mathrm{Pr}(\omega_j),$$

where R is the riskless return over time T. Thus $W_j(0) = \mathrm{Pr}(\omega_j)/R$.

Arrow-Debreu spot price by backward induction

One way to compute $W_k(0)$ for a fixed state $0 \leq k \leq N$ at expiry is by the recursive application of Eq.3.18.

Suppose that $N > 0$ and a fixed state index $0 \leq k \leq N$ are given, along with the risk neutral up probabilities $\{p(n, j) : 0 \leq n < N\}$ and riskless returns $\{R(n, j) : 0 \leq j \leq n)\}$ for a recombining binomial model with N time steps. Then the Octave code below computes $W_k(0) = \mathtt{W(1,1)}$:

```
W=zeros(N+1,N+1); W(N+1,k+1)=1;
for n=N-1:-1:0
  for j=0:n
    pup=p(n+1,j+1); pdown=1-pup;
    Wup=W(n+2,j+2); Wdown=W(n+2,j+1);
    W(n+1,j+1)=(pup*Wup+pdown*Wdown)/R(n+1,j+1);
  end
end
```

This code must be run $N+1$ times to find all the spot prices $W_0(0), \ldots, W_N(0)$ needed for Arrow-Debreu expansions.

Remark. Array indices start at 1 in Octave, so (n, j) must be implemented as $(\texttt{n+1},\texttt{j+1})$, and so on, to avoid problems if $n = 0$ or $j = 0$.

Computational complexity and profiling

Large discrete models may require large amounts of arithmetic and thus long run times. There are several methods of judging these costs:

FLOP count. One *FLOP*, or *floating-point operation*, is a loosely defined unit of computer arithmetic. It is typically a combination of operations needed for linear algebra: one multiplication and one addition. In some cases, reading and writing to internal memory are also included. FLOP counts may be estimated from program source code. A single backward induction computation of $W_k(0)$ above requires approximately

$$(N + 1)^2 + 1 + \sum_{n=0}^{N-1} \sum_{j=0}^{n} (10) \approx 5(N + 1)^2, \quad \text{for large } N,$$

total FLOPs, counting assignments, additions, multiplications, and divisions equally as one FLOP each and ignoring the index calculations. Another way to estimate this is to count the number of computed elements in triangular array W, which is $N(N + 1)/2$, and multiply that by the number of FLOPs (about 10) per output element.

Order of complexity. This is an upper bound on the FLOP count and is usually expressed in big-Oh notation. Adding two $N \times N$ matrices is said to cost $O(N^2)$ FLOPs, as $N \to \infty$, since there are N^2 sums in total. The notation means that there exists a constant C such that the actual number of FLOPs, by whatever definition, is no more than CN^2 for all sufficiently large N. By comparison, multiplying two such matrices costs $O(N^3)$ FLOPs as $N \to \infty$. The larger power of N indicates that the time required for multiplication will grow faster as N increases than the time required for addition.

Run time. *Profiling* an implementation is available in Octave and other software systems to measure the actual time used by a function. This depends on the host computer, so it is most useful for comparing programs on the same computer. It can also estimate the order of complexity if used on successive calculations with increasing N.

For example, in Octave there is a `profile` command for this purpose:

```
N=200; M1N=rand(N,N); M5N=rand(5*N,5*N); % two big matrices
profile off; profile clear; profshow % show the reset timer
profile on; M1N*M1N; profile off; profshow; profile clear
profile on; M5N*M5N; profile off; profshow; profile clear
```

This measures how much time was spent by the costliest arithmetic operators or functions while multiplying two $N \times N$ matrices. Repeating the computation with $5N \times 5N$ matrices shows the $O(N^3)$ order of complexity: the ratio of times should be approximately $125 = 5^3$.

When these tools are applied to the algorithm of individual backward induction for computing all Arrow-Debreu spot prices $\{W_k(0) : k = 0, 1, \dots, N\}$ at a fixed time step N, they show that it will cost $O(N^3)$ FLOPS.

3.3.4 Jamshidian's Forward Induction Formula

A recombining binomial tree can find the spot prices of all the Arrow-Debreu unit portfolios for all time steps and states at once.

Let $\lambda(n, j)$ be the price at time $t = 0$ of the Arrow-Debreu security that, at time step n, has price 1 in state j and price 0 in every other state $j' \neq j$. These prices may be arranged in a recombining binomial tree with N time steps and available states $0, 1, \dots, n$ at time step n.

The key observation is that the Arrow-Debreu unit portfolio that pays 1 at n, j has a spot price that may be computed by a tree with n time steps and known values in the bottom row n:

$$W(n, j') = \begin{cases} 1, & j' = j; \\ 0, & j' \neq j. \end{cases}$$

$$W(n', j') = \frac{p(n', j')W(n' + 1, j' + 1) + [1 - p(n', j')]W(n' + 1, j')}{R(n', j')},$$

for $n' = n - 1, \dots, 1, 0$ and $j' = 0, 1, \dots, n$. But since there is only one nonzero term in row n, there are only two nonzero terms in row $n - 1$, namely

$$W(n - 1, j') = \begin{cases} 0, & j' > j; \\ \dfrac{1 - p(n - 1, j)}{R(n - 1, j)}, & j' = j; \\ \dfrac{p(n - 1, j - 1)}{R(n - 1, j - 1)}, & j' = j - 1; \\ 0, & j' < j - 1. \end{cases}$$

Portfolio W decomposes into a linear combination of the two Arrow-Debreu securities that pay 1 at states $(n - 1, j)$ and $(n - 1, j - 1)$. Evaluate[6] $W(0, 0) =$

---

[6] Apply the One Price Theorem.

$\lambda(n, j)$ in terms of their spot prices $\lambda(n - 1, j)$ and $\lambda(n - 1, j - 1)$:

$$\lambda(n, j) = \frac{1 - p(n - 1, j)}{R(n - 1, j)} \lambda(n - 1, j) + \frac{p(n - 1, j - 1)}{R(n - 1, j - 1)} \lambda(n - 1, j - 1). \quad (3.21)$$

This gives a forward recursion as $n = 1, 2, \ldots, N$ with $j = 0, 1, \ldots, n$. Together with the initial condition $\lambda(0, 0) = 1$, Eq.3.21 is called *Jamshidian's forward induction formula*.

Note that $\lambda(0, 0) = 1$ follows from the definition of λ. Similarly, it follows that $\lambda(n, j) = 0$ if $j < 0$ or $j > n$ since there are no future states in which these securities have nonzero value.

The cost to compute $\Lambda \overset{\text{def}}{=} \{\lambda(n, j) : 0 \le n \le N, 0 \le j \le n\}$ by Jamshidian's induction is proportional to the number of prices that must be computed. That is because each price is found by combining just two or fewer previous prices with a fixed amount of arithmetic. But there are $(N+1)(N+2)/2$ prices in this triangular array, a polynomial in N that is dominated by $N^2/2$ as N gets large. In the "big Oh" notation, computing Λ has $O(N^2)$ complexity.

Arrow-Debreu security $\lambda(n, j)$ may be priced individually, of course, using a tree with $(n + 1)(n + 2)/2$ entries of which only the one at the top, with $(n', j') = (0, 0)$, is needed. There are $(N + 1)(N + 2)/2 > N^2/2$ trees needed to find all of Λ, and more than half of them have $N/2$ or more levels, so the average cost is at least $(N/2)^2/2 = N^2/8$. Thus, it takes at least $N^4/16$ computations, or $O(N^4)$ complexity, to use the individual tree algorithm. Jamshidian's induction computes only the final values and has far smaller $O(N^2)$ complexity.

Octave implementation of Jamshdian's induction

To build a tree of Arrow-Debreu prices to depth N, the risk neutral probabilities $p(n, j)$ and the riskless returns $R(n, j)$ must be specified to depth $N - 1$.

```
1   function L = AD(p, R, N)
2   % Octave/MATLAB function to compute Arrow-Debreu
3   % prices by Jamshidian's forward induction.
4   % INPUTS:                          (Example)
5   %    p =  NxN matrix of up probs.    (all 0.5)
6   %    R =  NxN matrix of riskless Rs  (all 1.002)
7   %    N =  binomial tree height, must be >1   (9)
8   % OUTPUTS:
9   %    L = Arrow-Debreu prices lambda(n,j).
10  % EXAMPLE:
11  %    L = AD(0.5*ones(9,9), 1.002*ones(9,9), 9)
12  %
13      L=zeros(N+1,N+1); % allocate the output matrix
14      L(1,1)=1; % special case: lambda(n=0,j=0)=1
15      L(2,1)=(1-p(1,1))/R(1,1); % (1,0) is special
16      L(2,2)=p(1,1)/R(1,1);    % (1,1) is special
```

```
17    for n=2:N  % rows n>1 need inner j loops:
18      L(n+1,1)=(1-p(n,1))*L(n,1)/R(n,1);  % (n,0)
19      L(n+1,n+1)=p(n,n)*L(n,n)/R(n,n);    % (n,n)
20      for j=1:n-1  % (n,j), 0<j<n
21        L(n+1,j+1)=( p(n,j)*L(n,j)/R(n,j)
22          + (1-p(n,j+1))*L(n,j+1)/R(n,j+1) );
23      end
24    end
25    return; % L(n+1,j+1) now holds lambda(n,j)
26  end
```

Remark. Because $\lambda(n,j) = 0$ if $j < 0$ or $j > n$, the recursion formula in Eq.3.21 has a single term for $\lambda(n,0)$ and $\lambda(n,n)$:

$$\lambda(n,0) = (1 - p(n-1,0))\lambda(n-1,0)/R(n-1,0);$$
$$\lambda(n,n) = p(n-1,n-1)\lambda(n-1,n-1)/R(n-1,n-1).$$

The code handles these two special endpoint cases before the inner loop over $j = 1, \ldots, n-1$.

Also, if $n = 1$, there are only the endpoint cases $j = 0$ and $j = 1$, with no inner j loop. Thus row $n = 1$ gets its own special formula like the initial row $n = 0$:

$$\lambda(0,0) = 1;$$
$$\lambda(1,0) = (1 - p(0,0))\lambda(0,0)/R(0,0) = (1 - p(0,0))/R(0,0);$$
$$\lambda(1,1) = p(0,0)\lambda(0,0)/R(0,0) = p(0,0)/R(0,0).$$

Arrow-Debreu prices and state probabilities

Arrow-Debreu spot prices behave like probabilities in computing prices with expansions like Eq.3.20. One way to visualize them is to plot a row, such as the bottom row, of the output matrix of AD() with the Ocave/MATLAB commands that produced Figure 3.1:

```
pkg load statistics  % contains function binopdf()
N=10; p=0.5; R=1.002; Mat=ones(N,N);
L=AD(p*Mat,R*Mat,N); binNp=binopdf(0:N,N,p);
subplot(1,2,1); plot(L(N+1,:)); xlabel("Arrow-Debreu");
subplot(1,2,2); plot(binNp); xlabel("Binomial");
```

Note that the first two inputs to AD(p*Mat,R*Mat,N) must be $N \times N$ matrices even if the parameters are constant. Also, row $\lambda(N,)$ is in L(N+1,:) of the output matrix because 0 is not a valid index in Octave or MATLAB.

Figure 3.1 suggests that for constant p and R, the Arrow-Debreu prices have a (discounted) binomial distribution. This will be proven for the CRR model below, where R and p are independent of time and state and there are explicit formulas for the Arrow-Debreu prices.

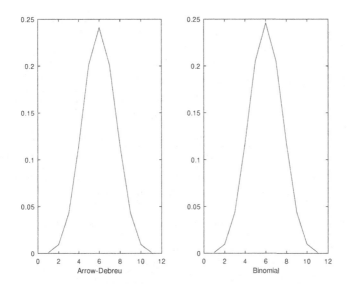

FIGURE 3.1
Arrow-Debreu prices (left) versus binomial distribution (right).

3.3.5   Zero-Coupon Bonds and Interest Rate Constraints

Suppose that $\{R(n,j) : 0 \le n < N, 0 \le j \le n\}$ is a recombining binomial tree of riskless returns. Then interest rates can be modeled using zero-coupon bond prices. This will expose some constraints on such models.

Recall the notation used in Section 1.1.5: $Z(t,T)$ is the market price, at time $t \in [0,T]$, of a zero-coupon bond that pays 1 at maturity $t = T$. Now fix $N > 0$ and let $D(n,j)$ be the discount factor, at time step n in state j, of a zero-coupon bond that expires at time step N. This may also be regarded as the price of such a bond if its face value is 1. If the maturity time is $t = T$ and the present time is $t = 0$, then

$$D(n,j) = Z(t_n, T), \qquad \text{if the state is } j \text{ at time step } n,$$

where $t_n = \frac{n}{N}T$ is the nth time step.

The terminal condition for D is evidently

$$D(N,j) = 1, \qquad j = 0, 1, \dots, N, \tag{3.22}$$

since such a bond pays its face value at N regardless of state j.

No-arbitrage discount factors $D(n,j)$ may be computed from riskless rates $R(n,j)$ and risk neutral probabilities p using the generalized backward pricing formula:

$$D(n,j) = \frac{p(n,j)D(n+1,j+1) + (1 - p(n,j))D(n+1,j)}{R(n,j)}, \qquad (3.23)$$

for $n = N - 1, \ldots, 1, 0$ and $j = 0, \ldots, n$.

Notice that at $n = N - 1$, in any state j,

$$
\begin{aligned}
D(N - 1, j) &= \frac{p(N - 1, j)D(N, j + 1) + (1 - p(N - 1, j))D(N, j)}{R(N - 1, j)} \\
&= \frac{p(N - 1, j) + (1 - p(N - 1, j))}{R(N - 1, j)} = \frac{1}{R(N - 1, j)},
\end{aligned}
$$

since $D(N, j) = D(N, j + 1) = 1$ for all j, so the backward induction formula for D does not require a row N for R, nor row N for p.

Octave computation of discount rates from riskless returns

One way to test a model for R is to compute the predicted zero-coupon bond discounts and then compare them with market prices. This may be done with an Octave/MATLAB function based on Eqs.3.22 and 3.23.

```
1   function D = ZCB(p, R, N)
2   % Octave/MATLAB function to compute Zero Coupon
3   % Bond discounts from recombining binomial trees of
4   % risk neutral probabilities and riskless returns.
5   % INPUTS:                          (Example)
6   %    p =  NxN matrix of up probs.     (all 0.5)
7   %    R =  NxN matrix of riskless Rs   (all 1.002)
8   %    N =  binomial tree height, must be >1    (9)
9   % OUTPUTS:
10  %    Z =  (N+1) rows of zero-coupon bond discounts
11  % EXAMPLE:
12  %    D = ZCB(0.5*ones(9,9), 1.002*ones(9,9), 9)
13  %
14    D = zeros(N+1,N+1); % initialize output matrix
15    for(j=0:N)           % level N
16      D(N+1,j+1) = 1;
17    end
18    for(j=0:N-1)         % level N-1
19      D(N,j+1) = 1/R(N,j+1);
20    end
21    for(n=N-2:-1:0)   % levels N-2,...,1,0
22      for(j=0:n)    % Compute D by backward induction
23        D(n+1,j+1) = ( p(n+1,j+1)*D(n+2,j+2) +
24          (1-p(n+1,j+1))*D(n+2,j+1) )/R(n+1,j+1);
25      end
26    end
27    return;  % D has rows 1:N+1
28  end
```

3.4 The Cox-Ross-Rubinstein Model

In the Cox-Ross-Rubinstein (CRR) simplification of the general multistep binomial model, u, d, R, p and $1-p$ are assumed to be constant and to depend on the more basic parameters σ, or *volatility*, and r, the *riskless interest rate*. With the time to expiry T divided into N equal steps, these formulas are:

$$u(n,j) = u \overset{\text{def}}{=} \exp(\sigma\sqrt{T/N}); \tag{3.24}$$

$$d(n,j) = d \overset{\text{def}}{=} \exp(-\sigma\sqrt{T/N}) = 1/u;$$

$$R(n,j) = R \overset{\text{def}}{=} \exp(rT/N);$$

$$p(n,j) = p \overset{\text{def}}{=} (R-d)/(u-d);$$

$$1 - p(n,j) = 1-p \overset{\text{def}}{=} (u-R)/(u-d);$$

for all n,j with $0 \le n \le N$ and $0 \le j \le n$.

Octave computation of CRR parameters

Eqs.3.24 are used so often that it pays to implement them separately.

```
1  function [pu,up,R] = CRRparams(T,r,v,N)
2  % Octave/MATLAB function to compute parameters
3  % for the Cox-Ross-Rubinstein (CRR) binomial tree.
4  % INPUTS:                              (Example)
5  %    T =   time to expiry              (1 year)
6  %    r =   riskless APR                (0.02)
7  %    v =   volatility; must be >0      (0.15)
8  %    N =   height of the binomial tree   (10)
9  % OUTPUT:
10 %    pu  =  risk-neutral up probability.
11 %    up  =  up factor
12 %    R   =  riskless return over dt
13 % EXAMPLE:
14 %    [pu,up,R] = CRRparams(1,0.02,0.15,10)
15 %
16    dt = T/N;         % one time step of N in [0,T]
17    up = exp(v*sqrt(dt)); % up factor, will be >1
18    down = 1/up;      % down factor, will be <1
19    R = exp(r*dt);    % riskless return over dt
20    pu = (R-down)/(up-down); % risk-neutral up prob.
21    return; % Parameters are all computed
22 end
```

There are constraints on the parameters σ, r, N, T imposed by the no-arbitrage axiom. In particular, the risk neutral probability p must satisfy $0 < p < 1$. This requires $d < R < p$, which implies

$$\exp(-\sigma\sqrt{T/N}) < \exp(rT/N) < \exp(\sigma\sqrt{T/N}) \implies -\sigma < r\sqrt{T/N} < \sigma.$$

Since $\sigma > 0$ in all markets with risky assets, this pair of inequalities will hold for all sufficiently large N, namely $N > r^2 T / \sigma^2$. In practice, r is smaller[7] than σ so that any $N \geq T$ is sufficiently large.

Octave computation of $S(n,j)$ using CRR assumptions

In the CRR model, the asset price $S(n,j) = S(0,0)u^j d^{n-j} = S(0,0)u^{2j-n}$ is completely determined for all n, j by the constant factor $u = 1/d$ and the spot price $S(0,0)$.

For a contingent claim whose value is determined by the price of S at expiry, namely $\{S(N,j) : 0 \leq j \leq N\}$, only the bottom row of prices is needed and it may be computed in $O(N)$ arithmetic operations. However, more exotic derivatives may need intermediate prices as well, and then it is useful to compute and store the whole tree of $O(N^2)$ prices in a matrix, using a separate function:

```
1  function S = StreeCRR(S0,up,N)
2  % Octave/MATLAB function to compute the recombining
3  % binomial tree S of underlying asset prices in the
4  % Cox-Ross-Rubinstein (CRR) model.
5  % INPUTS:                                  (Example)
6  %    S0 = asset spot price                   (100)
7  %    up = up factor, must be > 0             (1.2)
8  %    N = binomial tree height, must be >=0     (3)
9  % OUTPUT:
10 %    S =  (N+1)x(N+1) matrix S(n,j) of asset prices.
11 % EXAMPLE:
12 %    S = StreeCRR(100,1.2,3)
13 %
14    S = zeros(N+1,N+1); % allocate the output matrix
15    for n=0:N          % all time steps up to N
16      for j=0:n        % state indices at time n
17        S(n+1,j+1)=S0*up^(2*j-n);
18      end
19    end
20    return
21 end
```

3.4.1 Arrow-Debreu Decomposition in CRR

Under the CRR assumptions, it is possible to get a closed formula for Arrow-Debreu securities prices $\{\lambda(N,j) : j = 0, \ldots, N\}$ using Jamshidian's forward induction.

---

[7] Also, r is positive except in unusual negative interest rate scenarios. In those cases, the constraint is $|r| < \sigma$.

Let p be the risk neutral probability of an up transition and R the riskless return over one time subinterval. Then Eq.3.21 simplifies to

$$\lambda(n,j) = \frac{1-p}{R}\lambda(n-1,j) + \frac{p}{R}\lambda(n-1,j-1). \tag{3.25}$$

Lemma 3.5 *Eq.3.25, with the initial condition $\lambda(0,0) = 1$, has the unique solution*

$$\lambda(n,j) = \binom{n}{j}\frac{p^j(1-p)^{n-j}}{R^n}. \tag{3.26}$$

for all $n = 0, 1, 2, \ldots$ and all $j = 0, \ldots, n$.

Proof: Use induction on n.

For $n = 0$ there is only one state $j = 0$ and the initial condition guarantees equality:

$$\lambda(0,0) = 1 = \binom{0}{0}\frac{p^0(1-p)^{0-0}}{R^0}.$$

Now suppose that Eq.3.26 is true for $n-1$ and all $0 \leq j \leq n-1$. Fix $0 \leq j \leq n$ and apply Eq.3.25 to compute

$$
\begin{aligned}
\lambda(n,j) &= \frac{1-p}{R}\lambda(n-1,j) + \frac{p}{R}\lambda(n-1,j-1) \\
&= \frac{1-p}{R}\binom{n-1}{j}\frac{p^j(1-p)^{n-1-j}}{R^{n-1}} + \frac{p}{R}\binom{n-1}{j-1}\frac{p^{j-1}(1-p)^{n-j}}{R^{n-1}} \\
&= \binom{n-1}{j}\frac{p^j(1-p)^{n-j}}{R^n} + \binom{n-1}{j-1}\frac{p^j(1-p)^{n-j}}{R^n} \\
&= \left[\binom{n-1}{j} + \binom{n-1}{j-1}\right]\frac{p^j(1-p)^{n-j}}{R^n} \\
&= \binom{n}{j}\frac{p^j(1-p)^{n-j}}{R^n}
\end{aligned}
$$

as claimed. The last step follows from the identity

$$
\begin{aligned}
\binom{n-1}{j} + \binom{n-1}{j-1} &= \frac{(n-1)!}{j!(n-1-j)!} + \frac{(n-1)!}{(j-1)!(n-j)!} \\
&= \frac{(n-1)!(n-j)}{j!(n-j)!} + \frac{(j)(n-1)!}{j!(n-j)!} \\
&= \frac{(n-1)!(n-j+j)}{j!(n-j)!} = \frac{n!}{j!(n-j)!} = \binom{n}{j},
\end{aligned}
$$

which is known as *Pascal's triangle*. $\qquad\square$

Lemma 3.5 implies that, in the CRR model, the no-arbitrage prices of Arrow-Debreu securities with known values at expiry are just the binomial probabilities with parameters N and p, discounted by $e^{-rT} = R^{-N}$:

$$\lambda(N,j) = R^{-N}\binom{N}{j}p^j(1-p)^{N-j} = R^{-N}b(j\,|\,N,p), \tag{3.27}$$

where $b(j \mid N, p) \overset{\text{def}}{=} \binom{N}{j} p^j (1-p)^{N-j}$ is the binomial probability function, the probability of j successes in N independent trials, each with success probability p. Combined with the Arrow-Debreau decomposition formula of Eq.3.20, the CRR pricing of European-style options simplifies into a familiar formula:

Theorem 3.6 *In the N-step CRR model for European-style options, the no-arbitrage premiums are*

$$C(0) = \sum_{j=0}^{N} C(N,j)\lambda(N,j) = e^{-rT} \mathrm{E}(C(T)),$$

$$P(0) = \sum_{j=0}^{N} P(N,j)\lambda(N,j) = e^{-rT} \mathrm{E}(P(T)),$$

where expectation is computed with the binomial probability function.   □

Example: European-style Call and Put premiums in CRR

Theorem 3.6, combined with Lemma 3.5 and Eq.3.27, may be implemented in Octave to compute European-style option premiums with an N-step binomial model in $O(N)$ FLOPS. Using CRR model parameters, the expiry values are

$$C(N,j) = \left[S(N,j) - K\right]^{+} = \left[S_0 u^{2j-N} - K\right]^{+},$$

$$P(N,j) = \left[K - S(N,j)\right]^{+} = \left[K - S_0 u^{2j-N}\right]^{+},$$

where $S_0 = S(0,0)$ is the spot price, K is the strike price, and $u > 1$ is the up factor over one of the N time steps to expiry T.

```
1  function [C0, P0] = CRReurAD(T,S0,K,r,v,N)
2  % Octave/MATLAB function to price European Call
3  % and Put options by Arrow−Debreu expansion
4  % in the Cox−Ross−Rubinstein (CRR) model.
5  % INPUTS:                                  (Example)
6  %     T =   expiration time in years        (1)
7  %     S0 =   spot stock price               (90)
8  %     K =   strike price                    (95)
9  %     r =   riskless yield per year         (0.02)
10 %     v =   volatility; must be >0          (0.15)
11 %     N =   height of the binomial tree     (10)
12 % OUTPUTS:
13 %     C0 =   price of the Call option at t=0.
14 %     P0 =   price of the Put option at t=0.
15 % EXAMPLE:
16 %     [C0, P0]=CRReurAD(1,90,95,0.02,0.15,10);
17 %
18     [pu,up,R] = CRRparams(T,r,v,N); % Use CRR values
19     Njs=0:N;  % all j values for n=N (expiry t=T)
20     SNj=S0*up.^(2*Njs−N);  % Terminal prices S(N,j)
```

```
21    lAD=binopdf(Njs,N,pu)*exp(-r*T); % A-D prices
22    C0=max(SNj-K,0)*lAD'; % i.p. with Call payoff
23    P0=max(K-SNj,0)*lAD'; % i.p. with Put payoff
24    return; % Premiums C0 and P0 by A-D expansion
25  end
```

Remark. Since `binopdf()` is not in base Octave, it may be necessary to install and then load its package (`statistics`), which should be done securely from a trusted source like `https://octave.sourceforge.io/`, as follows:

```
pkg install -forge statistics % from Octave Forge
pkg load statistics   % contains function binopdf()
```

The sum in Theorem 3.6, which is evidently an inner product, may be implemented as matrix multiplication ($1 \times N$ times $N \times 1$) in Octave. Also, Octave functions `max()` and `binopdf()` accept matrix inputs and return matrix output of the same shape. Finally, Octave allows componentwise power operations such as the formula for `SNj`, and properly subtracts a scalar K from each component of a matrix in expressions such as `SNj-K`, eliminating the need for some loops.

Arrow-Debreu formula for option premiums in CRR

Lemma 3.5 and Theorem 3.6 lead to a formula for European-style option prices in terms of the CRR model parameters, in particular as a function of N. The limit as $N \to \infty$ then gives the continuous Black-Scholes model.

To find this formula, first consider the European-style Call option. Since

$$\left[S_0 u^{2j-N} - K\right]^+ = \begin{cases} 0, & S_0 u^{2j-N} < K, \\ S_0 u^{2j-N} - K, & S_0 u^{2j-N} \geq K, \end{cases}$$

only states $\{j : S_0 u^{2j-N} > K\}$ will contribute to the sum in Lemma 3.6. Those satisfy the inequality

$$S_0 u^{2j-N} > K \iff 2j - N > \frac{\log \frac{K}{S_0}}{\log u} \iff j > \frac{N}{2} + \frac{\log \frac{K}{S_0}}{2\sigma\sqrt{T/N}} \overset{\text{def}}{=} a,$$

since $u = \exp(\sigma\sqrt{T/N})$ in the CRR model at volatility σ. Hence the CRR model with N times steps produces

$$C(0,0) = R^{-N} \sum_{a < j \leq N} \binom{N}{j} p^j (1-p)^{N-j} \left(S_0 u^{2j-N} - K\right).$$

After some simplification, this may be rewritten in terms of the *complementary binomial distribution function*,

$$\mathcal{B}(x \mid N, p) \overset{\text{def}}{=} \sum_{x < j \leq N} \binom{N}{j} p^j (1-p)^{N-j}. \tag{3.28}$$

Begin by splitting the sum into two parts:

$$R^{-N} \sum_{a < j \leq N} \binom{N}{j} p^j (1-p)^{N-j} K = R^{-N} K \mathcal{B}(a \mid N, p),$$

and

$$R^{-N} \sum_{a < j \leq N} \binom{N}{j} p^j (1-p)^{N-j} S_0 u^{2j-N} = S_0 \sum_{a < j \leq N} \binom{N}{j} \left(\frac{pu}{R}\right)^j \left(\frac{1-p}{Ru}\right)^{N-j}$$

which may be further simplified by putting

$$q \stackrel{\text{def}}{=} \frac{pu}{R} = \left(\frac{R - \frac{1}{u}}{u - \frac{1}{u}}\right) \frac{u}{R} = \frac{u - \frac{1}{R}}{u - \frac{1}{u}}, \tag{3.29}$$

so that

$$1 - q = \frac{\left(u - \frac{1}{u}\right) - \left(u - \frac{1}{R}\right)}{u - \frac{1}{u}} = \frac{\frac{1}{R} - \frac{1}{u}}{u - \frac{1}{u}} = \frac{1}{Ru}\left(\frac{u - R}{u - \frac{1}{u}}\right) = \frac{1-p}{Ru}.$$

Thus the second term is

$$S_0 \sum_{a < j \leq N} \binom{N}{j} q^j (1-q)^{N-j} = S_0 \mathcal{B}(a \mid N, q).$$

Combining the two terms gives the formula

$$C(0,0) = S_0 \mathcal{B}(a \mid N, q) - R^{-N} K \mathcal{B}(a \mid N, p), \tag{3.30}$$

where

$$p = \frac{R - \frac{1}{u}}{u - \frac{1}{u}} = \frac{\exp(rT/N) - \exp(-\sigma\sqrt{T/N})}{\exp(\sigma\sqrt{T/N}) - \exp(-\sigma\sqrt{T/N})},$$

$$q = \frac{u - \frac{1}{R}}{u - \frac{1}{u}} = \frac{\exp(\sigma\sqrt{T/N}) - \exp(-rT/N)}{\exp(\sigma\sqrt{T/N}) - \exp(-\sigma\sqrt{T/N})},$$

$$\text{and} \quad a = \frac{N}{2} + \frac{\log \frac{K}{S_0}}{2\sigma\sqrt{T/N}}, \quad \text{as before.}$$

3.4.2   Limit of CRR as $N \to \infty$

To compute the limit[8] of the discrete CRR formula in Eq.3.30 as $N \to \infty$, use the Taylor expansions of degrees 2 and 1 about $x = 0$ for the exponential function, namely

$$\exp(x) = 1 + x + \frac{x^2}{2} + O(x^3) = 1 + x + O(x^2),$$

---

[8]Spoiler: the limit is the Black-Scholes Call formula, Eq.3.38 below.

to approximate p and q for large N, namely for small T/N, getting

$$p = \frac{[1 + \frac{rT}{N} + O\left(\left[\frac{T}{N}\right]^2\right)] - [1 - \sigma\sqrt{\frac{T}{N}} + \frac{\sigma^2}{2}\frac{T}{N} + O\left(\sqrt{\frac{T}{N}}^3\right)]}{[1 + \sigma\sqrt{\frac{T}{N}} + \frac{\sigma^2}{2}\frac{T}{N} + O\left(\sqrt{\frac{T}{N}}^3\right)] - [1 - \sigma\sqrt{\frac{T}{N}} + \frac{\sigma^2}{2}\frac{T}{N} + O\left(\sqrt{\frac{T}{N}}^3\right)]}$$

$$= \frac{\sigma\sqrt{\frac{T}{N}} + \left(r - \frac{\sigma^2}{2}\right)\frac{T}{N} + O\left(\sqrt{\frac{T}{N}}^3\right) + O\left(\left[\frac{T}{N}\right]^2\right)}{2\sigma\sqrt{\frac{T}{N}} + O\left(\sqrt{\frac{T}{N}}^3\right)}$$

$$= \frac{1}{2} + \frac{r - \frac{\sigma^2}{2}}{2\sigma}\sqrt{\frac{T}{N}} + O\left(\frac{T}{N}\right), \quad \text{as } N \to \infty, \tag{3.31}$$

and similarly,[9]

$$q = \frac{1}{2} + \frac{r + \frac{\sigma^2}{2}}{2\sigma}\sqrt{\frac{T}{N}} + O\left(\frac{T}{N}\right), \quad \text{as } N \to \infty. \tag{3.32}$$

These approximations suffice to prove that for fixed T, r, and $\sigma > 0$,

$$\lim_{N \to \infty} p = \frac{1}{2}, \qquad \lim_{N \to \infty} q = \frac{1}{2}. \tag{3.33}$$

It remains to compute

$$\lim_{N \to \infty} \mathcal{B}(a \mid N, p), \qquad \lim_{N \to \infty} \mathcal{B}(a \mid N, q).$$

This may be done with the following version of the *Central Limit Theorem*, together with the formulas relating a, p, q and N.

Theorem 3.7 (Berry-Esséen) *Suppose that* $\{X_n : n = 1, \ldots, N\}$ *is a sequence of independent, identically distributed random variables satisfying* $\mathrm{E}(X_n) = 0$, $\mathrm{E}(X_n^2) = v^2 > 0$, *and* $\mathrm{E}(|X_n|^3) = w < \infty$ *for all* $n = 1, \ldots, N$. *Let* $\mathcal{F}_N$ *be the cumulative distribution function of the normalized sum*

$$S_N \stackrel{\text{def}}{=} \frac{X_1 + \cdots + X_N}{v\sqrt{N}}; \qquad \mathcal{F}_N(x) \stackrel{\text{def}}{=} \Pr(S_N \le x).$$

Let Φ *denote the cumulative distribution function of the standard normal random variable* $\mathcal{N}(0, 1)$, *namely*

$$\Phi(x) = \Pr(\mathcal{N}(0, 1) \le x) = \frac{1}{\sqrt{2\pi}} \int_{-\infty}^{x} e^{-t^2/2} \, dt.$$

Then

$$|\mathcal{F}_N(x) - \Phi(x)| < \frac{3w}{v^3\sqrt{N}},$$

for all x.

---

[9]This is Exercise 8 at the end of this chapter.

Proof: See Feller, *An Introduction to Probability Theory and Its Applications*, Volume II, p.542, in Further Reading below.   □

To apply Theorem 3.7, suppose that $0 < s < 1$ is given and let Y be the random variable[10] taking only the values 0 and 1, with probabilities

$$\Pr(Y = 0) = 1 - s, \qquad \Pr(Y = 1) = s.$$

Let $X \stackrel{\text{def}}{=} Y - s$. This X is just the random variable Y with its mean s subtracted. Then

$$
\begin{aligned}
\mathrm{E}(X) &= (-s)(1-s) + (1-s)(s) = 0, \\
v^2 = \mathrm{E}(X^2) &= (-s)^2(1-s) + (1-s)^2(s) = s(1-s) > 0, \\
w = \mathrm{E}(|X|^3) &= |-s|^3(1-s) + |1-s|^3(s) = s(1-s)(s^2 + [1-s]^2) < \infty.
\end{aligned}
$$

Now let $X_1, \ldots, X_N$ be independent copies of X, corresponding to independent copies $Y_1, \ldots, Y_N$ of Y, and write the normalized sum

$$S_N = \frac{X_1 + \cdots + X_N}{v\sqrt{N}} = \frac{Y_1 + \cdots + Y_N - Ns}{v\sqrt{N}} = \frac{1}{v\sqrt{N}}(Z_{N,s} - Ns),$$

where $Z_{N,s} \sim \text{Binomial}(N, s)$ is the binomial random variable counting the number of successes in N independent trials, each with success probability s. The cumulative distribution function $\mathcal{F}_N$ of S_N may be written in terms of the complementary binomial distribution function $\mathcal{B}$ as follows:

$$
\begin{aligned}
\mathcal{F}_N(x) &= \Pr(S_N \leq x) \\
&= \Pr\left(\frac{1}{v\sqrt{N}}(Z_{N,s} - Ns) \leq x\right) \\
&= \Pr\left(Z_{N,s} \leq xv\sqrt{N} + Ns\right) \\
&= 1 - \Pr\left(Z_{N,s} > xv\sqrt{N} + Ns\right) \\
&= 1 - \mathcal{B}\left(xv\sqrt{N} + Ns \mid N, s\right).
\end{aligned}
$$

Now use Theorem 3.7 to estimate $\mathcal{B}$:

$$
\begin{aligned}
\left|1 - \mathcal{B}\left(xv\sqrt{N} + Ns \mid N, s\right) - \Phi(x)\right| &= |\mathcal{F}_N(x) - \Phi(x)| \\
&< \frac{w}{v^3\sqrt{N}} \\
&= \frac{s(1-s)(s^2 + [1-s]^2)}{s^{3/2}(1-s)^{3/2}\sqrt{N}} \\
&= \frac{s^2 + [1-s]^2}{\sqrt{s(1-s)N}}
\end{aligned}
$$

---

[10]Such Y is called a *Bernoulli random variable*.

This inequality holds for all x, so substitute $y \stackrel{\text{def}}{=} xv\sqrt{N} + Ns$, which satisfies

$$x = \frac{y - Ns}{v\sqrt{N}} = \frac{y - Ns}{\sqrt{s(1-s)N}},$$

and thereby gives

$$\left| 1 - \mathcal{B}(y \mid N, s) - \Phi\left(\frac{y - Ns}{\sqrt{s(1-s)N}}\right) \right| < \frac{s^2 + [1 - s]^2}{\sqrt{s(1-s)N}} \tag{3.34}$$

The inequality simplifies since $1 - \Phi(z) = \Phi(-z)$ for every z. Thus

$$\left| \mathcal{B}(y \mid N, s) - \Phi\left(\frac{Ns - y}{\sqrt{s(1-s)N}}\right) \right| < \frac{s^2 + [1 - s]^2}{\sqrt{s(1-s)N}} \tag{3.35}$$

To use these estimates in the CRR model, put

$$y = a = \frac{N}{2} + \frac{\log \frac{K}{S_0}}{2\sigma\sqrt{T/N}}$$

and start with $s = p$ in Eq.3.35 to get:

$$
\begin{aligned}
\frac{Ns - y}{\sqrt{s(1-s)N}} &= \frac{Np - a}{\sqrt{p(1-p)N}} \\[2mm]
&= \frac{Np - \frac{N}{2} - \frac{\log \frac{K}{S_0}}{2\sigma\sqrt{T/N}}}{\sqrt{p(1-p)N}} \\[2mm]
&= \frac{N\left(\frac{1}{2} + \frac{r - \frac{\sigma^2}{2}}{2\sigma}\sqrt{\frac{T}{N}} + O\left(\frac{T}{N}\right)\right) - \frac{N}{2} - \frac{\log \frac{K}{S_0}}{2\sigma\sqrt{T/N}}}{\sqrt{p(1-p)N}} \\[2mm]
&= \frac{\frac{r - \frac{\sigma^2}{2}}{2\sigma}\sqrt{T} - \frac{\log \frac{K}{S_0}}{2\sigma\sqrt{T}}}{\sqrt{p(1-p)}} + O\left(\frac{1}{\sqrt{N}}\right), \quad \text{as } N \to \infty.
\end{aligned}
$$

Now use the fact that $\lim_{N \to \infty} p = \frac{1}{2}$ to compute

$$\lim_{N \to \infty} \frac{Np - a}{\sqrt{p(1-p)N}} = \frac{r - \frac{\sigma^2}{2}}{\sigma}\sqrt{T} - \frac{\log \frac{K}{S_0}}{\sigma\sqrt{T}} \stackrel{\text{def}}{=} z_2. \tag{3.36}$$

By a similar argument using the fact that $\lim_{N \to \infty} q = \frac{1}{2}$ as well, compute

$$\lim_{N \to \infty} \frac{Nq - a}{\sqrt{q(1-q)N}} = \frac{r + \frac{\sigma^2}{2}}{\sigma}\sqrt{T} - \frac{\log \frac{K}{S_0}}{\sigma\sqrt{T}} \stackrel{\text{def}}{=} z_1. \tag{3.37}$$

Finally, for both $s = p$ and $s = q$, the right-hand side of Eq.3.35 tends to zero as $N \to \infty$, so conclude that

$$\lim_{N\to\infty} \mathcal{B}(a \mid N, q) = \Phi(z_1),$$
$$\lim_{N\to\infty} \mathcal{B}(a \mid N, p) = \Phi(z_2),$$
$$\lim_{N\to\infty} R^{-N} = e^{-rT}.$$

Thus, taking the limit as $N \to \infty$ in Eq.3.30 and using Eqs.3.36 and 3.37 gives the *Black-Scholes formula* for pricing European-style Call options:

$$C(0) = S_0 \Phi(z_1) - e^{-rT} K \Phi(z_2) \qquad (3.38)$$

3.4.3   CRR Greeks

Though the discrete Cox-Ross-Rubinstein option price modeled with N time steps converges to the Black-Scholes price as $N \to \infty$, the same is not true for Deltas or Gammas, the derivatives with respect to S. The problem is that neither $[S - K]^+$ nor $[K - S]^+$ is differentiable with respect to S at $S = K$. They are in fact piecewise linear functions of S with slope jumping from 0 to 1 (or -1 to 0) at K. This is evident in Figure 1.2 on p.6.

Replacing S by its bottom level CRR price $S_0 u^{2j-N}$ in state j shows that neither $\left[S_0 u^{2j-N} - K\right]^+$ nor $\left[K - S_0 u^{2j-N}\right]^+$ is differentiable with respect to S_0 at K/u^{2j-N}. Because the CRR option premium is a linear combination of those payoff functions, its graph[11] (as a function of S_0) will be piecewise linear with joints at K/u^{2j-N}. As $N \to \infty$, the joints will get closer together and the graph will appear smoother, but for any finite N there will be jumps in Delta at all those joints. Those jumps are small for large N, but they invalidate the centered difference formula for Delta in Eq.2.40. In particular, it is not true that the error is $O(h^2)$ as $h \to 0$.

Gammas computed from CRR models with second derivative centered differences, as in Eq.2.42, have even worse behavior. Because the second derivatives of the piecewise linear CRR approximation can be infinite at joints, they may have no relation to the Black-Scholes values. In this case, an alternative numerical differentiation method is needed. A quadratic polynomial $q(s)$ is interpolated through the option values at three prices near $s = S_0$. Then $q'(S_0)$ approximates Delta, while the second derivative $q''(S_0)$ of the quadratic polynomial approximates Gamma.

It is necessary to use sampled prices that are on different segments of the piecewise linear graph. Otherwise, the interpolation points could be collinear. The critical spacing is the distance between the joints nearest $S_0 \approx K/u^{2j-N}$, namely

$$\left(\frac{K}{u^{2(j+1)-N}}, \frac{K}{u^{2j-N}}, \frac{K}{u^{2(j-1)-N}} \right) \approx \left(d^2 S_0, S_0, u^2 S_0 \right), \qquad (3.39)$$

---

[11]It is left as an exercise to plot `CRReurAD()` with small N, in terms of S_0, to see this.

where $d = 1/u$. Note that the critical spacing is approximately

$$h_0 \stackrel{\text{def}}{=} \frac{K}{u^{2(j-1)-N}} - \frac{K}{u^{2j-N}} \approx S_0(u^2 - 1) \approx 2S_0 v\sqrt{T/N}, \qquad (3.40)$$

since $u^2 = e^{2v\sqrt{T/N}}$ and $2v\sqrt{T/N}$ is small when N is large. The three points in Eq.3.39 should be used in the quadratic interpolation method, described on p.37, as implemented by the following Octave code:

```
T=1; S0=100; K=100; r=0.02; v=0.10; N=100;
h0=2*S0*v*sqrt(T/N);  % critical h
u2=exp(2*v*sqrt(T/N)); % squared up factor
x=[S0/u2, S0, S0*u2]-S0; % shifted abscissas
C0=CRReurAD(T,S0,K,r,v,N); % C(0) at S0
Cu=CRReurAD(T,S0*u2,K,r,v,N); % ...at S0*u^2
Cd=CRReur(T,S0/u2,K,r,v,N); % ...at S0/u^2
y=[Cd, C0, Cu]; % ordinates for interpolation
p=polyfit(x,y,2); DeltaC=p(2), GammaC=2*p(1)
```

Output vector p=[a b c] is the coefficients of the shifted quadratic interpolating polynomial

$$q(S_0 + x) = ax^2 + bx + c,$$

so $\Delta_C \approx q'(S_0) = b = $ p(2) and $\Gamma_C \approx q''(S_0) = 2a = 2*$p(1). The approximate values DeltaC=0.59809 and GammaC=0.038568 agree to more than two decimal places with the Black-Scholes Greeks computed on p.37.

Remark. Coefficient b from quadratic interpolation is given by the *divided differences* formula[12]

$$b = \frac{1}{x_u - x_d} \left(\frac{f(x_u) - f(x_0)}{x_u - x_0} - \frac{f(x_0) - f(x_d)}{x_0 - x_d} \right),$$

where $x_0 = S_0$, $x_d = d^2 S_0$, $x_u = u^2 S_0$, and $f(x)$ is the option premium at spot price x. This is the formula derived in the paper by Pelsser and Vorst in Further Reading below. Note that the denominators are unchanged by the shift $x \leftarrow x - S_0$.

The Black-Scholes Call premium may be computed by Octave and compared with the $N = 100$ CRR approximation as follows:

```
C=BS(T,S0,K,r,v); C, C0  % C(0) from BS vs. CRR
```

This produces C=5.017 and C0=5.0069, which also have a little more than two decimal places of agreement. The accuracy of the quadratic interpolation approximation to Gamma from CRR will be roughly the difference between the CRR approximation and its Black-Scholes limit. This difference is $O(1/\sqrt{N})$, by inequalities in the proof of the Berry-Esséen Theorem 3.7.

---

[12]See Wickerhauser, p.110, cited in Further Reading after Chapter 2.

Theorem 3.8 *The derivatives* DeltaC *and* GammaC, *computed as above from CRR values with N time steps using quadratic interpolation through the points of Eq.3.39, converge to the Black-Scholes values in Eqs.2.28 and 2.30, respectively, as $N \to \infty$. The same is true for the corresponding Put formulas.* □

The other Greeks, Theta, Rho, and Vega/Kappa, come from differentiation with respect to time T, riskless rate r, and volatility v, respectively. These parameters all contribute smooth, many-times differentiable effects on the option price. For them, the centered difference formula in Eq.2.41 has $O(h^2)$ error and agrees closely with the corresponding Black-Scholes Greeks for small h and large N.

Theorem 3.9 *The derivatives Θ_C, κ_C, and ρ_C, computed by the centered difference formula in Eq.2.41 above from CRR values with N time steps and $h = 1/N$, converge to the Black-Scholes values in Eqs.2.32, 2.35, and 2.37, respectively, using $h = 1/N$, as $N \to \infty$. The same result holds for the corresponding Put formulas.* □

The proofs of Theorem 3.8 and Theorem 3.9, along with estimates of the rate of convergence, are left as an exercise.

3.5 Exercises

1. Suppose that $S(t, \omega)$, $0 \le t \le T$ is the price of a risky asset S, and that the riskless return over time $[0, T]$ is R. Model the future at time $t = T$ using $\Omega = \{\uparrow, \downarrow\}$ and assume that $S(T, \downarrow) < S(T, \uparrow)$.

   (a) Use the no-arbitrage Axiom 1 to conclude that

   $$S(T, \downarrow) < RS(0) < S(T, \uparrow).$$

   (b) Use the Fair Price Theorem 1.4 to prove the same inequalities.

2. In Exercise 1 above, model the future at time $t = T$ using the N-step binomial model $\Omega = \{\omega_0, \omega_1, \ldots, \omega_N\}$ and assume that

   $$S(T, \omega_k) = S(0)u^k d^{N-k},$$

   where $S(0) > 0$ is the spot price and $0 < d < u$ are the up factor and down factor, respectively, over one time step T/N.

   (a) Use the no-arbitrage Axiom 1 to conclude that

   $$d < R^{1/N} < u.$$

   (b) Use the Fair Price Theorem 1.4 to prove the same inequalities.

3. Suppose that a portfolio X contains risky stock S and riskless bond B in amounts h_0, h_1:

$$X(t, \omega) = h_0 B(t, \omega) + h_1 S(t, \omega).$$

Model the future at time $t = T$ using $\Omega = \{\uparrow, \downarrow\}$, assuming only that $S(T, \uparrow) \neq S(T, \downarrow)$ and that $B(T, \uparrow) = B(T, \downarrow) = R$. Compute h_0 and h_1 in terms of all the other quantities. (Hint: use Macsyma to derive Eq.3.1.)

4. In Exercise 3 above, suppose that X is a European-style Call option for S with expiry T and strike price K. Use the payoff formula $X(T) = [S(T) - K]^+$ in the equation for h_1 to prove that

$$0 \leq h_1 \leq 1.$$

Conclude that, in this model of the future, a European-style Call option for S is equivalent to a portfolio containing part of a share of S plus or minus some cash.

5. In Exercise 3 above, suppose that X is a European-style Put option for S with expiry T and strike price K. Use the payoff formula $X(T) = [K - S(T)]^+$ in the equation for h_1 to prove that

$$-1 \leq h_1 \leq 0.$$

Conclude that, in this model of the future, a European-style Put option for S is equivalent to a portfolio containing part of a share of S sold short plus or minus some cash.

6. Suppose that $C(0)$ and $P(0)$ are the premiums for European-style Call and Put options, respectively, on an asset S with the following parameters: expiry at $T = 1$ year, spot price $S(0) = 90$, strike price $K = 95$. Assume that the riskless annual percentage rate is $r = 0.02$, and the volatility for S is $\sigma = 0.15$, and that these will remain constant from now until expiry.

(a) Use a LibreOffice Calc spreadsheet to implement the Cox-Ross-Rubinstein (CRR) model to compute $C(0)$ and $P(0)$ with $N = 10$ time steps, using the backward pricing formula in Eq.3.18. (Hint: compare output with CRReurAD() to check for bugs.)

(b) Use the Octave function CRReurAD() with $N = 10$, $N = 100$, and $N = 1000$ time steps to compute $C(0)$ and $P(0)$.

(c) Repeat part (b) with the Octave function CRReur() on p.88, again using $N = 10$, $N = 100$, and $N = 1000$ time steps to compute $C(0)$ and $P(0)$. Profile the time required to compute them, and compare the time and the output with that of CRReurAD().

(d) Compare the prices from parts (b) and (c). Is it justified to use $N = 1000$? Is $N = 10$ sufficiently accurate?

7. Compare the prices from parts (a) and (b) of previous Exercise 6 with the Black-Scholes prices computed using Eqs.2.25 and 2.26. Plot the logarithm of the differences against $\log N$ to estimate the rate of convergence. (Hint: Use the programs in Chapter 2, Section 2.4.)

8. Derive Eq.3.32 on p.79:

$$ q = \frac{1}{2} + \frac{r + \frac{\sigma^2}{2}}{2\sigma} \sqrt{\frac{T}{N}} + O\left(\frac{T}{N}\right). $$

9. Use the CRR approximation with $N = 4$ to compute the European-style Call option premiums at several hundred equally spaced spot prices $75 \leq S_0 \leq 115$, with expiry $T = 1$, strike $K = 95$, $r = 0.02$, and $\sigma = 0.15$.

   (a) Plot the values against S_0.

   (b) At what values of S_0 in that range does the graph appear to be nonsmooth?

   (c) Compute the points of nondifferentiability for S_0 in $[75, 115]$.

10. Compute the CRR option premiums and Greeks for European-style Call and Put options on a risky asset with the following parameters: spot price \$90, strike price \$95, expiry in 1 year, annual riskless rate 2%, and volatility 15%. Use $N = 100$ steps. Justify the method used.

3.6   Further Reading

- Fischer Black, Emanuel Derman, and William Toy. "A One Factor Model of Interest Rates and its Application to Treasury Bond Options." *Financial Analysts Journal* 46 (1990), pp.33–39.

- John C. Cox, Stephen A. Ross, and Mark Rubinstein. "Option Pricing: A Simplified Approach." *Journal of Financial Economics* 7:3 (1979), pp.229–263.

- William Feller. *An Introduction to Probability Theory and Its Applications, Volume II (2nd Edition)*. John Wiley & Sons, Inc., New York (1971).

- Thomas S. Y. Ho and Sang-Bin Lee. "Term Structure Movements and Pricing Interest Rate Contingent Claims." *Journal of Finance* 41 (1986), pp.1011–1029.

- Antoon Pelsser and Ton Vorst. "The Binomial Model and the Greeks." *Journal of Derivatives* 1 (1994), pp.45–49.

4

Exotic Options

Contingent claims other than European-style Call and Put options are sometimes called *exotic*, like flavors of ice cream in contrast to *vanilla*. They are constructed by financial institutions to appeal to investors with specific goals, but in such a way that they can be hedged. In this chapter, the CRR model will be used to price various exotic options that are currently marketed.

4.1   Recombining Binomial Tree Prices

Unlike the Black-Scholes formula for European-style option prices, CRR does not provide a simple exact formula that can be evaluated in a fixed number $O(1)$ of FLOPs.

The fastest discrete algorithm is an Arrow-Debreu expansion for spot prices in CRR, using the binomial p.d.f. on N states at expiry as in AD(). It has very low $O(N)$ complexity.

The backward induction algorithm for CRR options pricing is not quite as fast. Counting the number of quantities needed to fill a recombining binomial tree with N steps, there are about $N^2/2$ prices to compute. Each price requires a fixed amount of arithmetic with the stored values u, d, R, p, so the total number of operations is $O(N^2)$. This is costlier than $O(N)$ but is still considered *low computational complexity*.

Computations using the whole model are needed to price exotic contingent claims affected by events at times t between purchase $(t = 0)$ and expiry $(t = T)$. The backward induction algorithm can be modified to accommodate early exercise, dividend payments, barrier price encounters, or other features, justifying the higher order of complexity.

Some exotic contingent claims are *path dependent*, affected by intermediate states as well as the terminal state at expiry. In general, this requires a binary tree model which, for N time steps, contains $O(2^N)$ prices. This severely limits N in practice but otherwise offers the greatest flexibility of all.

4.1.1   European-Style Options in CRR

In the Cox-Ross-Rubinstein model, prices are computed by backward induction from the bottom row:

$$C(N, j) = \left[S_0 u^j d^{N-j} - K\right]^+ = \left[S_0 u^{2j-N} - K\right]^+, \quad 0 \le j \le N;$$

$$C(n, j) = \frac{pC(n+1, j+1) + [1-p]C(n+1, j)}{R}, \quad n = N-1, \ldots, 0, \ 0 \le j \le n.$$

Likewise, for the European-style Put option,

$$P(N, j) = \left[K - S_0 u^j d^{N-j}\right]^+ = \left[K - S_0 u^{2j-N}\right]^+, \quad 0 \le j \le N;$$

$$P(n, j) = \frac{pP(n+1, j+1) + [1-p]P(n+1, j)}{R}, \quad n = N-1, \ldots, 0, \ 0 \le j \le n.$$

The inductive step is the same for both, only the bottom row of the tree gets different terminal values. It is reasonable to perform both with the same loop.

Octave implementation of CRR for European-style options

Since array indices in Octave (and MATLAB, and Fortran) start at 1, the Call and Put premiums at time 0 are stored at C(1,1) and P(1,1), respectively. In general, the prices at nodes $C(n, j)$ and $P(n, j)$ are stored at C(n+1,j+1) and P(n+1,j+1), respectively.

```
1   function [C, P] = CRReur(T, S0, K, r, v, N)
2   % Octave/MATLAB function to price European Call
3   % and Put options using the Cox−Ross−Rubinstein
4   % (CRR) binomial pricing model.
5   % INPUTS:                                        (Example)
6   %      T =  expiration time in years             (1)
7   %      S0 = spot stock price                     (90)
8   %      K =  strike price                         (95)
9   %      r =  riskless yield per year              (0.02)
10  %      v =  volatility; must be >0               (0.15)
11  %      N =  height of the binomial tree          (10)
12  % OUTPUTS:
13  %      C =  price of the Call option at all (n,j).
14  %      P =  price of the Put option at all (n,j).
15  % EXAMPLE:
16  %      [C, P] = CRReur(1, 90, 95, 0.02, 0.15, 10);
17  %      C(1,1),P(1,1)  % to get just C(0) and P(0)
18  %
19      [pu,up,R] = CRRparams(T,r,v,N); % Use CRR values
20      C=zeros(N+1,N+1); P=zeros(N+1,N+1); % Initial C,P
21      for j=0:N % Set terminal values at time step N
22      SNj=S0*up^(2*j−N);          % Expiry price S(N,j)
23      C(N+1,j+1)=max(SNj−K, 0); % Plus part of (SNj−K)
24      P(N+1,j+1)=max(K−SNj, 0); % Plus part of (K−SNj)
25      end
```

```
26    for n=N−1:−1:0   % Price earlier grid values
27      for j=0:n % ...with the backward pricing formula
28        C(n+1,j+1)=(pu*C(n+2,j+2)+(1−pu)*C(n+2,j+1))/R;
29        P(n+1,j+1)=(pu*P(n+2,j+2)+(1−pu)*P(n+2,j+1))/R;
30      end
31    end
32    return; % Prices in C and P are now fully defined.
33  end
```

The cost of CRR in arithmetic over Black-Scholes formula evaluations is justified by the much greater flexibility of CRR in modeling exotic options. A few examples are described and implemented in the next sections.

4.1.2 American-Style Options in CRR

An American-style option can be exercised at any time up to and including its expiry. This additional right does not affect its price at expiry time T, when the Call option is worth $[S(T) - K]^+$ and the Put option is worth $[K - S(T)]^+$ just like the European-style option. However, the right of early exercise affects its price at intermediate times. In the CRR model, this changes the backward induction to include the early exercise value.

For the American-style Call option,

$$
\begin{aligned}
C(N,j) &= \left[S_0 u^j d^{N-j} - K\right]^+ = \left[S_0 u^{2j-N} - K\right]^+, \quad 0 \le j \le N; \\
C_{\text{Exercise}} &= \left[S_0 u^j d^{n-j} - K\right]^+ = \left[S_0 u^{2j-n} - K\right]^+, \\
C_{\text{Binomial}} &= \frac{pC(n+1,j+1) + [1-p]C(n+1,j)}{R}, \\
C(n,j) &= \max\left(C_{\text{Binomial}}, C_{\text{Exercise}}\right), \quad n = N-1,\ldots,0,\ 0 \le j \le n.
\end{aligned}
$$

Likewise, for the American-style Put option,

$$
\begin{aligned}
P(N,j) &= \left[K - S_0 u^j d^{N-j}\right]^+ = \left[K - S_0 u^{2j-N}\right]^+, \quad 0 \le j \le N; \\
P_{\text{Exercise}} &= \left[K - S_0 u^j d^{n-j}\right]^+ = \left[K - S_0 u^{2j-n}\right]^+, \\
P_{\text{Binomial}} &= \frac{pP(n+1,j+1) + [1-p]P(n+1,j)}{R}, \\
P(n,j) &= \max\left(P_{\text{Binomial}}, P_{\text{Exercise}}\right), \quad n = N-1,\ldots,0,\ 0 \le j \le n.
\end{aligned}
$$

These computations are similar enough to be combined into one function.

Octave implementation of CRR for American-style options

```
1  function [C, P] = CRRa(T, S0, K, r, v, N)
2  % Octave/MATLAB function to price American Call
3  % and Put options using the Cox−Ross−Rubinstein
```

```
4   % (CRR) binomial pricing model.
5   % INPUTS:                                    (Example)
6   %     T =   expiration time                  (1 year)
7   %     S0 =  spot stock price                    (90)
8   %     K =   strike price                        (95)
9   %     r =   risk-free yield                    (0.02)
10  %     v =   volatility; must be >0            (0.15)
11  %     N =   height of the binomial tree         (10)
12  % OUTPUT:
13  %     C =   price of the Call option at all (n,j).
14  %     P =   price of the Put option at all (n,j).
15  % EXAMPLE:
16  %     [C,P] = CRRa(1, 90, 95, 0.02, 0.15, 10);
17  %     C(1,1),P(1,1)  % to get just C(0) and P(0)
18  %
19      [pu,up,R] = CRRparams(T,r,v,N); % Use CRR values
20      C=zeros(N+1,N+1); P=zeros(N+1,N+1); % Initial C,P
21      for j = 0:N % Set terminal values at time step N
22        V = S0*up^(2*j-N)-K;   % S(N,j) - K
23        C(N+1,j+1)=max(V,0); P(N+1,j+1) = max(-V,0);
24      end
25      for n = (N-1):(-1):0 % Price earlier grid values
26        for j = 0:n % states j={0,1,...,n} at time n
27          % Binomial pricing model value
28          bC=(pu*C(n+2,j+2) + (1-pu)*C(n+2,j+1))/R;
29          bP=(pu*P(n+2,j+2) + (1-pu)*P(n+2,j+1))/R;
30          % Exercise value: X for Call, -X for Put
31          X = S0*up^(2*j-n)-K;   % S(n,j) - K
32          % Price at this node is the larger:
33          C(n+1,j+1)=max(bC,X); P(n+1,j+1)=max(bP,-X);
34        end
35      end
36      return; % Prices are in matrices C and P.
37  end
```

4.1.3 Binary Options in CRR

A binary option is like a European-style option on a stock S with strike price K but with a fixed payoff B at expiry T if it finishes "in the money." That means $S(T) > K$ for a Call, or $S(T) < K$ for a Put.

Prices are first set at expiry, in the bottom row N:

$$C(N,i) = \begin{cases} B, & S_0 u^i d^{N-i} > K, \\ 0, & \text{otherwise}; \end{cases} \qquad P(N,i) = \begin{cases} B, & S_0 u^i d^{N-i} < K, \\ 0, & \text{otherwise}, \end{cases}$$

for $i = 0, 1, \ldots, N$. All other prices are computed by backward induction.

Remark.   Prices $C(n,j)$ and $P(n,j)$ are proportional to B, so it is sufficient to compute them for $B_0 = 1$ and then multiply the results by any desired B.

Octave implementation of CRR for binary options

These are also similar enough to be combined into one function. In this version, the fixed payoff B is retained as an input parameter.

```octave
 1  function [C, P] = CRRbin(T, S0, K, B, r, v, N)
 2  % Octave/MATLAB function to price binary Call and Put
 3  % options using the Cox-Ross-Rubinstein (CRR) model.
 4  %  INPUTS:                                    (Example)
 5  %     T =  expiration time in years           (1)
 6  %     S0 = spot stock price                   (90)
 7  %     K =  strike price                       (95)
 8  %     B =  payoff at expiry                   (20)
 9  %     r =  risk-free yield per year           (0.02)
10  %     v =  volatility; must be >0             (0.15)
11  %     N =  height of the binomial tree        (10)
12  %  OUTPUTS:
13  %     C = Call option price at all gridpoints (n,j).
14  %     P = Put option price at all gridpoints (n,j).
15  %  EXAMPLE:
16  %     [C, P] = CRRbin(1, 90, 95, 20, 0.02, 0.15, 10);
17  %     C(1,1),P(1,1)  % to get just C(0) and P(0)
18  %
19     [pu,up,R] = CRRparams(T,r,v,N); % Use CRR values
20     C=zeros(N+1,N+1); P=zeros(N+1,N+1); % Initial C,P
21     for j = 0:N % Set terminal values at time step N
22       SNj = S0 * up^(2*j-N);   % Asset price S(N,j)
23       if( SNj > K )
24         C(N+1,j+1) = B; % ...trigger the Call payoff
25       end
26       if( SNj < K )
27         P(N+1,j+1) = B; % ...trigger the Put payoff
28       end
29     end  % ...otherwise the payoff is zero.
30     for n = N-1:-1:0   % Price earlier grid values
31       for j = 0:n  % ...by backward pricing formula
32         C(n+1,j+1)=(pu*C(n+2,j+2)+(1-pu)*C(n+2,j+1))/R;
33         P(n+1,j+1)=(pu*P(n+2,j+2)+(1-pu)*P(n+2,j+1))/R;
34       end
35     end
36     return; % Prices in matrices C and P are all defined.
37  end
```

Remark. The premiums for a binary Call and Put are related by

$$C(0) + P(0) = B/R,$$

since a portfolio containing one of each will be worth a riskless B at expiry time T. Such a portfolio, to avoid an arbitrage, must have a price equal to the present value of B, which is B/R if the riskless return over time T is R.

4.1.4   Compound Options in CRR

A compound option is a European-style Call with strike price L and expiry at a fixed future time T_1. What makes it exotic is that the asset possibly to be purchased is itself an option. Thus there is another underlying risky asset S and some additional parameters: spot price S_0 at time 0, strike price K for S, and final expiry T, with $0 < T_1 < T$.

For example, suppose that the option to be purchased is a European-style Call. Denote its price at time t to be $C(t)$ for $0 \le t \le T$. Then at T_1 its value will be $C(T_1)$.

Denote the price of the compound Call option at time t by $W(t)$. Its value at time T_1 is then $[C(T_1) - L]^+$, since it would only be exercised if it allowed the cheaper purchase of C at price L. The price $W(0)$ may then be computed using backward induction from these known values $W(T_1)$.

Two multistep binomial trees are needed to price the compound Call. One is needed to find $C(T_1)$ from the known prices $C(T) = [C(T) - K]^+$. The second, smaller tree will then find $W(0)$ by backward induction from $W(T_1) = [C(T_1) - L]^+$.

The first induction formula is thus

$$C(N,j) = \left[S_0 u^j d^{N-j} - K\right]^+ = \left[S_0 u^{2j-N} - K\right]^+;$$
$$C(n,j) = \frac{pC(n+1,j+1) + [1-p]C(n+1,j)}{R},$$

where $n = N-1, \ldots, M$ and $j = 0, 1, \ldots, n$. Here $M = \lfloor T_1 N/T \rfloor$.

The second induction formula is

$$W(M,i) = [C(M,i) - L]^+, \quad i = 0, 1, \ldots, M,$$
$$W(n,j) = \frac{pW(n+1,j+1) + [1-p]W(n+1,j)}{R},$$

where $n = M-1, \ldots, 2, 1, 0$ and $j = 0, 1, \ldots, n$. This extracts $W(0,0)$ from a CRR tree of depth N:

Note that the first tree does not need to be filled all the way back to time 0, nor does the other option need to be priced with its own tree. However, the extra cost of filling the top of the tree is outweighed by the benefit of reusing existing software `CRReur()` rather than writing and debugging a custom function.

Octave implementation of CRR for compound Call options

```
1  function  [W,C]  =  CRRcc(T,T1,S0,K,L,r,v,N)
2  %   Octave/MATLAB  function  to  price  a  compound  Call  option
3  %   using  the  Cox-Ross-Rubinstein  (CRR)  binomial  model.
4  % INPUTS:                                            (Example)
5  %    T  =  expiration  time  in  years               (1)
6  %    T1 =  choice  time;  must  have  0<T1<T          (0.5)
```

```
 7  %      S0 =  spot  stock  price                        (90)
 8  %      K =   stock  strike  price  at  expiry  T        (95)
 9  %      L =   option  strike  price  at  time  T1       (4.50)
10  %      r =   risk−free  yield                          (0.02)
11  %      v =   volatility ;  must  be  >0                (0.15)
12  %      N =   height  of  the  binomial  tree            (10)
13  % OUTPUT:
14  %      W =   price  of  the  compound  Call  option
15  %      C  =  price  of  the  vanilla  European  Call  option
16  % EXAMPLE:
17  %      [W,C]  =  CRRcc ( 1 ,0.5 ,90 ,95 ,4.50 ,0.02 ,0.15 ,10 );
18  %      W( 1 ,1 ) ,C( 1 ,1 )   % to  get  just  W(0)  and  C(0)
19  %
20         [ pu , up ,R]=CRRparams (T, r , v ,N );  % Use  CRR  values
21         [C,P]=CRReur (T, S0 ,K, r , v ,N );  % vanilla  options  at  T
22         M=round (T1∗N/T );  % number  of  time  steps  to  time  T1
23         W=zeros (M+1,M+1 );    % smaller  output  matrix
24         for  j=0:M  % Set  terminal  values  at  (M,0 ) ,... ,(M,M) :
25           W(M+1,j +1) = max(C(M+1,j +1)−L,0 );  % positive  part
26         end
27         for  n=M−1:−1:0    % Backward  induction
28           for  j=0:n  % Binomial  pricing  model  value
29             W(n+1,j +1)=(pu∗W(n+2,j +2)+(1−pu )∗W(n+2,j +1)) /R;
30           end
31         end
32         return   % Prices  in  matrices  W and  C  are  defined .
33       end
```

4.1.5   Chooser Options in CRR

A chooser option is a contract granting the right to choose either a European-style Call option or Put option at strike price K, with the choice to be made at a fixed future time T_1 but before expiry time T. Thus $0 < T_1 < T$. At T_1 the value of the chooser option will be the greater of the Call and Put prices with expiries $T - T_1$. Thus the chooser option's premium may be computed by backward induction from expiry prices $\max(C(T_1), P(T_1))$:

$$
\begin{aligned}
W(M, i) &= \max(C(M, i), P(M, i)), \quad i = 0, 1, \ldots, M, \\
W(n, j) &= \frac{pW(n + 1, j + 1) + [1 - p]W(n + 1, j)}{R},
\end{aligned}
$$

where $n = M - 1, \ldots, 2, 1, 0$ and $j = 0, 1, \ldots, n$, and $M = \lfloor T_1 N / T \rfloor$.

Octave implementation of CRR for chooser options

Three multistep binomial trees are used to price the chooser option. Two of them find the European-style Call and Put option prices at time T_1 for options with expiry T, and they may be computed with the previous implementation:

[C,P]=CRReur(T,S0,K,r,v,N). The third requires its own backward induction loop from max(C,T) at bottom row M.

```
1  function [W, C, P] = CRRcho(T, T1, S0, K, r, v, N)
2  % Octave/MATLAB function to price a chooser option
3  % with the Cox−Ross−Rubinstein (CRR) binomial model.
4  % INPUTS:                                    (Example)
5  %    T  =  expiry time in years              (1)
6  %    T1 =  choice time; must have 0<T1<T      (0.5)
7  %    S0 =  spot stock price                  (90)
8  %    K  =  strike price                      (95)
9  %    r  =  risk−free annual yield            (0.02)
10 %    v  =  volatility; must be >0            (0.15)
11 %    N  =  height of the binomial tree       (10)
12 % OUTPUT:
13 %    W  =  price of the chooser option
14 %    C  =  price of the vanilla European Call option
15 %    P  =  price of the vanilla European Put option
16 % EXAMPLE:
17 %    [W, C, P] = CRRcho(1, 0.5, 90, 95, 0.02, 0.15, 10);
18 %    W(1,1),C(1,1),P(1,1) % get just W(0),C(0), and P(0)
19 %
20    [pu,up,R] = CRRparams(T,r,v,N); % Use CRR values
21    [C,P]=CRReur(T,S0,K,r,v,N); % price C,P from expiry T
22    M=floor(T1*N/T);  % number of time steps before T1
23    W=zeros(M+1,M+1); % smaller output matrix W
24    for j = 0:M      % terminal values at (M,0),...,(M,M)
25      W(M+1,j+1) = max(C(M+1,j+1), P(M+1,j+1)); % choose
26    end  % Call and Put prices at time T1 initialize W
27    for n = M−1:−1:0  % start at terminal time M
28      for j = 0:n % Binomial pricing model value
29        W(n+1,j+1)=(pu*W(n+2,j+2)+(1−pu)*W(n+2,j+1))/R;
30      end  % ...backward induction fills W.
31    end
32    return; % Prices in matrices W, C, and P are defined.
33 end
```

4.1.6   Forward Start Options in CRR

A forward start option is an exotic derivative that is equivalent to a vanilla option with expiry T but with a strike price equal to the underlying stock price at a fixed future time T_1, where $0 < T_1 < T$. There are two types: a forward start Call and a forward start Put, each with its premium that may be computed in the CRR model by backward induction.

Multiple multistep binomial trees are needed to price each forward start option. All but one of them are needed to find the vanilla Call or Put option prices at time T_1 for options with expiry T and different at-the-money strike prices. The final tree will price the forward start option by backward induction from a bottom row filled with the different $C(T_1)$ or $P(T_1)$. The algorithm separates naturally into three parts.

Let $M = \lfloor T_1 N/T \rfloor$ be the grid time corresponding to T_1 in an N-step binomial tree. First compute the strike prices, which are the modeled stock prices $S(M, j)$ for $j = 0, 1, \ldots, M$:

$$K(j) \stackrel{\text{def}}{=} S(M, j) = S_0 u^j d^{M-j} = S_0 u^{2j-M}.$$

Second, model the vanilla option prices vC or vP at time T_1 for each of the strike prices $K(j)$, $j = 0, 1, \ldots, M$. This may be done for both with $M + 1$ uses of CRReur(): for each $j = 0, 1, \ldots, M$, compute

$$
\begin{aligned}
[C, P] &= \mathrm{CRReur}(T, S_0, K(j), r, v, N); \\
vC(j) &\stackrel{\text{def}}{=} C(M, j), \\
vP(j) &\stackrel{\text{def}}{=} P(M, j).
\end{aligned}
$$

(Note that both $vC(j)$ and $vP(j)$ are at-the-money premiums for options expiring $T - T_1$ after time T_1.)

Third and finally, use backward induction in an M-level binomial tree to compute the forward start Call or Put option prices, fC or fP, respectively. For Calls, this will be

$$
\begin{aligned}
fC(M, i) &= vC(i), \quad i = 0, 1, \ldots, M; \\
fC(n, j) &= \frac{p fC(n+1, j+1) + [1 - p] fC(n+1, j)}{R},
\end{aligned}
$$

and for Puts, it will be

$$
\begin{aligned}
fP(M, i) &= vP(i), \quad i = 0, 1, \ldots, M; \\
fP(n, j) &= \frac{p fP(n+1, j+1) + [1 - p] fP(n+1, j)}{R},
\end{aligned}
$$

where $n = M - 1, \ldots, 2, 1, 0$ and $j = 0, 1, \ldots, n$. This leaves the forward start Call and Put premiums in $fC(0, 0)$ and $fP(0, 0)$, respectively.

Octave implementation of CRR for forward start options

```
1  function [C, P] = CRRfws(T, S0, T1, r, v, N)
2  %   Octave/MATLAB function to price forward start
3  %   Call and Put options using the Cox-Ross-Rubinstein
4  %   (CRR) binomial pricing model.
5  % INPUTS:                                    (Example)
6  %    T  =  expiration time in years           (1)
7  %    S0 =  spot stock spot price              (90)
8  %    T1 =  strike price set time; 0<T1<T      (0.5)
```

```
 9   %   r  =  risk-free  annual  yield           (0.02)
10   %   v  =  volatility;  must  be  >0          (0.15)
11   %   N  =  height  of  the  binomial  tree     (10)
12   % OUTPUT:
13   %   C =  price  of  the  Call  option  at  gridpoints  (n,j).
14   %   P =  price  of  the  put  option  at  gridpoints  (n,j).
15   % EXAMPLE:
16   %   [C,P] = CRRfws(1,  90,  0.5,  0.02,  0.15,  10);
17   %   C(1,1),P(1,1)   % get  just  C(0),P(0)  as  output
18   %
19     [pu,up,R] = CRRparams(T,r,v,N); % Use  CRR  values
20     M = round(T1*N/T); % set  strike  price  grid  time
21     for (j=0:M)
22       SMj(j+1)= S0*up^(2*j-M); % modeled  S(M,j)
23     end
24     for(j=0:M) % price  at-the-money  vanilla  options
25       [C,P]=CRReur(T,S0,SMj(j+1),r,v,N); % vanilla
26       CMj(j+1)=C(M+1,j+1);   % Call  price  at  (M,j)
27       PMj(j+1)=P(M+1,j+1);   % Put  price  at  (M,j)
28     end
29     % Reinitialize  smaller  forward  start  C  and  P:
30     C=zeros(M+1,M+1); P=zeros(M+1,M+1);
31     for j=0:M  % set  terminal  values  at  time  M
32       C(M+1,j+1)=CMj(j+1); P(M+1,j+1)=PMj(j+1);
33     end
34     for n = M-1:-1:0  % backward  induction
35       for j = 0:n      % binomial  pricing
36         C(n+1,j+1)=(pu*C(n+2,j+2)+(1-pu)*C(n+2,j+1))/R;
37         P(n+1,j+1)=(pu*P(n+2,j+2)+(1-pu)*P(n+2,j+1))/R;
38       end
39     end
40     return; % Prices  in  C  and  P  are  all  defined.
41   end
```

4.1.7 Barrier Options

Barrier options are a class of exotic derivatives engineered to have a lower premium than their vanilla versions by removing some extreme world states from consideration. The "barrier" is a price for the underlying asset that, if reached, either

- voids the option, for *knock-out* barrier options, or

- activates the option, for *knock-in* barrier options.

There are two *up-and-in*, and two *down-and-in*, knock-in barrier options, Call and Put. Similarly, there are four knock-out barrier options, *up-and-out* and *down-and-out* Calls and Puts. The barrier may be above or below the strike price K, and may likewise be above or below the spot price $S(0)$ of the underlying asset.

Barrier monitoring is done at specified times, usually when the market is closed after the end of a trading day. The closing price of the underlying asset is compared with the barrier price, and any change in status of the option is published before the next trading session begins. Continuous barrier monitoring is also possible but with high computational cost, and likewise a discrete binomial model of such an option with frequent or continuous barrier monitoring requires a large number of time steps for accurate pricing.

In all cases, the premium to charge for such exotic options is heavily influenced by the model of S prices. Several examples are priced below using the CRR model for simplicity.

Octave implementation of CRR Up-and-In barrier Call

This Call option is activated when the underlying asset price equals or exceeds the barrier price B, after which it behaves like a vanilla European Call. To price it in the CRR binomial model, first price the asset S and the vanilla Call V in their own binomial trees. The backward induction from level N has this terminal condition: for $j = 0, 1, \ldots, N$,

$$C^i(N, j) \overset{\text{def}}{=} \begin{cases} V(N, j) = [S(N, j) - K]^+, & S(N, j) \geq B, \\ 0, & S(N, j) < B. \end{cases}$$

The backward pricing formula also tests the barrier: for $j = 0, 1, \ldots, n$,

$$C^i(n, j) = \begin{cases} V(n, j), & S(n, j) \geq B, \\ [p\, C^i(n+1, j+1) + (1-p)\, C^i(n+1, j)]/R, & S(n, j) < B, \end{cases}$$

where p is the risk neutral up probability and R is the riskless return over one time step, both computed from the CRR parameters.

```
1  function [Exotic, Vanilla] = CRRuiC(T,S0,K,B,r,v,N)
2  %   Octave/MATLAB function to price up-and-in barrier
3  %   Call options using the Cox-Ross-Rubinstein (CRR)
4  %   binomial pricing model.
5  %   INPUTS:                              (Example)
6  %      T =   expiration time             (1 year)
7  %      S0 = spot stock price             (100)
8  %      K =   strike price                (80)
9  %      B =   barrier price               (120)
10 %      r =   risk-free yield             (0.05)
11 %      v =   volatility; must be >0      (0.20)
12 %      N =   height of the binomial tree (40)
13 %   OUTPUT:
14 %      Exotic =   price of the option at all (n,j).
15 %      Vanilla =  price of European Call at (n,j).
16 %   EXAMPLE:
```

```
17 %     [uiCa, eCa]  = CRRuiC(1,100,80,120,0.05,0.20,40);
18 %
19      [pu,up,R] = CRRparams(T,r,v,N); % Use CRR values
20      Exotic=zeros(N+1,N+1);  Vanilla=zeros(N+1,N+1);
21      for j = 0:N  % set values at expiry step N
22        SNj = S0*up^(2*j-N); % asset price S(N,j)
23        Vanilla(N+1,j+1)=max(SNj-K, 0); % plain Eur. Call
24        if(SNj < B) % barrier not exceeded, so no Exotic
25          Exotic(N+1,j+1) = 0;
26        else  % barrier exceeded, so Exotic is created
27          Exotic(N+1,j+1) = Vanilla(N+1,j+1);
28        end
29      end  % Now price earlier values by induction:
30      for n = (N-1):(-1):0  % grid times n={N-1,...,0}
31        for j = 0:n % states j={0,1,...,n} at time n
32          Vanilla(n+1,j+1)=( pu*Vanilla(n+2,j+2)...
33            + (1-pu)*Vanilla(n+2,j+1))/R; % all cases
34          % Test the barrier
35          Snj = S0*up^(2*j-n); % asset price S(n,j)
36          if(Snj<B) % barrier not exceeded, so use Exotic
37            Exotic(n+1,j+1)=(pu*Exotic(n+2,j+2)...
38              +(1-pu)*Exotic(n+2,j+1) )/R; % bin. price
39          else   % barrier exceeded, so use Vanilla
40            Exotic(n+1,j+1) = Vanilla(n+1,j+1);
41          end
42        end
43      end
44      return % Prices are in matrices Exotic and Vanilla.
45    end
```

Octave implementation of CRR Up-and-Out barrier Call

Here the Call option is voided when the underlying asset price hits the barrier price B. To price it in the CRR binomial model, first price the asset S and the vanilla Call V in their own binomial trees. The backward induction from level N reverses the terminal conditions of the up-and-in Call: for $j = 0, 1, \ldots, N$,

$$C^o(N,j) = \begin{cases} 0, & S(N,j) \geq B, \\ V(N,j) = [S(N,j) - K]^+, & S(N,j) < B. \end{cases}$$

The backward pricing formula likewise tests the barrier but returns 0 if S equals or exceeds it: for $j = 0, 1, \ldots, n$,

$$C^o(n,j) = \begin{cases} 0, & S(n,j) \geq B, \\ [p\,C^o(n+1,j+1) + (1-p)\,C^o(n+1,j)]/R, & S(n,j) < B, \end{cases}$$

where p is the risk neutral up probability and R is the riskless return over one time step, both computed from the CRR parameters.

```
1  function Call = CRRuoC(T,S0,K,B,r,v,N)
2  % Octave/MATLAB function to price up-and-out barrier
3  % Call options using the Cox-Ross-Rubinstein (CRR)
4  % binomial pricing model.
5  % INPUTS:                                  (Example)
6  %      T =  expiration time                (1 year)
7  %      S0 = spot stock price               (100)
8  %      K =  strike price                   (80)
9  %      B =  barrier price                  (120)
10 %      r =  risk-free annual yield         (0.05)
11 %      v =  volatility; must be >0         (0.20)
12 %      N =  height of the binomial tree    (40)
13 % OUTPUT:
14 %      Call =  price of the option at all (n,j).
15 % EXAMPLE:
16 %      uoCa = CRRuoC(1,100,80,120,0.05,0.20,40);
17 %
18    [pu,up,R] = CRRparams(T,r,v,N); % Use CRR values
19    Call=zeros(N+1,N+1);  % allocate output matrix
20    for j=0:N % set the values at expiry step N
21      SNj=S0*up^(2*j-N);      % asset price S(N,j)
22      if(SNj < B) % then the barrier is not exceeded
23        Call(N+1,j+1) = max( SNj - K, 0 );
24      else     % barrier exceeded, so option vanishes
25        Call(N+1,j+1) = 0;
26      end
27    end % use backward induction to price the rest
28    for n=(N-1):(-1):0 % grid times n={N-1,..., 0}
29      for j=0:n % test the barrier in all states j
30        Snj = S0*up^(2*j-n); % asset price S(n,j)
31        if(Snj < B) % then barrier is not exceeded
32          Call(n+1,j+1)=( pu*Call(n+2,j+2)...
33            +(1-pu)*Call(n+2,j+1) )/R; % bin. price
34        else % barrier exceeded, so option vanishes
35          Call(n+1,j+1) = 0;
36        end
37      end
38    end
39    return   % Prices in Call matrix are now defined.
40 end
```

Octave implementation of CRR Partial Up-and-Out barrier Call

A *partial barrier option* is one of the eight mentioned types, with barrier monitoring occurring only after some specified time between now and expiry.

```
1  function C = CRRpuoC(T,S0,K,B,r,v,N,N0)
2  % Octave/MATLAB function to price partial up-and-out
```

```
3  % barrier Call options using the Cox-Ross-Rubinstein
4  % (CRR) binomial pricing model.
5  % INPUTS:                                          (Example)
6  %    T =   expiration time                         (1 year)
7  %    S0 =  spot stock price                          (100)
8  %    K =   strike price                              (80)
9  %    B =   barrier price                            (120)
10 %    r =   risk-free yield                         (0.05)
11 %    v =   volatility; must be >0                  (0.20)
12 %    N =   height of the binomial tree              (4)
13 %    N0 =  first barrier test time;                 (3)
14 %              must satisfy 0 < N0 < N
15 % OUTPUT:
16 %    C =   price at all gridpoints (n,j).
17 % EXAMPLE:
18 %    puoCa = CRRpuoC(1,100,80,120,0.05,0.20,4,3);
19 %
20    [pu,up,R] = CRRparams(T,r,v,N); % Use CRR values
21    C=zeros(N+1,N+1); % allocate output matrix
22    for j=0:N % set values at expiry step N
23      SNj=S0*up^(2*j-N); % asset price S(N,j)
24      if(SNj < B) % then the barrier is not exceeded
25        C(N+1,j+1) = max( SNj - K, 0 );
26      else    % barrier exceeded, so option vanishes
27        C(N+1,j+1) = 0;
28      end
29    end
30    % Use backward induction to price earlier values:
31    for n = (N-1):(-1):0  % grid times n={N-1,..., 0}
32      for j = 0:n    % states j={0,1,...,n} at time n
33        % Test the barrier
34        Snj = S0*up^(2*j-n);   % asset price S(n,j)
35        if(n < N0 || Snj < B) % too early, or else
36                              % barrier not exceeded
37        % ...so use the binomial pricing model value
38          C(n+1,j+1)=(pu*C(n+2,j+2)+(1-pu)*C(n+2,j+1))/R;
39        else % time after N0 and barrier exceeded,
40              % so option vanishes
41          C(n+1,j+1) = 0;
42        end
43      end
44    end
45    return; % Prices in C are now defined.
46  end
```

Octave implementation of CRR Rebate Up-and-Out barrier Call

An additional feature of knock-out options is a *rebate* of the premium if the option is voided. To calculate its price, let $W(x)$ be the price of the knock-out option which returns x if it is voided but otherwise behaves like the vanilla option. Then the price of the rebate knock-out option is the solution to the fixed point problem

$$x = W(x),$$

for that will have a price equal to its rebate. For example, the up-and-out Call has this implementation of $W(x)$:

```octave
1   function C = CRRruoC(T,S0,K,B,refund,r,v,N)
2   % Octave/MATLAB function to price up-and-out barrier
3   % Call options with a refund at knock-out using the
4   % Cox-Ross-Rubinstein (CRR) binomial pricing model.
5   % INPUTS:                                    (Example)
6   %    T =   expiration time                   (1 year)
7   %    S0 =  spot stock price                  (100)
8   %    K =   strike price                      (80)
9   %    B =   barrier price                     (120)
10  %    refund =  rebate paid if the            (17)
11  %              option knocks out at
12  %              the barrier price
13  %    r =   risk-free annual yield            (0.05)
14  %    v =   volatility; must be >0            (0.20)
15  %    N =   height of the binomial tree       (4)
16  % OUTPUT:
17  %    C =   price of the option at all (n,j).
18  % EXAMPLE:
19  %    uorCa = CRRruoC(1,100,80,120,17,0.05,0.20,4);
20  %
21     [pu,up,R] = CRRparams(T,r,v,N); % Use CRR values
22     C=zeros(N+1,N+1); % allocate the output matrix
23     for j = 0:N % set values at expiry step N
24       SNj = S0*up^(2*j-N); % asset price S(N,j)
25       if(SNj < B) % then the barrier is not exceeded
26         C(N+1,j+1) = max( SNj-K, 0 );
27       else    % barrier exceeded, so option vanishes
28         C(N+1,j+1) = refund; % ...with a rebate
29       end
30     end % Use backward induction to price the rest:
31     for n = (N-1):(-1):0 % grid times n={N-1,..., 0}
32       for j = 0:n % states j={0,1,...,n} at time n
33         % Test the barrier
34         Snj = S0*up^(2*j-n); % asset price S(n,j)
35         if(Snj < B) % then barrier is not exceeded
36           % ...so use the binomial pricing model value
37           C(n+1,j+1)=(pu*C(n+2,j+2)+(1-pu)*C(n+2,j+1))/R;
38         else % barrier exceeded, so option vanishes
```

```
39          C(n+1,j+1) = refund; % ...with a rebate
40        end
41      end
42    end
43    return  % Prices in C are now fully defined.
44  end
```

To price the option with a full rebate, start with an initial return $x = 0$. Now iterate, replacing x with the price of the option, until the price of the option with refund x equals x:

```
T=1; S=100; K=80; B=120;r=0.05; v=0.20; N=4;
X=[0];   % initial guess: zero price, zero refund
X=CRRruoC(T,S,K,B,X(1,1),r,v,N); X(1,1) % 6.2558
for i=1:10
  X=CRRruoC(T,S,K,B,X(1,1),r,v,N); % fixed-point iteration
end
X=CRRruoC(T,S,K,B,X(1,1),r,v,N); X(1,1) % 10.765
X=CRRruoC(T,S,K,B,X(1,1),r,v,N); X(1,1) % 10.765
```

Convergence seems to have occured after 12 iterations. It is left as an exercise to reimplement the loop so that it checks for convergence by measuring the change between iterations.

Another exotic twist is a partial rebate of a portion α, with $0 < \alpha < 1$, of the exotic option premium. This reuses the function CRRruoC(), but the iteration is slightly different:

```
T=1; S=100; K=80; B=120;r=0.05; v=0.20; N=4; % as before
alpha=0.50;   % partial rebate, 50 percent of the premium
X=[0];        % initial guess is zero price, zero refund
X=CRRruoC(T,S,K,B,alpha*X(1,1),r,v,N); X(1,1) % 6.2558
for i=1:10
  X=CRRruoC(T,S,K,B,alpha*X(1,1),r,v,N); % iteration with alpha
end
X=CRRruoC(T,S,K,B,alpha*X(1,1),r,v,N); X(1,1) % 7.9131
X=CRRruoC(T,S,K,B,alpha*X(1,1),r,v,N); X(1,1) % 7.9131
```

This also seems to have converged after 12 iterations. The premium is lower, obviously, than for the full rebate barrier option.

4.1.8 Booster Options

This is a variation on barrier options for investors who expect to profit from low volatility. The option pays an amount proportional to the time (in $[0, T]$) that the underlying asset spends with its price between two bounds, the upper barrier U and the lower barrier L, where of course $L < U$.

The price of the booster option decreases as volatility increases since, with increasing volatility, the underlying asset price has a shorter expected time spent inside the interval $[L, U]$ and thus has lower expected payoff.

A booster option can augment a Call at strike U and a Put with strike L. It will pay off in case neither the Call nor the Put yield a profit.

Octave implementation of CRR of a Booster option

```
 1  function B = CRRboo(T,S0,L,H,r,v,N)
 2  % Octave/MATLAB function to price booster options
 3  % using the Cox-Ross-Rubinstein (CRR) model.
 4  % INPUTS:                           (Example)
 5  %     T =  expiration time          (1 year)
 6  %     S =  stock price              ($100)
 7  %     L =  low barrier price        ($85)
 8  %     H =  high barrier price       ($115)
 9  %     r =  risk-free yield          (0.05)
10  %     v =  volatility, >0           (0.20)
11  %     N =  tree height              (4)
12  % OUTPUT:
13  %     B =  price of the option at all (n,j).
14  % EXAMPLE:
15  %     boo = CRRboo(1,100,85,115,0.05,0.20,4);
16  %
17      [pu,up,R] = CRRparams(T,r,v,N); % Use CRR values
18      B=zeros(N+1,N+1); % allocate the output matrix
19      for j=0:N % Set the values at expiry step N
20        SNj=S0*up^(2*j-N); % asset price S(N,j)
21        if(SNj < H)             % less than high barrier
22          if(L < SNj )
23            B(N+1,j+1) = N; % time spent between L,H
24          end
25        end
26      end % Now test for barrier crossing before n=N
27      for n=1:N-1              % n=0 case is unneeded
28        for j=0:n % test barriers at (n,0),...,(n,n):
29          Snj=S0*up^(2*j-n); % asset price S(n,j)
30          if(Snj >= H || Snj <= L )
31            B(n+1,j+1) = n;
32          end
33        end
34      end
35      % Backward induction prices any zero grid values:
36      for n=(N-1):(-1):0  % grid times n={N-1,..., 0}
37        for j=0:n    % states j={0,1,...,n} at time n
38          if (B(n+1,j+1)==0) % then use backwarding
39            B(n+1,j+1)=(pu*B(n+2,j+2)+(1-pu)*B(n+2,j+1))/R;
40          end
41        end
42      end
43      return;    % Prices in B are now defined.
44  end
```

4.2   Path Dependent Prices

For some assets A, the price $A(T, \omega)$ may depend on the path by which state ω was reached at time T. To compute such a price in a discrete model with path dependent states, the successor states that follow from (n, k) require their own unique labels as they are not assumed to be related to any other predecessor.

In a model where there is one up state and one down state after each (n, k), successors may be labeled uniquely by $(n + 1, 2k)$ and $(n + 1, 2k + 1)$, respectively. This difference from the recombining tree is evident after two time steps when there are four rather than three states:

$$A(0,0)$$

$$A(1,0) \qquad\qquad\qquad A(1,1)$$

$$A(2,0) \quad A(2,1) \qquad\qquad A(2,2) \quad A(2,3)$$

With such labeling, the number of distinct states doubles at each time step. There will be 2^n different states at intermediate time step n and 2^N possible states at final time N. With such rapid growth, the practical limit is $N \approx 30$ on contemporary computers.

4.2.1   Efficient Data Structures

Let $k \in \{0, 1, \ldots, 2^n - 1\}$ be one of the path dependent states at time step n. Its unique predecessor will then be labeled $(n - 1, \lfloor k/2 \rfloor)$, where $\lfloor x \rfloor$ is the *floor* function that returns the greatest integer less than or equal to x. Thus, for integer $k \geq 0$,

$$\lfloor k/2 \rfloor = \begin{cases} k/2, & k \text{ even,} \\ (k-1)/2, & k \text{ odd.} \end{cases} \tag{4.1}$$

Putting $A(n, k)$ into an $N \times 2^N$ matrix gives the following:

$$\boxed{A(n,k)} \qquad\qquad\qquad k \quad \rightarrow$$

$$
\begin{array}{c}
A(0,0) \\
A(1,0) \quad A(1,1) \\
n \qquad A(2,0) \quad A(2,1) \quad A(2,2) \quad A(2,3) \\
\downarrow \qquad \vdots \qquad \vdots \qquad \vdots \qquad \ddots \\
A(N,0) \quad A(N,1) \quad A(N,2) \quad \cdots \quad A(N, 2^N - 1)
\end{array}
$$

For large N it is inefficient to depict such a tree as part of a mostly empty $2^N \times 2^N$ matrix. It is better to use a linear array $\{\bar{A}(m) : 1 \leq m < 2^{N+1}\}$ and an indexing function

$$m(n, k) \stackrel{\text{def}}{=} 2^n + k, \qquad 0 \leq n \leq N, \ 0 \leq k < 2^n. \tag{4.2}$$

Note that $m = 2^n + k$ determines n, k uniquely by the formulas

$$n = \lfloor \log_2(m) \rfloor, \qquad k = m - 2^n. \tag{4.3}$$

Alternatively, expressing m in binary notation reveals n as the place of $m_n = 1$, the most significant *bit*, or base-2 digit of m, and k as the binary number represented by the remaining bits:

$$m = m_n m_{n-1} \cdots m_1 m_0 \text{ (base 2)}, \qquad k = m_{n-1} \cdots m_1 m_0 \text{ (base 2)}. \tag{4.4}$$

Then $A(n, k) = \bar{A}(m(n, k))$ is the path dependent price of the asset at time step n reached along the path indexed by k. If $m = 2^n + k$ corresponds to the current time and state indices, then $2m = 2^{n+1} + 2k$ and $2m + 1 = 2^{n+1} + 2k + 1$ index the subsequent down state and up state, respectively. Conversely, the unique predecessor of state m has index $\lfloor m/2 \rfloor$.

$\bar{A}$ may be arranged as a tree in rows by time step:

$$\boxed{\bar{A}(2^n + k)} \qquad\qquad\qquad k \quad \rightarrow$$

$$
\begin{array}{c}
n \\
\downarrow
\end{array}
\quad
\begin{array}{cccc}
\bar{A}(1) & & & \\
\bar{A}(2) & \bar{A}(3) & & \\
\bar{A}(4) & \bar{A}(5) & \bar{A}(6) & \bar{A}(7) \\
\vdots & \vdots & \vdots & \ddots \\
\bar{A}(2^N) & \bar{A}(2^N + 1) & \bar{A}(2^N + 2) & \cdots \quad \bar{A}(2^N + 2^N - 1)
\end{array}
$$

The no-arbitrage assumption implies that the future value of an asset price (at riskless rate R, which may depend on time and state) must lie between its two successor prices. Namely,

$$A(n + 1, 2k) < R(n, k)A(n, k) < A(n + 1, 2k + 1), \quad 0 \le n < N, \ 0 \le k < 2^n,$$

or equivalently, in the other indexing convention,

$$\bar{A}(2m) < \bar{R}(m)\bar{A}(m) < \bar{A}(2m + 1), \quad 1 \le m < 2^N,$$

where $\bar{R}(m) = \bar{R}(2^n + k) \overset{\text{def}}{=} R(n, k)$.

Indexing examples

The unique path to any state (n, k) at time step n is determined by the binary digits of k, or equivalently from those of the *terminus index* $m = 2^n + k$ using Eq.4.4. For example,

$$
\begin{aligned}
m = 51 &= 32 + 16 + 2 + 1 = 2^5 + 2^4 + 2^1 + 2^0 = 110011 \text{ (base 2)} \\
\implies k = 19 &= 16 + 2 + 1 = 2^4 + 2^1 + 2^0 = 10011 \text{ (base 2)},
\end{aligned}
$$

which defines $k = m_4 m_3 m_2 m_1 m_0$ (base 2) $= 10011$ (base 2).

Now denote by $k(i)$ the (non-recombining) state reached at time step i along the unique path to state $m = 2^n + k$. Then k is defined recursively by

$$
\begin{aligned}
k(0) &= 0, \quad \text{since } (0,0) \text{ is the unique initial state,} \\
k(i) &= 2k(i-1) + m_{n-i}, \quad i = 1, \ldots, n,
\end{aligned}
$$

where $m_{n-i} = 1$ indicates an up transition and $m_{n-i} = 0$ indicates a down transition to time step i. For example, state $k = 19$, of the $2^N = 32$ states at the bottom of the $N = 5$ time-step tree, defines

$$
\begin{aligned}
k(0) &= 0, \\
k(1) &= 2k(0) + m_4 = 1, \quad (m_4 = 1) \\
k(2) &= 2k(1) + m_3 = 2, \quad (m_3 = 0) \\
k(3) &= 2k(2) + m_2 = 4, \quad (m_2 = 0) \\
k(4) &= 2k(3) + m_1 = 9, \quad (m_1 = 1) \\
k(5) &= 2k(4) + m_0 = 19 = k. \quad (m_0 = 1)
\end{aligned}
$$

The path's terminus index $m = 2^N + k = 32 + 19 = 51$, in the linear array, thus similarly defines a unique path in the recombining tree.

Prices from up and down steps

The modeled asset prices in a non-recombining binomial tree define up and down factors:

$$
\begin{aligned}
d(n,k) &\stackrel{\text{def}}{=} A(n+1, 2k)/A(n,k); \\
u(n,k) &\stackrel{\text{def}}{=} A(n+1, 2k+1)/A(n,k).
\end{aligned}
$$

Alternatively, in the m notation,

$$
\begin{aligned}
\bar{d}(m) &\stackrel{\text{def}}{=} \bar{A}(2m)/\bar{A}(m); \\
\bar{u}(m) &\stackrel{\text{def}}{=} \bar{A}(2m+1)/\bar{A}(m),
\end{aligned}
$$

and so $\bar{u}(m(n,k)) = u(n,k)$ and $\bar{d}(m(n,k)) = d(n,k)$.

Conversely, given non-recombining trees u, d (or $\bar{u}, \bar{d}$) of up and down steps, a non-recombining tree of asset prices may be computed by forward induction from the spot price A_0:

$$
\begin{aligned}
A(0,0) &= A_0; \\
A(n+1, 2k) &= d(n,k)A(n,k); \\
A(n+1, 2k+1) &= u(n,k)A(n,k), \quad 0 \le n < N; \quad 0 \le k < 2^n.
\end{aligned}
$$

In the m notation this is

$$
\begin{aligned}
\bar{A}(1) &= A_0; \\
\bar{A}(2m) &= \bar{d}(m)\bar{A}(m), \\
\bar{A}(2m+1) &= \bar{u}(m)\bar{A}(m), \quad 1 \le m < 2^N.
\end{aligned}
$$

The price $A(N,k)$, or equivalently $\bar{A}(m) = \bar{A}(2^N + k)$, is the spot price $A(0,0) = \bar{A}(1)$ times the product of factors $s_1, s_2, \ldots, s_N$ where

$$s_i = \begin{cases} u(i-1, k(i-1)), & k_{N-i} = 1, \\ d(i-1, k(i-1)), & k_{N-i} = 0. \end{cases} \quad \text{(in } A(N,k) \text{ indexing)},$$

$$= \begin{cases} \bar{u}(m(i-1)), & m_{N-i} = 1, \\ \bar{d}(m(i-1)), & m_{N-i} = 0, \end{cases} \quad \text{(in the } \bar{A}(m) \text{ convention)}.$$

For the example above, the price of path dependent asset A in state $(N,k) = (5,19)$ is therefore

$$A(5,19) = u(0,0)d(1,1)d(2,2)u(3,4)u(4,9)A(0,0).$$

Equivalently, using $\bar{A}$ and state index $m = 2^N + k = 51$, the price is

$$\bar{A}(m) = \bar{A}(2^N + k) = \bar{A}(51) = \bar{u}(1)\bar{d}(3)\bar{d}(6)\bar{u}(12)\bar{u}(25)\bar{A}(1).$$

Observe how the factors *udduu* correspond to the bits of $k = 10011$ (base 2).

Non-recombining binary trees with constant factors from CRR

In the Cox-Ross-Rubinstein (CRR) model with constant u, d and riskless return R, all the prices in the non-recombining tree come from the recombining CRR tree but are injected into multiple places. The no-arbitrage axiom requires that

$$0 < d < R < u,$$

and then the risk neutral probabilities will be constant at each time step in each state:

$$p = \frac{R-d}{u-d}, \qquad 1-p = \frac{u-R}{u-d}.$$

A risky asset S with spot price S_0 may be modeled by a non-recombining tree $\bar{S}$ using just the constant up and down factors:

$$\bar{S}(1) = S_0; \qquad \bar{S}(2m) = d\bar{S}(m); \qquad \bar{S}(2m+1) = u\bar{S}(m). \quad (4.5)$$

Eq.4.5 is implemented by the following Octave function:

```
1  function Sbar = NRTCRR(S0, up, down, N)
2  % Octave/MATLAB function to compute the binary non-
3  % recombining tree of underlying asset prices from a
4  % spot price and constant up and down factors.
5  % INPUTS:                              (Example)
6  %    S0 = asset spot price               (100)
7  %    up = up factor                      (1.2)
8  %    down = down factor                  (0.8)
9  %    N = binomial tree height, must be >=0  (3)
10 % OUTPUT:
```

```
11  %      Sbar = tree as linear array Sbar(m), 0<m<2*2^N
12  % EXAMPLE:
13  %      Sbar = NRTCRR(100,1.2,0.8,3)
14  %
15     Sbar = zeros(1,2*2^N-1);  % allocate the output matrix
16     Sbar(1) = S0;     % initialize with the spot price
17     for m=1:2^N-1     % all future times up to N-1
18       Sbar(2*m) = down*Sbar(m);    % down descendent
19       Sbar(2*m+1) = up*Sbar(m);    % up descendent
20     end
21     return
22  end
```

The output is a linear array containing the non-recombining depth-N binary tree of prices $\bar{S}(m)$, defined for $0 < m < 2 \times 2^N$.

4.2.2 Paths in Recombining Trees

Every path $(0,0) \to (1, j_1) \to \cdots \to (n, j_n)$ to level n in a recombining binomial tree corresponds to a unique terminus index $m = m(j_1, \ldots, j_n)$ in a non-recombining tree of depth n. All such path terminus indices form the set

$$\mathcal{M}(n) \stackrel{\text{def}}{=} \{2^n, 2^n + 1, \ldots, 2^n + (2^n - 1)\} \tag{4.6}$$

Each m in this set depends on more than just its last j variable: at time steps $n > 1$, there will be several indices m corresponding to some of the states (n, j_n) in the recombining tree. However, there is a function mapping m to (n, j_n). It may be defined using the binary bits of m, using Eq.4.4. Alternatively, a non-recombining tree may be built to identify the terminal state number j_n from a path dependent terminus index m:

$$j(m) = j_n, \qquad 2^n \leq m \leq 2^n + 2^n - 1.$$

Notice that j_n is the number of up steps on the path to (n, j_n). These may be counted by the following forward induction:

$$
\begin{aligned}
j(1) &= 0; & &(4.7) \\
j(2m) &= j(m), & m &\geq 1; \\
j(2m + 1) &= j(m) + 1, & m &\geq 1.
\end{aligned}
$$

The recursion increments the up count by one every time a path turns upwards. Similarly, the time step n may be identified from m by the recursion

$$
\begin{aligned}
n(1) &= 0; & &(4.8) \\
n(2m) &= n(m) + 1, & m &\geq 1; \\
n(2m + 1) &= n(m) + 1, & m &\geq 1.
\end{aligned}
$$

These two indexing functions are easily implemented in Octave:

```
1  function [nm,jm] = NJfromM(N)
2  % Octave/MATLAB function to compute indexes (n,j)
3  % in a recombining binomial tree from the index m
4  % in a non−recombining tree, by induction.
5  % INPUTS:                              (Example)
6  %    N = tree height, must be >0            (3)
7  % OUTPUTS:
8  %    nm = n(m), time step
9  %    jm = j(m), recombining state at time step n(m)
10 % EXAMPLE:
11 %    [nm,jm] = NJfromM(3);
12 %
13    % Initialize output arrays with zeros
14    nm = zeros(1,2*2^N−1); %... so  nm(1)==0;
15    jm = zeros(1,2*2^N−1); %... so  jm(1)==0;
16    for m=1:2^N−1 % all m indices before the last row
17       nm(2*m)=nm(m)+1;  nm(2*m+1)=nm(m)+1; % increment
18       jm(2*m)=jm(m);    jm(2*m+1)=jm(m)+1; % bifurcate
19    end % nm(m) and jm(m) now hold (n(m),j(m)), all m
20    return
21 end
```

Example: recombining into non-recombining

One application of the mappings in Eqs.4.7 and 4.8 is the injection of prices from a recombining tree into a non-recombining tree. Suppose that risky prices $S(n,j)$ and riskless returns $R(n,j)$ are two such trees, for $0 \leq n \leq N$ and $0 \leq j \leq n$. They may be placed into non-recombining trees by

$$\bar{S}(m) \stackrel{\text{def}}{=} S(n(m), j(m)), \tag{4.9}$$

$$\bar{R}(m) \stackrel{\text{def}}{=} R(n(m), j(m)), \tag{4.10}$$

for $m = 1, 2, \ldots, 2^{N+1} - 1$. Similarly, the up factors, down factors, and risk neutral up probabilities from a recombining tree may be mapped into non-recombining trees as follows:

$$\bar{u}(m) \stackrel{\text{def}}{=} u(n(m), j(m)), \tag{4.11}$$

$$\bar{d}(m) \stackrel{\text{def}}{=} d(n(m), j(m)), \tag{4.12}$$

$$\bar{p}(m) \stackrel{\text{def}}{=} p(n(m), j(m)), \tag{4.13}$$

A single function can perform this injection from a given recombining tree matrix into a non-recombining linear array:

```
1  function Xbar = NRTinject(X,N)
2  % Octave/MATLAB function to inject values from a
3  % recombining binomial tree X with depth N, stored
```

```
4  % in an (N+1)x(N+1) matrix, into a non-recombining
5  % tree stored in an array indexed by m=1:2*2^N-1.
6  % INPUTS:                                      (Example)
7  %    X = recombining tree matrix               (4x4 lower)
8  %    N = tree height, must be >0                   (3)
9  % OUTPUTS:
10 %    Xbar = output array, Xbar(m)=X(n(m)+1,j(m)+1)
11 % EXAMPLE:
12 %    X = [1,0,0,0; 2,3,0,0;4,5,6,0;7,8,9,10];
13 %    Xbar = NRTinject(X,3);
14 %
15    [nm,jm] = NJfromM(N); % get indexing functions
16    for m=1:2*2^N-1  % all m indices including last row
17       Xbar(m) = X(nm(m)+1,jm(m)+1); % no 0 index: add 1
18    end
19    return
20 end
```

Example: risk neutral path probabilities

Another application is computing the probability of individual paths through a recombining binomial tree. Let $\Pr(m)$ be the probability of path m. Using $\bar{p}(m)$ as defined in Eq.4.13, compute Pr by the following induction:

$$
\begin{aligned}
\Pr(1) &= 1; & &\text{(4.14)}\\
\Pr(2m) &= (1-\bar{p}(m))\Pr(m), & 1 \le m < 2^N,\\
\Pr(2m+1) &= \bar{p}(m)\Pr(m), & 1 \le m < 2^N.
\end{aligned}
$$

Note that the 2^N longest paths that end at time step N correspond to the 2^N terminus indices $m \in \mathcal{M}(N) = \{2^N, 2^N+1, \ldots, 2^N+(2^N-1)\}$ from Eq.4.6.

```
1  function Pr = PathPr(pu,N)
2  % Octave/MATLAB function to compute the probability
3  % of a path in a recombining binomial tree X with
4  % depth N, given up probabilities pu stored in an
5  % NxN matrix, and store them in a non-recombining
6  % tree indexed by m=1:2*2^N-1.
7  % INPUTS:                                      (Example)
8  %    pu = recombining tree matrix              (3x3 lower)
9  %    N = tree height, must be >0                   (3)
10 % OUTPUTS:
11 %    Pr = probability of each m-path to (n(m),j(m))
12 % EXAMPLE:
13 %    pu = [.51,0,0; .52,.53,0;.54,.55,.56];
14 %    Pr = PathPr(pu,3);
15 % Visualize the probabilities of the longest paths:
16 %    plot(Pr(8:15))
17 %
18    pubar = NRTinject(pu,N-1); % inject rows 1:(N-1)
```

```
19    Pr = ones(1,2*2^N−1);  %... so Pr(1)=1
20    for m=1:2^N−1  % all m indices before row N
21      Pr(2*m)   = (1−pubar(m))*Pr(m);  % down state
22      Pr(2*m+1) = pubar(m)*Pr(m);  % up state
23    end
24    return
25  end
```

4.2.3 Path Dependent Arrow-Debreu Securities

Recall that the spot price of an asset W may be computed with an Arrow-Debreu expansion:

$$W(0,0) = \sum_k W(N,k)\lambda(N,k),$$

where $\lambda(N,k)$ is the spot price of the unit security that, at time N, pays 1 in state k but 0 in all other states $k' \neq k$. This applies to path dependent assets where k indexes each path from time 0 to time N.

As with the recombining binomial tree model, there is an efficient forward induction for computing all the path dependent Arrow-Debreu spot prices with a single non-recombining tree. It follows the construction of Eq.3.21.

The linear array "m" indexing notation will be used throughout.

Fix a time step n, $0 \leq n \leq N$, and a state k, $0 \leq k < 2^n$, and put $m = 2^n + k$. Let $\bar{\lambda}(m) = \bar{\lambda}(2^n + k)$ be the spot price at time $t = 0$ of the Arrow-Debreu portfolio that, at time step n, is worth 1 in state k and 0 in every other state $k' \neq k$. These path dependent prices fill a non-recombining tree of depth N.

The initial observation is that each $\bar{\lambda}(m)$ may be computed by backward induction on its own non-recombining tree $\bar{W}$ of depth n. The terminal condition applies to row n, which corresponds to terminus indices $2^n \leq m' < 2^{n+1}$:

$$\bar{W}(m') = \bar{W}(2^n + k') = \begin{cases} 1, & k' = k; \\ 0, & k' \neq k, \end{cases} \qquad 0 \leq k' < 2^n. \qquad (4.15)$$

The backward induction step is

$$\bar{W}(m) = \frac{\bar{p}(m)\bar{W}(2m+1) + [1 - \bar{p}(m)]W(2m)}{\bar{R}(m)}, \qquad (4.16)$$

for $m = 2^n - 1, 2^n - 2\ldots, 1$. Here the risk neutral probabilities $\bar{p}$ and riskless returns $\bar{R}$ are taken from their own non-recombining trees into which they were injected from recombining binomial trees via Eqs.4.13 and 4.10. But since there is only one nonzero term in row n (at $m = 2^n + k$), there is only one nonzero term in row $n - 1$, at predecessor state $\lfloor m/2 \rfloor$. Its value depends on whether k, and thus m, is odd or even:

$$\bar{W}(\lfloor m/2 \rfloor) = \begin{cases} \dfrac{1 - \bar{p}(\lfloor m/2 \rfloor)}{\bar{R}(\lfloor m/2 \rfloor)}, & m \text{ even,} \\[3mm] \dfrac{\bar{p}(\lfloor m/2 \rfloor)}{\bar{R}(\lfloor m/2 \rfloor)}, & m \text{ odd.} \end{cases}$$

This portfolio is a multiple of the Arrow-Debreu portfolio that pays 1 at state $\lfloor m/2 \rfloor$, so the One Price Theorem implies that $\bar{W}(1) = \bar{\lambda}(m)$ is a multiple of the spot price $\bar{\lambda}(\lfloor m/2 \rfloor)$:

$$\bar{\lambda}(m) = \lambda(\lfloor m/2 \rfloor) \times \begin{cases} \dfrac{1 - \bar{p}(\lfloor m/2 \rfloor)}{\bar{R}(\lfloor m/2 \rfloor)}, & m \text{ even,} \\[3mm] \dfrac{\bar{p}(\lfloor m/2 \rfloor)}{\bar{R}(\lfloor m/2 \rfloor)}, & m \text{ odd.} \end{cases}$$

Now substitute $m \leftarrow 2m$ and $m \leftarrow 2m + 1$ to get the forward induction formulas for the down and up descendents, respectively, to conclude:

Theorem 4.1 *Path dependent Arrow-Debreu securities satisfy*

$$\bar{\lambda}(1) = 1, \qquad \bar{\lambda}(2m) = \frac{1 - \bar{p}(m)}{\bar{R}(m)}\bar{\lambda}(m), \qquad \bar{\lambda}(2m + 1) = \frac{\bar{p}(m)}{\bar{R}(m)}\bar{\lambda}(m),$$

for all $m = 1, 2, \ldots, 2^N - 1$. $\qquad\qquad\qquad\qquad\qquad\qquad\qquad\qquad\quad\square$

The Octave function to find $\bar{\lambda}$ is very similar to that for Pr:

```
 1  function Lbar = PathAD(pu, R, N)
 2  % Octave/MATLAB function to compute spot prices
 3  % for path−dependent Arrow−Debreu securities by
 4  % forward induction.
 5  % INPUTS:                                    (Example)
 6  %   pu = NxN matrix of up probabilities  (3x3 lower)
 7  %   R  = NxN matrix of riskless returns  (3x3 lower)
 8  %   N = tree height, must be >0                  (3)
 9  % OUTPUTS:
10  %     Lbar = array of Arrow−Debreu prices barlambda(m).
11  % EXAMPLE:
12  %     pu = [.51,0,0; .52,.53,0; .54,.55,.56];
13  %     R = [1.01,0,0; 1.02,1.03,0; 1.04,1.05,1.06];
14  %     Lbar = PathAD(pu,R,3)
15  %
16      Lbar = ones(1,2*2^N−1); %... so Lbar(1)=1=lambda(1)
17      pbar = NRTinject(pu,N−1); % inject up probs.
18      Rbar = NRTinject(R,N−1); % inject riskless returns
```

```
19      for m=1:2^N-1  % all indices before the last row
20          Lbar(2*m)=(1-pbar(m))*Lbar(m)/Rbar(m); % down
21          Lbar(2*m+1) = pbar(m)*Lbar(m)/Rbar(m); % up
22      end
23      return; % Lbar(m) == barlambda(m), 0<m<2*2^N
24  end
```

Corollary 4.2 *The recursion in Theorem 4.1 is solved by*

$$\bar{\lambda}(m) = \Pr(m) / \prod_{i=0}^{n} \bar{R}(\lfloor m/2^i \rfloor), \tag{4.17}$$

where $n = \lfloor \log_2 m \rfloor$ *is the highest power of 2 in* m.

Proof: This is evident since the path probability $\Pr(m)$ in Eq.4.14 satisfies the same recursion as $\bar{\lambda}(m)$ only without the denominator factors $\bar{R}(m)$. □

4.2.4 Asian-Style Options

An *average rate option*, sometimes called an Asian-style option, is like a European-style option on a stock S in that it can only be exercised at expiry T, but its payoff depends on the average price of the underlying asset over the period $[0, T]$. Thus its Call and Put payoffs, for strike price K, are respectively

$$\overline{C}(T) = [A(T) - K]^+; \qquad \overline{P}(T) = [K - A(T)]^+, \tag{4.18}$$

where the average price $A(T)$ is computed from samples at specific times $0 = t_0 < t_1 < \cdots < t_N = T$:

$$A(T) \stackrel{\text{def}}{=} \frac{1}{N+1} \sum_{i=0}^{N} S(t_i).$$

If the option is exercised, the holder is paid a cash settlement by the counterparty equal to the difference between the strike price and the average price. This is a useful risk-reduction tool for volatile assets that are vulnerable to price manipulation.

It is reasonable to use a discrete-time model to price such options since the specific times are usually equispaced. The underlying asset S prices may be modeled by a recombining binary tree such as in CRR. However, the averages will be path dependent in all but the one-step case. To see this, suppose that u, d are up and down factors, respectively, in a CRR model of depth $N > 1$. At time $n = 2$ there is a single modeled price for $S(2, 1)$:

$$S(2, 1) = S(0, 0)ud = S(0, 0)du,$$

since $ud = du$. However, there are two paths from $(0,0)$ to $(2,1)$ through the recombining tree, namely

$$(0,0) \overset{d}{\to} (1,0) \overset{u}{\to} (2,1) \quad \Longrightarrow \quad A_{ud} = \frac{S(0,0)[1 + d + ud]}{3},$$

$$(0,0) \overset{u}{\to} (1,1) \overset{d}{\to} (2,1) \quad \Longrightarrow \quad A_{du} = \frac{S(0,0)[1 + u + du]}{3},$$

and they must differ because $d \neq u$. Hence the averages must be stored in a non-recombining tree of path dependent prices.

Call-Put parity for Asian-style options

The payoffs for an average rate Call and Put are related by

$$\overline{C}(T) - \overline{P}(T) = [A(T) - K]^+ - [K - A(T)]^+ = A(T) - K,$$

using Eq.1.17. Then Fair Price Theorem 1.4 gives a parity formula for the premiums at time $t = 0$:

$$\overline{C}(0) - \overline{P}(0) = \exp(-rT)\mathrm{E}(A(T) - K),$$

where r is the riskless interest rate per year. Now $\mathrm{E}(A(T) - K) = \mathrm{E}(A(T)) - K$, and $\mathrm{E}(A(T))$ is computed using path probabilities, which requires a model of the future:

Theorem 4.3 *In the N-step CRR model for S with annual riskless rate r and riskless return $R = \exp(rT/N)$, the averages $A(T)$ of S along all paths from 0 to expiry T at time step N satisfy*

$$\mathrm{E}(A(T)) = \frac{1 + R + \cdots + R^N}{N + 1} S_0,$$

where $S_0 = S(0,0)$ is the spot price of S at time 0.

Proof: Denote by $U(N, m)$ the sum of S along the path m from 0 to N:

$$U(N, m) \overset{\mathrm{def}}{=} \sum_{n=0}^{N} S(n, j_n(m))$$

using the notation j_n from Eq.4.7 for intermediate states along the path m. Then the average of S along path m may be denoted by

$$A(N, m) \overset{\mathrm{def}}{=} \frac{1}{N + 1} U(N, m),$$

since there are $N + 1$ summands in $U(N, m)$. Hence, computing expectations with path probabilities in the N-step CRR model gives

$$(N + 1)\mathrm{E}(A(T)) = (N + 1) \sum_{m \in \mathcal{M}(N)} A(N, m)\mathrm{Pr}(m)$$

$$= \sum_{m \in \mathcal{M}(N)} U(N, m)\mathrm{Pr}(m) \overset{\mathrm{def}}{=} \mathrm{E}(U(N)),$$

where $\Pr(m)$ is defined recursively by Eq.4.14, and

$$\mathcal{M}(N) = \{2^N, 2^N + 1, \ldots, 2^N + (2^N - 1)\}$$

is the set of m-indices of all N-step paths through the S tree.

Now use induction on N to establish that

$$E(U(N)) = (1 + R + \cdots + R^N)S_0. \qquad (4.19)$$

The base case $N = 0$ holds since $U(0,0) = S_0$ is the one-element sum of just the spot price.

For the inductive step, suppose that Eq.4.19 holds for N and consider

$$\begin{aligned} E(U(N+1)) &= \sum_{m \in \mathcal{M}(N+1)} U(N+1, m)\Pr(m) \\ &= \sum_{m \in \mathcal{M}(N)} \Big[U(N+1, 2m)\Pr(2m) \\ &\qquad\qquad + U(N+1, 2m+1)\Pr(2m+1) \Big], \end{aligned}$$

breaking the sum into two parts by odd and even descendent from step N. But

$$\begin{aligned} U(N+1, 2m) &= U(N, m) + \bar{S}(2m), \\ U(N+1, 2m+1) &= U(N, m) + \bar{S}(2m+1), \\ \Pr(2m) &= (1-p)\Pr(m) \\ \Pr(2m+1) &= p\Pr(m), \end{aligned}$$

where p and $1-p$ are the risk neutral up and down probabilities from the CRR model for S, and $\bar{S}$ is the price from the S tree injected into an m-indexed tree by Eq.4.9. Use these to evaluate

$$\begin{aligned} \sum_{m \in \mathcal{M}(N)} U(N+1, 2m)\Pr(2m) &= \sum_{m \in \mathcal{M}(N)} [U(N, m) + \bar{S}(2m)]\Pr(2m) \\ &= (1-p) \sum_{m \in \mathcal{M}(N)} U(N, m)\Pr(m) \\ &\qquad + \sum_{m \in \mathcal{M}(N)} \bar{S}(2m)\Pr(2m), \end{aligned}$$

and likewise

$$\begin{aligned} \sum_{m \in \mathcal{M}(N)} U(N+1, 2m+1)\Pr(2m+1) &= p \sum_{m \in \mathcal{M}(N)} U(N, m)\Pr(m) \\ &\qquad + \sum_{m \in \mathcal{M}(N)} \bar{S}(2m+1)\Pr(2m+1). \end{aligned}$$

Combining these two parts gives

$$
\begin{aligned}
\mathrm{E}(U(N{+}1)) &= \sum_{m\in\mathcal{M}(N)} U(N,m)\mathrm{Pr}(m) + \sum_{m\in\mathcal{M}(N+1)} \bar{S}(m)\mathrm{Pr}(m) \\
&= \mathrm{E}(U(N)) + \mathrm{E}(S(N+1)) \\
&= (1 + R + \cdots + R^N)S_0 + R^{N+1}S_0,
\end{aligned}
$$

using the inductive hypothesis and the fair price $\mathrm{E}(S(N+1)) = R^{N+1}S_0$ from the CRR model with $N+1$ time steps and riskless return R per step. This establishes Eq.4.19 for all N and completes the proof. $\qquad\square$

Corollary 4.4 *As $N \to \infty$ in the CRR model with N steps to expiry T,*

$$
\mathrm{E}(A(T)) \to \frac{\exp(rT) - 1}{rT} S_0
$$

Proof: Apply the geometric sum formula to Theorem 4.3:

$$
\mathrm{E}(A(T)) = \frac{1}{N+1}(1 + R + \cdots + R^N) = \frac{1}{N+1}\frac{R^{N+1} - 1}{R - 1}
$$

Substitute $R = \exp(rt/N)$ and use the Taylor approximation $e^x - 1 \approx x$, valid for small x, to find the limit of the denominator:

$$
(N+1)(R-1) = (N+1)(\exp(rT/N) - 1) \approx (N+1)rT/N \quad \to \quad rT,
$$

as $N \to \infty$. Likewise, the numerator limit is

$$
R^{N+1} - 1 = \exp(rt(N+1)/N) - 1 \quad \to \quad \exp(rT) - 1,
$$

as $N \to \infty$. $\qquad\square$

With Corollary 4.4, the Call-Put parity formula for average rate options may be stated in a model-independent manner:

$$
\overline{C}(0) - \overline{P}(0) = \frac{1 - \exp(-rT)}{rT} S_0 - \exp(-rT)K. \tag{4.20}
$$

Octave implementation of partial sums along paths

Let $\bar{U}(m) \overset{\text{def}}{=} U(n(m), m)$ be the partial sum along path m as used in the proof of Theorem 4.3. This may be computed from the non-recombining tree $\bar{S}$ by the following recursion: put $\bar{U}(1) = \bar{S}(1)$, and then for $m = 1, 2, \ldots, 2^N - 1$, compute

$$
\bar{U}(2m) = \bar{U}(m) + \bar{S}(2m); \qquad \bar{U}(2m{+}1) = \bar{U}(m) + \bar{S}(2m{+}1). \tag{4.21}
$$

Eq.4.21 is easily implemented in Octave:

```
1  function Ubar = NRTpsums(Sbar, N)
2  % Octave/MATLAB function to compute partial sums
3  % along paths in a non-recombining tree (NRT).
4  % INPUTS:                              (Example)
5  %    Sbar = NRT array of length 2*2^N-1     (1:15)
6  %    N = tree depth, must be >=0            (3)
7  % OUTPUT:
8  %    Ubar = NRT of partial sums, same size as Sbar
9  % EXAMPLE:
10 %   Ubar = NRTpsums(1:15,3)
11 %
12    Ubar=zeros(size(Sbar)); % allocate the output
13    Ubar(1)=Sbar(1); % trivial first partial sum
14    for m=1:2^N-1  % all future times up to N-1
15    Ubar(2*m)=Ubar(m)+Sbar(2*m);      % down
16    Ubar(2*m+1)=Ubar(m)+Sbar(2*m+1); % up
17    end
18    return
19 end
```

Calculating path averages from these partial sums requires division by the number of time steps on the path indexed by m. This may be done with a few more lines:

```
Avgs=Ubar; % initialze an NRT array for path averages
for n=1:N % all time steps after now, to expiry
  mn=2^n:(2*2^n-1) % all m indices for time step n
  Avgs(mn)=Ubar(mn)/(n+1);
end
```

The loop is superfluous if only the terminal averages (to $n = N$) are needed.

Octave implementation of average rate options in CRR

From partial sums it is easy to compute averages, so one application is to price Asian-style average rate options. This implementation uses the simplifying CRR assumptions:

```
1  function [C, P] = CRRaro(T, S0, K, r, v, N)
2  % Octave/MATLAB function to price Average Rate Options
3  % (ARO, or Asian Options) using the Cox-Ross-Rubinstein
4  % (CRR) binomial pricing model.
5  % INPUTS:                              (Example)
6  %    T  = expiration time in years      (0.5)
7  %    S0 = stock price                   (100)
8  %    K  = strike price                  (95)
9  %    r  = riskless yield per year       (0.05)
10 %    v  = volatility; must be >0        (0.15)
11 %    N  = height of the binomial tree   (5)
12 % OUTPUTS:
```

```
13  %      [C,P] = Call and Put option prices, all m
14  % EXAMPLE:
15  %      [C,P]=CRRaro(0.5,100,95,0.05,0.15,5); C(1),P(1)
16  %
17    [pu,up,R]=CRRparams(T,r,v,N);
18    Sbar=NRTCRR(S0,up,1/up,N); % expanded S tree
19    Ubar=NRTpsums(Sbar,N); % sums along each path
20    C=zeros(size(Sbar)); P=zeros(size(Sbar));
21    for m=2^N:(2*2^N -1) % m indices at expiry
22      Avg = Ubar(m)/(N+1); % N+1 summands for these
23      C(m) = max(0,Avg-K); % Call payoff at expiry
24      P(m) = max(0,K-Avg); % Put payoff at expiry
25    end % ...prices set at expiry.
26    for m=(2^N-1):-1:1 % recursive previous indices
27      C(m) = (pu*C(2*m+1) + (1-pu)*C(2*m))/R;
28      P(m) = (pu*P(2*m+1) + (1-pu)*P(2*m))/R;
29    end % ...all prices set by backward induction.
30    return
31  end
```

Average rate options may also be priced in the CRR model using path dependent Arrow-Debreu securities. The implementation `CRRaroAD()` differs from `CRRaro()` in three ways:

- It returns just the option premiums $C(0), P(0)$, rather than two complete non-recombining trees used to price the options.

- It uses `PathAD(pu*ones(N,N), R*ones(N,N), N)` to compute the path dependent Arrow-Debreu spot prices. Note that the recombining up probability and riskless return trees needed by `PathAD()` are entered as constant $N \times N$ matrices.

- It replaces the backward recursion of `CRRaro()` that fills the non-recombining C and R trees with two Arrow-Debreu expansions, the inner products

$$C_0 = \sum_{m \in \mathcal{M}} \bar{\lambda}(m)C(m), \quad P_0 = \sum_{m \in \mathcal{M}} \bar{\lambda}(m)P(m),$$

where $C(m)$ and $P(m)$ are the Call and Put option payoffs at expiry in state $m \in \mathcal{M} = \{2^N, \dots, 2^N + 2^N - 1\}$.

```
1  function [C0, P0] = CRRaroAD(T,S0,K,r,v,N)
2  % Octave/MATLAB function to price Average Rate
3  % Options (ARO, or Asian Options) using pathwise
4  % Arrow-Debreau expansions with the Cox-Ross-
5  % Rubinstein (CRR) binomial pricing model.
6  % INPUTS:                              (Example)
7  %    T = expiration time in years      (0.5)
```

```
8    %      S0 = stock price              (100)
9    %      K  = strike price             (95)
10   %      r  = riskless yield per year  (0.05)
11   %      v  = volatility; must be >0   (0.15)
12   %      N  = height of the binomial tree  (5)
13   % OUTPUTS:
14   %      [C0,P0] =   Call and Put option premiums
15   % EXAMPLE:
16   %      [C0,P0]=CRRaroAD(0.5,100,95,0.05,0.15,5)
17   %
18      [pu,up,R]=CRRparams(T,r,v,N);
19      Sbar=NRTCRR(S0,up,1/up,N); % expanded S tree
20      Ubar=NRTpsums(Sbar,N); % sums along each path
21      Lbar=PathAD(pu*ones(N,N),R*ones(N,N),N);
22      mN=2^N:(2*2^N-1); % all m-indices at expiry
23      AvgN=Ubar(mN)/(N+1); % S averages at expiry
24      C0=max(0,AvgN-K)*Lbar(mN)'; % Call payoff i.p.
25      P0=max(0,K-AvgN)*Lbar(mN)'; % Put payoff i.p.
26      return % A-D expansion of C0,P0 is completed.
27   end
```

Compare (and test for bugs) by running both functions on the same suggested example inputs:

```
[C0,P0]=CRRaroAD(0.5,100,95,0.05,0.15,5)
[C,P]=CRRaro(0.5,100,95,0.05,0.15,5); C(1),P(1)
```

The outputs agree: `C0=6.5124=C(1)` and `P0=0.40545=P(1)`.

Octave implementation of CRR for geometric rate options

A variant of average rate options, which use arithmetic means to compute $A(T)$, is the *geometric mean* average rate option, in which the payoff depends on the difference between the strike price K and the geometric mean $G(T)$ of sampled prices for S on $t \in [0, T]$:

$$G(T) = \sqrt[N+1]{\prod_{i=0}^{N} S(t_i)}.$$

The geometric mean is more easily calculated using logarithms. This has the added benefit of avoiding huge numbers when S and N are large:

$$\log G(T) = \frac{1}{N+1} \sum_{i=0}^{N} \log S(t_i).$$

At expiry, the values in $\log G$ are exponentiated before comparison with K to give the Call and Put payoffs:

$$C(T) = [\exp(\log G(T)) - K]^+, \qquad P(T) = [K - \exp(\log G(T))]^+.$$

The premiums $C(0), P(0)$ are thereafter computed by backward induction.

```
1  function [C, P] = CRRgro(T, S0, K, r, v, N)
2  % Octave/MATLAB function to price Geometric Rate Options
3  % (GRO, or geometric mean Asian-style Options) with the
4  % Cox-Ross-Rubinstein (CRR) binomial pricing model.
5  % INPUTS:                                    (Example)
6  %    T  = expiration time in years         (0.5)
7  %    S0 = stock price                      (100)
8  %    K  = strike price                     (95)
9  %    r  = riskless yield per year          (0.05)
10 %    v  = volatility; must be >0           (0.15)
11 %    N  = height of the binomial tree      (5)
12 % OUTPUTS:
13 %    [C,P] = Call and Put option prices, all m
14 % EXAMPLE:
15 %    [C,P]=CRRgro(0.5,100,95,0.05,0.15,5); C(1),P(1)
16 %
17   [pu,up,R]=CRRparams(T,r,v,N);
18   Sbar=NRTCRR(S0,up,1/up,N); % expanded S tree
19   lU=NRTpsums(log(Sbar),N); % sums of logarithms
20   C=zeros(size(Sbar));  P=zeros(size(Sbar));
21   for m=2^N:(2*2^N -1) % m indices at expiry
22     Geom = exp(lU(m)/(N+1));  % geometric mean
23     C(m) = max(0,Geom-K); % Call payoff at expiry
24     P(m) = max(0,K-Geom); % Put payoff at expiry
25   end % ...prices set at expiry.
26   for m=(2^N-1):-1:1 % recursive previous indices
27     C(m) = (pu*C(2*m+1) + (1-pu)*C(2*m))/R;
28     P(m) = (pu*P(2*m+1) + (1-pu)*P(2*m))/R;
29   end % ...all prices set by backward induction.
30   return
31 end
```

Geometric rate options may also be priced using path dependent Arrow-Debreu securities. The modifications needed to do this are left as an exercise.

4.2.5 Floating Strike Options

In this exotic variation the strike price is set at expiry T to be the average price of the underlying asset over the period $[0, T]$. Thus the Call and Put options repectively confer the right, but not the obligation, to buy or sell the underlying asset at its average market price. The Call and Put payoffs, therefore, are respectively

$$\widehat{C}(T) = [S(T) - A(T)]^{+}; \qquad \widehat{P}(T) = [A(T) - S(T)]^{+}, \qquad (4.22)$$

where the average price $A(T)$ is computed as before from samples at specific times $0 = t_0 < t_1 < \cdots < t_N = T$:

$$A(T) \stackrel{\text{def}}{=} \frac{1}{N+1} \sum_{i=0}^{N} S(t_i).$$

Call-Put parity for floating strike options

The payoffs for a floating strike Call and Put are related by

$$\widehat{C}(T) - \widehat{P}(T) = [S(T) - A(T)]^+ - [A(T) - S(T)]^+ = S(T) - A(T),$$

using Eq.1.17. Then Fair Price Theorem 1.4 gives a parity formula for the premiums at time $t = 0$:

$$\widehat{C}(0) - \widehat{P}(0) = \exp(-rT)\mathrm{E}(S(T) - A(T)) = S_0 - \exp(-rT)\mathrm{E}(A(T)),$$

where r is the riskless interest rate per year. In the CRR model of the future, $\mathrm{E}(A(T))$ is given by Theorem 4.3:

$$\widehat{C}(0) - \widehat{P}(0) = S_0 - \exp(-rT)\frac{1 + R + \cdots + R^N}{N+1}S_0.$$

Letting $N \to \infty$ and applying Corollary 4.4 then gives

$$\widehat{C}(0) - \widehat{P}(0) = \frac{\exp(-rT) + rT - 1}{rT}S_0. \tag{4.23}$$

For small values of rT, where $\exp(-rT) \approx 1 - rT + (rT)^2/2$ by Taylor's theorem, this parity formula is well approximated by

$$\widehat{C}(0) - \widehat{P}(0) \approx \frac{rT}{2}S_0. \tag{4.24}$$

Octave implementation of CRR for floating strike options

```
1  function [C, P] = CRRflt(T, S0, r, v, N)
2  % Octave/MATLAB function to price floating strike
3  % Call and Put options using the Cox-Ross-Rubinstein
4  % (CRR) binomial pricing model.
5  % INPUTS:                              (Example)
6  %    T  =  expiration time in years    (1)
7  %    S0 =  spot stock price            (90)
8  %    r  =  riskless yield per year     (0.02)
9  %    v  =  volatility; must be >0      (0.20)
10 %    N  =  height of the tree          (4)
11 % OUTPUTS:
12 %    C = Call option NRT.
13 %    P = Put option NRT.
```

```
14  % EXAMPLE:
15  %      [C,P] = CRRflt(1,90,0.02,0.20,4);
16  % To get just the premiums at t=0, use
17  %      C(1),P(1)
18  %
19      [pu,up,R]=CRRparams(T,r,v,N);
20      Sbar=NRTCRR(S0,up,1/up,N); % expanded S tree
21      Ubar=NRTpsums(Sbar,N); % path partial sums
22      C=zeros(size(Sbar));   P=zeros(size(Sbar));
23      for m=2^N:(2*2^N −1) % m indices at expiry
24        Avg = Ubar(m)/(N+1);        % N+1 summands
25        C(m) = max(0,Sbar(m)−Avg); % Call payoff
26        P(m) = max(0,Avg−Sbar(m)); % Put payoff
27      end % ...prices set at expiry.
28      for m=(2^N−1):−1:1 % recursive previous indices
29        C(m) = (pu*C(2*m+1) + (1−pu)*C(2*m))/R;
30        P(m) = (pu*P(2*m+1) + (1−pu)*P(2*m))/R;
31      end % ...all prices set by backward induction.
32      return
33  end
```

It is left as an exercise to implement floating strike option pricing using geometric means rather than arithmetic means.

4.2.6 Lookback Options

These exotic derivatives may be called *extremal*. They grant the right, at time T, to buy or sell the underlying asset at its best price over the period $[0, T]$. Thus the lookback Call and Put payoffs are respectively

$$\widetilde{C}(T) = S(T) - S_{\min}(T); \qquad \widetilde{P}(T) = S_{\max}(T) - S(T), \qquad (4.25)$$

where *minimum* and *maximum* stochastic processes derived from S are defined, respectively, by

$$S_{\min}(T) \stackrel{\text{def}}{=} \min\{S(t) : 0 \le t \le T\}; \qquad S_{\max}(T) \stackrel{\text{def}}{=} \max\{S(t) : 0 \le t \le T\}.$$

(The state variable ω is omitted for simplicity.) Note that the plus-part is not needed in these payoff formulas as the arguments are always nonnegative.

In practice, these maximum and minimum prices are computed as before from samples at specific times $0 = t_0 < t_1 < \cdots < t_N = T$. Abusing notation:

$$S_{\min}(N) \stackrel{\text{def}}{=} \min\{S(t_i) : 0 \le i \le N\}; \qquad S_{\max}(N) \stackrel{\text{def}}{=} \max\{S(t_i) : 0 \le i \le N\}.$$

A multistep discrete model for S is therefore appropriate. But even if S is modeled by a recombining tree, the extremal prices $S_{\min}, S_{\max}$ are path dependent and require a non-recombining tree. In that case, the omitted state variable ω describes the path, so the m notation is appropriate.

Octave program to find minimums along paths

Let $\bar{S}_{\min}(m)$ be the minimum value of $S(n, j)$ along the path indexed by m. It may be computed from the non-recombining tree $\bar{S}$ by the following recursion:

$$
\begin{aligned}
\bar{S}_{\min}(1) &= \bar{S}(1), \quad \text{and then for } m = 1, 2, \dots, 2^N - 1, \\
\bar{S}_{\min}(2m) &= \min\{\bar{S}_{\min}(m), \bar{S}(2m)\}, \\
\bar{S}_{\min}(2m+1) &= \max\{\bar{S}_{\min}(m), \bar{S}(2m+1)\}.
\end{aligned}
\tag{4.26}
$$

Eq.4.26 is easily implemented in Octave:

```
1  function Minb = NRTmin(Sbar, N)
2  % Octave/MATLAB function to compute minimums
3  % along paths in a non-recombining tree (NRT).
4  % INPUTS:                           (Example)
5  %    Sbar = NRT array of length 2*2^N-1   (1:15)
6  %    N = tree depth, must be >=0          (3)
7  % OUTPUT:
8  %    Minb = NRT of minimums, same size as Sbar
9  % EXAMPLE:
10 %  Minb = NRTmin(1:15,3)
11 %
12    Minb=zeros(size(Sbar)); % allocate the output
13    Minb(1)=Sbar(1); % min along the trivial path
14    for m=1:2^N-1   % all future times up to N-1
15      Minb(2*m)=min(Minb(m),Sbar(2*m));    % down
16      Minb(2*m+1)=min(Minb(m),Sbar(2*m+1)); % up
17    end
18    return
19 end
```

Implementation of the function `NRTmax()` to find maximums along all paths is left as an exercise.

Octave implementation of CRR for lookback options

Once the maximums and minimums along paths are computed, the lookback Put and Call options are found by backward recursion from terminal values:

```
1  function [C,P] = CRRlb(T,S0,r,v,N)
2  % Octave/MATLAB function to price Lookback Call and
3  % Put options using the Cox-Ross-Rubinstein (CRR)
4  % binomial pricing model.
5  % INPUTS:                           (Example)
6  %    T  = expiration time in years     (1)
7  %    S0 = spot stock price             (100)
8  %    r  = riskless yield per year      (0.05)
9  %    v  = volatility; must be >0       (0.20)
10 %    N  = height of the tree           (4)
11 % OUTPUTS:
```

```
12  %       C = Call option NRT
13  %       P = Put option NRT
14  % EXAMPLE:
15  %       [C,P]=CRRlb(1,100,0.05,0.20,4); C(1),P(1)
16  %
17     [pu,up,R]=CRRparams(T,r,v,N);
18     Sbar=NRTCRR(S0,up,1/up,N); % expanded S tree
19     MinS=NRTmin(Sbar,N); MaxS=NRTmax(Sbar,N);
20     C=zeros(size(Sbar)); P=zeros(size(Sbar));
21     for m=2^N:(2*2^N -1) % states at expiry N
22        C(m)=Sbar(m)-MinS(m); % Call payoff
23        P(m)=MaxS(m)-Sbar(m); % Put payoff
24     end % ...payoffs at expiry are set.
25     for m=(2^N-1):-1:1 % indices down to t=0
26        C(m) = (pu*C(2*m+1) + (1-pu)*C(2*m))/R;
27        P(m) = (pu*P(2*m+1) + (1-pu)*P(2*m))/R;
28     end % ... backward induction is complete
29     return % C(1),P(1) contain the premiums
30  end
```

Lookback options may also be priced using path dependent Arrow-Debreu securities. The modifications needed to do this are left as an exercise.

4.2.7 Ladder Options

Another exotic version of an extremal option is the *ladder option*. Here the payoff acquires a lower bound at expiry as soon as the extremal price crosses a rung on a ladder of in-the-money prices. The holder of such an option is thus assured a positive payoff even if the underlying asset price retraces back out of the money at expiry.

For example, a ladder Call option C on $S(t)$ has a specified strike price K and an increasing sequence $\{L_i\}$, or ladder, of in-the-money prices above K:

$$S(0) \leq \max\{S(0), K\} < L_1 < \cdots < L_k.$$

Let $M(T) = \max\{S(t) : 0 \leq t \leq T\}$ be the maximum price of S up to expiry. Then the ladder Call option payoff is

$$C(T) \stackrel{\text{def}}{=} \max\{[S(T) - K]^+, L_i - K\}, \tag{4.27}$$

where L_i is the highest rung of the ladder reached by $S(t)$ as recorded by $M(T)$:

$$L_i \leq M(T) < L_{i+1}.$$

Note that

$$C(T) \geq [S(T) - K]^+ \geq 0,$$

so this exotic option cannot cost less than a vanilla European Call with the same parameters.

The ladder for the corresponding Put option satisfies

$$L_k < \cdots < L_1 < \min\{S(0), K\} \le S(0),$$

and the very similar analysis and implementation is left as an exercise.

Octave implementation of CRR for ladder Call options

A discrete model for pricing ladder options requires a non-recombining tree of paths to hold the extremal prices. It is convenient to use a non-recombining tree for the underlying asset prices, too, even if they are modeled by a recombining binomial tree such as in CRR.

In this example with N time steps to expiry, both S and M are stored as linear arrays with m indexing. The paths of length N are indexed by $m \in \mathcal{M}(N) = \{2^N, \ldots, 2^N + (2^N - 1)\}$, so those are the indices in the non-recombining tree C where terminal values are set using Eq.4.27. Then the option premium is found by backward induction as usual:

$$C(m) = \frac{pC(2m+1) + (1-p)C(2m)}{R}, \quad m = 2^N - 1, 2^N - 2, \ldots, 2, 1,$$

where p is the risk neutral up probability from the CRR model, and R is the riskless return per time step.

```
1   function LadC = CRRladC(T,S0,K,L,r,v,N)
2   % Octave/MATLAB function to price a ladder Call
3   % option using a Cox-Ross-Rubinstein (CRR) model.
4   % INPUTS:                                    (Example)
5   %    T  = expiration time in years           (1)
6   %    S0 = spot price                         (50)
7   %    K  = strike price                       (55)
8   %    L  = increasing prices > S0,K           ([60,65,70])
9   %    r  = riskless yield per year            (0.05)
10  %    v  = volatility; must be >0             (0.20)
11  %    N  = height of the tree                 (4)
12  % OUTPUTS:
13  %    LadC = Ladder Call option price array.
14  % EXAMPLE:
15  %    LadC=CRRladC(1,50,55,[60,65,70],0.05,0.20,4)
16  %
17      [pu,up,R]=CRRparams(T,r,v,N);
18      Sbar=NRTCRR(S0,up,1/up,N); % S in an NRT
19      MaxS=NRTmax(Sbar,N); % non-recombining max tree
20      LadC=zeros(size(Sbar)); % Call prices NRT
21      % Initialize with the payoffs at expiry:
22      k=length(L);        % number of ladder levels, >1
23      for m=2^N:(2*2^N-1) % state indexes at expiry
24          if(MaxS(m)<L(1)) % below level L(1) is special
25              LadC(m)=max(Sbar(m)-K,0);
26          else % MaxS >= L(1)
```

```
27      if (MaxS(m)<L(k)) %  ... use  levels  L(2)<...<L(k)
28        for  l=2:k        % loop  to  find  the  level
29          if (MaxS(m)<L(l))
30            LadC(m)=max(max(Sbar(m)-K,L(l-1)-K),0);
31            break; % found  l:  L(l-1)=<MaxS<L(l),
32          end
33        end          % ... so  exit  the  "l"  loop
34      else % MaxS>=L(k),  another  special  case
35        LadC(m)=max(max(Sbar(m)-K,L(k)-K),0);
36      end
37    end
38  end
39  for  m=(2^N-1):-1:1 % backward  recursion
40    LadC(m)=(pu*LadC(2*m+1)+(1-pu)*LadC(2*m))/R;
41  end
42  return % LadC(1)  is  the  option  premium
43 end
```

An easy experiment with this function shows how much more the ladder Call costs than the vanilla European-style Call, using the same model for the underlying asset:

```
LadC=CRRladC(1,50,50,[55,60,65],0.05,0.20,4); LadC(1)
[eC,eP]=CRReur(1,50,50,0.05,0.20,4); eC(1,1)
```

The returned values are \$6.98 for the ladder Call versus \$4.99 for the vanilla European Call.

4.3   Exercises

1. Implement a compound Put option, the option to purchase a European-style Put option, with expiry T and strike price K, for price L at time T_1 satisfying $0 < T_1 < T$. Use your code to price such an option with parameters $(T, T_1, S_0, K, L, r, v, N) = (1, 0.5, 90, 95, 4.50, 0.02, 0.15, 20)$. (Hint: modify CRRcc.m.)

2. Let $\mathcal{M}(n)$ be the number of paths in a recombining binomial tree to depth n, as defined in Eq.4.6. Prove that

$$\mathcal{M}(n + 1) = [2\mathcal{M}(n)] \cup [2\mathcal{M}(n) + 1],$$

where $aX + b \stackrel{\text{def}}{=} \{ax + b : x \in X\}$ for sets X of numbers.

3. Implement floating strike option pricing in the CRR model using geometric means instead of arithmetic means, as in CRRgro versus CRRaro. Compare the results on the suggested example inputs.

4. Use `CRRaro` to compute the average-rate Call and Put premiums in the CRR model for $S_0 = K = 100$, $T = 1$, $r = 0.05$, $v = 0.15$, and different values of N. Compare $\overline{C}(0) - \overline{P}(0)$ with the limit value in Eq.4.20.

5. Implement floating strike option pricing in the CRR model using geometric means instead of arithmetic means, as in `CRRgro` versus `CRRaro`. Compare the results on the suggested example inputs.

6. Implement floating strike option pricing in the CRR model using path-dependent Arrow-Debreu securities. Check that the results agree with `CRRflt`.

7. Write an Octave program to compute the maximums along all paths in a non-recombining binary tree of depth N. (Hint: modify `NRTmin()`.)

8. Implement lookback option pricing in the CRR model using path-dependent Arrow-Debreu securities. Check that the results agree with `CRRlb`.

9. Implement ladder Put option pricing in the CRR model by modifying `CRRladC` appropriately. Check that the premium is at least as great as that for the vanilla European-style Put with the same parameters.

4.4   Further Reading

- Antoine Conze and Viswanathan, "Path Dependent Options: The Case of Lookback Options." *Journal of Finance* 46:5 (1991), pp.1893–1907.

- Espen Gaardner Haug, *The Complete Guide to Option Pricing Formulas.* McGraw-Hill, New York (1997).

- Naoto Kunitomo and Masayuki Ikeda. "Pricing options with curved boundaries." *Mathematical Finance* 2:4 (1992), pp.275–298.

- Israel Nelken (editor). *Handbook of Exotic Options.* McGraw-Hill, New York (1996).

- Antoon Pelsser. "Pricing double barrier options using Laplace transforms." *Finance and Stochastics* 4:95 (2000), pp.95–104.

- Peter Guangping Zhang. *Exotic Options: A Guide to the Second Generation Options (2nd Edition).* World Scientific, New York (1998).

5

Forwards and Futures

Forwards and Futures are contracts for transactions to be completed in the future. A strike price is negotiated between the parties, but no money is exchanged until the expiry time when the Long side is obligated to purchase the underlying asset from the Short side at that price. Models of the future are used to estimate this strike price, though ultimately it is set by market action in response to supply and demand.

The strike price does not represent a prediction of the future, but only a consensus on the probabilities of various future states. A simple example is the Forward contract described in Section 1.2.1. Its strike price is the spot price of the underlying asset, which balances current supply and demand, times the expected riskless return to expiry, which depends on a model of the future.

As time advances and expiry approaches, the riskless returns and the updated spot prices may change. Also, circumstances may arise that affect the ability or willingness of parties to a contract to fulfill their obligations, raising the risk of *default*. A Long Forward default occurs if the buyer does not pay the agreed price; a Short Forward default occurs if the seller does not deliver the specified underlying asset.

The Futures contract provides a means to adjust prices and protect both parties against default. It involves a supervising financial institution that holds some money from each party, monitors the market, and instantly adjusts balances in these *margin accounts* to match changes in the value of the contract. These adjustments mimic dynamic hedging of the Futures contract with a riskless asset, though the margin account typically holds only a fraction of the total contract amount. The result is that both sides have their profits and losses *marked to market*, following the market price of the underlying, so that either side can exit the contract without injuring the other.

5.1 Discrete Models for Forwards

Suppose that S is a risky asset with spot price S_0. The fair price at time $t = 0$ for a Forward contract to buy S at time $t = T$ is

$$F_0 \stackrel{\text{def}}{=} \frac{S_0}{Z(0,T)},$$

where $Z(0,T)$, as defined in Section 1.1.5, is the spot discount factor for a zero-coupon bond maturing at time T.

Now suppose that the price of S is $S(n,j)$ at time index n and state index j in a discrete model of the future for which $n = N$ represents time T. Then $S(0,0) = S_0$ is the spot price of S. The fair price at time 0 in the unique state 0 for a Forward contract to buy S at time step N is

$$F_0 = F(0,0) \stackrel{\text{def}}{=} \frac{S(0,0)}{D(0,0)},$$

where $D(n,j)$ is the modeled discount factor, at time step n in state j, for that zero-coupon bond maturing at time T, namely at time step N. Note that

$$D(N,j) = 1, \qquad j = 0,1,\ldots,N. \tag{5.1}$$

as the zero-coupon bond is worth its face value at maturity in any state.

As time advances, the fair price may change. In the discrete model, at time n in state j, the fair price for a Forward contract that expires at time N is

$$F(n,j) = \frac{S(n,j)}{D(n,j)}, \tag{5.2}$$

which is adjusted by the interest rates described by $D(n,j)$.

5.1.1 No-Arbitrage Forwards Values

Let $V(n,j)$ be the value at time n in state j of the Long Forward contract initiated at time 0 with maturity N at price $F(0,0)$. It may be computed by the hedging method as

$$V(n,j) = S(n,j) - F(0,0)D(n,j) = S(n,j) - \frac{D(n,j)}{D(0,0)}S(0,0).$$

The technique is to construct a portfolio whose value at time n in state j is $S(n,j) - F(0,0)D(n,j)$ and then show that it is a hedge for the Long Forward contract. This may be done as follows:

- Short-sell S for $S(n,j)$;

- Buy a zero-coupon bond maturing at N with face value $F(0,0)$ for the discounted price $F(0,0)D(n,j)$.

The net proceeds will be $S(n,j) - F(0,0)D(n,j)$. This could be positive or negative. To show that the portfolio is a hedge, evaluate it at expiry time N in any state:

- Redeem the bond for $F(0,0)$;

- Execute the Forward contract to buy S for $F(0,0)$;

- Return S to cover the short sale.

This covers all liabilities of the Long Forward, so it and the hedge portfolio must have the same price at time N in any state. Conclude by the One Price Theorem 1.2 that, to avoid an arbitrage opportunity, the value $V(n, j)$ of the Long Forward contract at time n in state j must equal $S(n, j) - F(0, 0)D(n, j)$ as claimed.

5.1.2 Binomial Models for Forwards Prices

Now suppose that the asset prices $\{S(n, j)\}$ and riskless returns $\{R(n, j)\}$ for $0 \le n \le N$ and $0 \le j \le n$ are tabulated in a recombining binomial tree. These tables may be used to model the no-arbitrage price $F(n, j)$ at times between now ($n = 0$, $j = 0$) and maturity ($n = N$, any $0 \le j \le N$).

It is evident that

$$F(N, j) = S(N, j)$$

for any state j at maturity, since otherwise there is an obvious arbitrage. This also follows from Eqs.5.1 and 5.2.

Since $S(n, j)$ is given, by Eq.5.2 it remains to model $D(n, j)$ using the generalized backward pricing formula of Section 3.3.2:

$$D(n, j) = \frac{p(n, j)D(n + 1, j + 1) + (1 - p(n, j))D(n + 1, j)}{R(n, j)},$$

where $p(n, j)$ is computed by Eq.3.16. The terminal condition from which the backward induction procedes is Eq.5.1. An immediate consequence is that

$$D(N - 1, j) = 1/R(N - 1, j), \qquad 0 \le j \le N - 1, \tag{5.3}$$

namely that the zero-coupon bond yields the riskless return $R(N - 1, j)$ if it is purchased in state j one time step before maturity. Combine this with Eq.5.2 to compute

$$F(N - 1, j) = S(N - 1, j)R(N - 1, j), \qquad 0 \le j \le N - 1, \tag{5.4}$$

which is the Forward price given by any discrete model with one time step.

5.2 Discrete Models for Futures

Both parties to Futures contract must have *margin accounts* at a supervisory financial institution, usually a brokerage firm, exchange, or bank. Denote these accounts by M_{long} for the long side (the future buyer of S) and M_{short} for the short side (the future seller). In a discrete model, the amounts in these accounts at time n in state j will be $M_{\text{long}}(n,j)$ and $M_{\text{short}}(n,j)$, respectively. They contain positive amounts above some minimum set by the institution, and they earn interest at the riskless rate on their contents.

Suppose that at the initiation of the contract at time 0 in unique state 0, the accounts hold $M_{\text{long}}(0,0)$ and $M_{\text{short}}(0,0)$, respectively. The Long party agrees to buy S at time N from the Short party for a price denoted by $G(0,0)$.

More generally, denote by $G(n,j)$ the price at time n in state j for a Futures contract also maturing at time N. This price is affected by S and R as derived below. It is evident that

$$G(N,j) = S(N,j) \tag{5.5}$$

for any state j at maturity, since otherwise there is an obvious arbitrage. For other n,j, use backward induction and risk neutral probabilities to find no-arbitrage prices $G(n,j)$.

The margin accounts are adjusted at specified intermediate times in response to price changes in the underlying assets. This is called *marking to market*, and it typically happens daily or weekly, after the exchange closes. In a discrete model, suppose that the world states at times n and $n+1$ are j and j', respectively. Then the margin adjustments at time $n+1$ are

$$M_{\text{long}}(n+1,j') = M_{\text{long}}(n,j)R(n,j) + [G(n+1,j') - G(n,j)] \tag{5.6}$$
$$M_{\text{short}}(n+1,j') = M_{\text{short}}(n,j)R(n,j) - [G(n+1,j') - G(n,j)] \tag{5.7}$$

Thus the Long side is paid immediately for increases in G while the Short side is debited its loss immediately. Decreases in G have the reverse effect.

Theorem 5.1 *Suppose that at times $n = 0,1,\ldots,N$, the price of S is given by $S(n,j(n))$ where $j(n)$ is one of the discrete possible states at time n. Assume that riskless interest rates are nonnegative and that all margin calls are covered. Then for all $0 \le n \le N$,*

$$M_{\text{long}}(n,j(n)) \ge M_{\text{long}}(0,0) + [G(n,j(n)) - G(0,0)];$$
$$M_{\text{short}}(n,j(n)) \ge M_{\text{short}}(0,0) - [G(n,j(n)) - G(0,0)].$$

Proof: This may be proved by induction on time n. First assume that there are no margin calls.

At time $n = 0$, since $j(0) = 0$, both inequalities evidently hold as equalities.

Now assume the inductive hypothesis that the inequalities hold at time n. Compare sides at time $n+1$ by applying the margin adjustments from Eqs.5.6 and 5.7 with $j = j(n)$ and $j' = j(n+1)$. This gives, for the Long side,

$$\begin{aligned} M_{\text{long}}(n+1,j') &= M_{\text{long}}(n,j)R(n,j) + [G(n+1,j') - G(n,j))] \\ &\ge M_{\text{long}}(n,j(n)) + [G(n+1,j') - G(n,j(n))] \\ &\ge M_{\text{long}}(0,0) + [G(n+1,j') - G(0,0)]. \end{aligned}$$

The middle inequality follows since $R(n, j) \geq 1$ for all states and times when riskless interest rates are positive. The last inequality follows from the inductive hypothesis.

The proof for the Short side is similar but for a change of sign.

Finally, if either side covers a margin call by adding cash at some $(n, j(n))$, then its margin account balance will be larger for all subsequent times, preserving the claimed inequality. □

Margin account requirements differ but, by Theorem 5.1, if $M_{\text{long}}(0, 0)$ contains enough money to pay $G(0, 0)$ at expiry N, it will contain enough money at time n to pay $G(n, j(n))$ at expiry N. Hence the Long party may withdraw from the contract by forfeiting $M_{\text{long}}(n, j(n))$ to the supervising institution which may then serve as a replacement Long side with no cost of its own.

Similarly, the Short party may withdraw by forfeiting $M_{\text{short}}(n, j(n))$ to the supervising institution which may then find a replacement Short side at the current Futures price $G(n, j(n))$.

5.2.1 Binomial Models for Futures Prices

The value of a Futures contract depends on both the underlying asset prices S and the interest rates R. That is because the margin account balance is affected by both. The margin account for a Futures contract may thus be considered a contingent claim on the underlying S. Modeling its balance yields formulas for Futures prices.

Suppose that both R and S are described by recombining binomial trees of N levels with states $j = 0, 1, \ldots, n$ at each time step n. Let

$$p(n, j) = \frac{R(n, j)S(n, j) - S(n+1, j)}{S(n+1, j+1) - S(n+1, j)}$$

be the risk neutral up probability for S at time n in state j. The backward pricing formula for contingent claim M_{long} in this model gives

$$
\begin{aligned}
M_{\text{long}}(n, j) &= \frac{p(n, j)M_{\text{long}}(n+1, j+1) + (1 - p(n, j))M_{\text{long}}(n+1, j)}{R(n, j)} \\
&= \frac{p(n, j)}{R(n, j)} \left[M_{\text{long}}(n, j)R(n, j) + [G(n+1, j+1) - G(n, j)] \right] \\
&\quad + \frac{1 - p(n, j)}{R(n, j)} \left[M_{\text{long}}(n, j)R(n, j) + [G(n+1, j) - G(n, j)] \right] \\
&= M_{\text{long}}(n, j) + \frac{p(n, j)}{R(n, j)} [G(n+1, j+1) - G(n, j)] \\
&\quad + \frac{1 - p(n, j)}{R(n, j)} [G(n+1, j) - G(n, j)].
\end{aligned}
$$

Solving for $G(n, j)$ in terms of future values gives the backward induction formula for G:

$$G(n, j) = p(n, j)G(n + 1, j + 1) + (1 - p(n, j))G(n + 1, j). \qquad (5.8)$$

The induction starts at N, as usual, with the terminal conditions from Eq.5.5 specifying $G(N, j)$ for $0 \le j \le N$.

Observe that p, and thus R, need only be given down to time $N - 1$. Notice too that there is no division by $R(n, j)$, which for other assets represents adjustment for present value. That is because in Futures contracts the present value is realized by adjustments to the margin account. These adjustments accumulate differently along different paths through the R model and thus may differ from the path independent discount encoded in $D(n, j)$. However, if the riskless return at each time n is not dependent on the state j, then the two adjustments will be the same, so Futures and Forwards will have equal fair prices:

Theorem 5.2 *If $R(n, j) = R(n, \cdot)$ is independent of j, then $F(n, j) = G(n, j)$ for all $0 \le n \le N$ and all $0 \le j \le n$.*

Proof: First note that at expiry,

$$G(N, j) = F(N, j) = S(N, j), \quad j = 0, 1, \ldots, N.$$

Next, show that F and G satisfy the same backward recursion and thus that F and G have the same price in all prior states. But

$$F(n, j) = \frac{S(n, j)}{D(n, j)}$$

and if $R(n, j)$ does not depend on j, then

$$D(n, j) = D(n, \cdot) \overset{\text{def}}{=} \frac{1}{R(n, \cdot)} \frac{1}{R(n + 1, \cdot)} \frac{1}{R(N - 1, \cdot)}$$

for every $0 \le n < N$, so

$$F(n, j) = R(n, \cdot)R(n + 1, \cdot) \cdots R(N - 1, \cdot)S(n, j)$$

However, the backward induction formula for S gives

$$
\begin{aligned}
S(n, j) &= \frac{1}{R(n, \cdot)} \left[p(n, j)S(n + 1, j + 1) + (1 - p(n, j))S(n + 1, j) \right] \\
&= \frac{D(n + 1, \cdot)}{R(n, \cdot)} \left[p(n, j)F(n + 1, j + 1) + (1 - p(n, j))F(n + 1, j) \right],
\end{aligned}
$$

which, in combination with the previous equations, gives

$$F(n, j) = p(n, j)F(n + 1, j + 1) + (1 - p(n, j))F(n + 1, j).$$

This is the same backward recursion as Eq.5.8, with the same terminal values. Conclude that $F(n,j) = G(n,j)$ for all n, j. □

Modeled Forwards and Futures prices may differ when interest rates are not independent of state. Here is a simple example with three time steps:

$$
S = \begin{matrix} 100 \\ 85 & 115 \\ 75 & 100 & 130 \\ 65 & 85 & 115 & 150 \end{matrix}
\qquad
R = \begin{matrix} 1.025 \\ 1.03 & 1.02 \\ 1.035 & 1.025 & 1.015 \end{matrix}
$$

With these inputs, at $t = 0$ the Forward contract price is $F(0) = \$107.46$, whereas the Futures contract price is $G(0) = \$107.25$.

Remark. Exercise 2 below gives an even simpler example where $F(0) \neq G(0)$, using only two time steps. Conversely, Exercise 1 shows that every one-step two-state model must have $F(0) = G(0)$.

Octave code to compute Forwards and Futures

```
1   function [F,G] = FwdFut(S, R, N)
2   % Octave/MATLAB function to compute Forward prices F
3   % and Futures prices G from recombining binomial trees
4   % S,R of underlying asset prices and riskless returns.
5   % INPUTS:                                        (Example)
6   %    S = asset price matrix              (N+1 x N+1)
7   %    R = riskless returns matrix             (N x N)
8   %    N = binomial tree height, must be >1        (3)
9   % OUTPUTS:
10  %    F =  (N+1) row matrix of Forward prices
11  %    G =  (N+1) row matrix of Futures prices
12  % EXAMPLE:
13  %    S=[100,0,0,0;87,115,0,0;75,100,133,0;65,87,115,152]
14  %    R=[1.025,0,0;1.03,1.02,0;1.035,1.025,1.01]
15  %    [F,G]= FwdFut(S,R,3); F(1,1),G(1,1)
16  %
17      p = RiskNeut(S,R,N); % Risk neutral up probabilities
18      D = ZCB(p,R,N);      % Zero coupon bond discounts
19      F = zeros(N+1,N+1);  % Forward fair prices
20      for(n=0:N)
21        for(j=0:n)
22          F(n+1,j+1) = S(n+1,j+1)/D(n+1,j+1);
23        end
24      end
25      G = zeros(N+1,N+1);  % Futures fair prices
26      for(j=0:N)
27        G(N+1,j+1) = S(N+1,j+1);
28      end
29      for(n=N-1:-1:0)
30        for(j=0:n)
```

```
31        G(n+1,j+1) = ( p(n+1,j+1)*G(n+2,j+2)
32                      + (1-p(n+1,j+1))*G(n+2,j+1) );
33      end
34    end
35    return
36  end
```

5.2.2   No-Arbitrage Futures Values

A Forward contract costs nothing to either side, but a Futures contract requires a deposit into a margin account. From the investor's viewpoint, its cost is equal to the margin balance minus the initial deposit and any additions required to satisfy margin calls. A Futures portfolio constructed from zero cash, using borrowed money, has value

$$M_f(t) - B(t), \qquad 0 \le t \le T,$$

where $B(t)$ is the time-evolving debt incurred to fund the margin account $M_f(t)$ which is adjusted according to changes in the Futures price for the underlying asset. Because of that margin account, whose balance is path dependent, the value of a Futures contract is path dependent.

The no-arbitrage Futures prices G have the same values for any margin account balance. They may be modeled by a recombining binomial tree just like the underlying asset prices. However, the margin account balance needs to be modeled by a non-recombining tree of paths using Eqs.5.6 and 5.7.

Octave code to compute margin balances

This function assumes a sufficiently large initial balance so that no additions are needed to satisfy any margin calls.

```
1   function [ML,MS] = MarFut(ML0, MS0, G, R, N)
2   % Octave/MATLAB function to find margin balances for
3   % the Long and Short side of a Futures contract with
4   % modeled prices G and riskless returns R.
5   % INPUTS:                                      (Example)
6   %     ML0 = Long-side initial margin               (200)
7   %     MS0 = Short-side initial margin              (200)
8   %     G = Futures price matrix               (N+1 x N+1)
9   %     R = riskless returns matrix                (N x N)
10  %     N = binomial tree height, must be >1          (3)
11  % OUTPUTS:
12  %     ML = Long margin, non-recombining tree array
13  %     MS = Short margin, non-recombining tree array
14  % EXAMPLE:
15  %   S=[100,0,0,0;87,115,0,0;75,100,133,0;65,87,115,152]
16  %   R=[1.025,0,0;1.03,1.02,0;1.035,1.025,1.01]
17  %   [F,G]= FwdFut(S,R,3);
18  %   [ML,MS]=MarFut(200,200,G,R,3)
```

```
19  %
20     ML = zeros(1,2*2^N-1);   ML(1) = ML0;
21     MS = zeros(1,2*2^N-1);   MS(1) = MS0;
22     [nm,jm] = NJfromM(N); % recombining index functions
23     for (m=1:2^N-1)
24        Rnj=R(nm(m)+1,jm(m)+1); Gnj=G(nm(m)+1,jm(m)+1);
25        Gup=G(nm(m)+2,jm(m)+2); Gdown=G(nm(m)+2,jm(m)+1);
26        dGup = Gup-Gnj;        dGdown = Gdown-Gnj;
27        ML(2*m)=ML(m)*Rnj+dGdown; ML(2*m+1)=ML(m)*Rnj+dGup;
28        MS(2*m)=MS(m)*Rnj-dGdown; MS(2*m+1)=MS(m)*Rnj-dGup;
29     end
30     return
31  end
```

Octave code to compute riskless balances

The cost of borrowing the initial margin deposit is also path dependent. It
is modeled by the riskless balance in an unadjusted account $B(t)$. It may be
computed by forward induction using the riskless return rate model R.

```
1   function Bal = PathBal(Bal0, R, N)
2   % Octave/MATLAB function to track an account balance
3   % along all paths in a recombining binomial tree R
4   % of riskless returns.
5   % INPUTS:                                  (Example)
6   %    Bal0 = initial deposit                    (200)
7   %    R = riskless returns matrix             (N x N)
8   %    N = binomial tree height, must be >1        (3)
9   % OUTPUTS:
10  %    Bal = non-recombining tree array
11  % EXAMPLE:
12  %   R=[1.025,0,0;1.03,1.02,0;1.035,1.025,1.01]
13  %   Bal=PathBal(200,R,3)
14  %
15     Bal = zeros(1,2^(N+1)-1);  Bal(1) = Bal0;
16     [nm,jm] = NJfromM(N); % recombining index functions
17     for (m=1:2^N-1)
18        Rnj=R(nm(m)+1,jm(m)+1);
19        Bal(2*m)=Bal(m)*Rnj; Bal(2*m+1)=Bal(m)*Rnj;
20     end
21     return
22  end
```

How does entering a Futures contract for S at strike price $G(0)$ with initial
margin deposit $M_f(0)$ compare with simply depositing $B(0) = M_f(0)$ into a
riskless account without the Futures contract? With `MarFut()` and `PathBal()`
it is possible to experiment and gain some intuition through discrete binomial

modeling. The fair price of the contract is computed with a path dependent
Arrow-Debreu expansion as returned by `PathAD(pu,R,N)`, given the risk neutral probabilities `pu=RiskNeut(S,R,N)` implied by asset prices S and riskless
returns R down to time step N:

```
S=[100,0,0,0;87,115,0,0;75,100,133,0;65,87,115,152]; N=3; M0=200;
R=[1.025,0,0;1.03,1.02,0;1.035,1.025,1.01]; ends=2^N:2*2^N-1;
[F,G]= FwdFut(S,R,N); [ML,MS]=MarFut(M0,M0,G,R,N);
Bal=PathBal(M0,R,N); pu=RiskNeut(S,R,N); Lbar=PathAD(pu,R,N)';
ML(ends)*Lbar(ends), MS(ends)*Lbar(ends), Bal(ends)*Lbar(ends)
```

All three values will be the same \$200. This experimental result suggests that
both the Long and Short Futures contracts with margin accounts at $t = 0$ are
worth the same as the margin deposit in a riskless account. But that result
holds in general for Futures contracts priced by the discrete binomial model:

Theorem 5.3 *The value of the Futures contract at its fair price is equal to
the value of the initial margin deposit without the contract.*

Proof: This will be proved by induction in a discrete model of underlying
asset prices S and riskless returns R, using recombining binomial trees but
with path dependent prices for the accounts and the Arrow-Debreu securities.
    Let $\bar{p}$ be the risk neutral probabilities corresponding to S and R, injected
into a non-recombining tree with the m notation. Likewise, let $\bar{R}$ be the riskless
returns injected into another such tree. Then the Futures margin account M_f
satisfies the backward induction

$$M_f(m) = \frac{\bar{p}(m)M_f(2m+1) + (1 - \bar{p}(m))M_f(2m)}{\bar{R}(m)}, \quad 0 < m < 2^N, \quad (5.9)$$

if expiry corresponds to time step N. Denote the initial deposit by $M_f(1)$.
    On the other hand, the path dependent value of the initial deposit without
margin adjustments satisfies the forward induction

$$B(2m) = B(m)\bar{R}(m); \quad B(2m+1) = B(m)\bar{R}(m), \quad 0 < m < 2^N,$$

with initial deposit $B(1) = M_f(1)$ by hypothesis.
    Now let $\bar{\lambda}(m)$ be the price of the path dependent Arrow-Debreu security
that pays 1 at the end of path m and 0 at the end of all other paths of the
same length. By Theorem 4.1, the set of such prices satisfies

$$\bar{\lambda}(1) = 1,$$
$$\bar{\lambda}(2m) = \frac{1 - \bar{p}(m)}{\bar{R}(m)}\bar{\lambda}(m),$$
$$\bar{\lambda}(2m+1) = \frac{\bar{p}(m)}{\bar{R}(m)}\bar{\lambda}(m), \quad m = 1, 2, \ldots.$$

Let the collection of paths of length N be indexed by

$$m \in \{2^N, 2^N + 1, \ldots, 2^N + 2^N - 1\} \overset{\text{def}}{=} \mathcal{M}(N),$$

as in the proof of Theorem 4.3. For the current theorem, it suffices to prove that

$$\sum_{m \in \mathcal{M}(N)} M_f(m)\bar{\lambda}(m) = \sum_{m \in \mathcal{M}(N)} B(m)\bar{\lambda}(m),$$

or equivalently, that for all $N = 0, 1, \ldots,$

$$\sum_{m \in \mathcal{M}(N)} \Big(M_f(m) - B(m) \Big) \bar{\lambda}(m) = 0. \tag{5.10}$$

This may be done by induction. The base case, $N = 0$, is evident since then $\mathcal{M}(0) = \{1\}$, and $B(1) = M_f(1)$ by hypothesis.

Now suppose that Eq.5.10 holds for N and consider the case $N + 1$:

$$\sum_{m \in \mathcal{M}(N+1)} \Big(M_f(m) - B(m) \Big) \bar{\lambda}(m) =$$

$$= \sum_{m \in \mathcal{M}(N)} \Big(M_f(2m) - B(2m) \Big) \bar{\lambda}(2m) +$$

$$+ \sum_{m \in \mathcal{M}(N)} \Big(M_f(2m+1) - B(2m+1) \Big) \bar{\lambda}(2m+1)$$

$$= \sum_{m \in \mathcal{M}(N)} \Big(M_f(2m) - B(m)\bar{R}(m) \Big) \frac{1 - \bar{p}(m)}{\bar{R}(m)} \bar{\lambda}(m) +$$

$$+ \sum_{m \in \mathcal{M}(N)} \Big(M_f(2m+1) - B(m)\bar{R}(m) \Big) \frac{\bar{p}(m)}{\bar{R}(m)} \bar{\lambda}(m)$$

$$= \sum_{m \in \mathcal{M}(N)} \Big(M_f(m) - B(m) \Big) \bar{\lambda}(m) = 0,$$

using Eq.5.9 and the inductive hypothesis. This completes the proof. $\qquad \square$

5.2.3 Margin Calls and Defaults

Either party, Long or Short, in a Futures contract may choose to sell its interest voluntarily before expiry. However, one side may be forced, by the supervising institution, to sell under certain circumstances. The precipitating event is a *margin call*, if the party's margin balance falls below the institution's minimum. Unless additional funds are added to restore the minimum, by a specified deadline, the contract falls into *default*. This happens without injury to the counterparty. The supervising institution markets the defaulted side of the contract to another trader at the current spot price, including all fluctuations between the margin call and the default date.

The defaulting party thereby covers the default with funds from its margin account. In extraordinary circumstances, that margin balance may fall below zero. In that case, the defaulting party will owe the supervising institution additional funds. This creates a risk for the institution since the defaulting party may not have the funds to pay this bill, so margin balance requirements must be set high enough so that falling below zero is vanishingly unlikely in the modeled future.

Margin account simulations in the CRR model

CRR offers one simple way to model margin accounts for Futures contracts. Since path-dependent trees are needed, the number N of time steps must be small so that the $O(2^N)$ non-recombining tree size stays manageable. This will also keep the recombining binomial model S and the riskless return model R small, at $O(N^2)$ in size.

Suppose that fixed CRR parameters are given: riskless annual rate r, volatility σ, expiry time T, and number of time steps N. Let the underlying asset S have spot price S_0, and let ML_0 and MS_0, respectively, be the initial deposits into the Long and Short margin accounts, respectively. Implement the model in Octave as follows:

```octave
function [ML,MS,Pr] = CRRmargin(T,S0,ML0,MS0,r,v,N)
% Octave/MATLAB function to find Margin balances for
% the Long and Short side of a Futures contract in
% the Cox-Ross-Rubinstein (CRR) model.
% INPUTS:                                   (Example)
%    T = time to expiry, in years                 (1)
%    S0 = Underlying asset spot price           (100)
%    ML0 = Long-side initial Margin             (200)
%    MS0 = Short-side initial Margin            (200)
%    r = riskless annual interest rate         (0.02)
%    v = volatility                            (0.15)
%    N = binomial tree height                    (10)
% OUTPUTS:
%    ML = Long margin, non-recombining tree array
%    MS = Short margin, non-recombining tree array
%    Pr = path probs., non-recombining tree array
% EXAMPLE:
%    [ML,MS,Pr]=CRRmargin(1,100,200,200,0.02,0.15,10);
%
   [pu,up,R1] = CRRparams(T,r,v,N); % key parameters
   S = StreeCRR(S0,up,N);   % CRR model S matrix
   R = R1*ones(N,N); % const. riskless returns matrix
   [F,G] = FwdFut(S, R, N); % Futures prices matrix G
   [ML,MS] = MarFut(ML0, MS0, G, R, N); % accounts
   Pr = PathPr(pu*ones(N,N), N); % path probabilities
   return
end
```

The returned linear arrays are the non-recombining trees of modeled Long and Short margin acccount balances and the path probabilities.

Probability of default in the CRR model

For simplicity, suppose that a margin call will occur if and only if the margin balance falls below 0. Further suppose that ML and MS have been computed and returned by CRRmargin(). If no element in either array is negative, then neither side will receive a margin call and there will be no defaults. That may be detected with the Octave commands

```
T=1; S0=100; ML0=50; MS0=30; r=0.02; v=0.15; N=10;
[ML,MS,Pr]=CRRmargin(T,S0,ML0,MS0,r,v,10);
sum(ML<0)  % ans = 0, so there are no Long margin calls
sum(MS<0)  % ans = 77, so there are Short margin calls
```

For this example, the Long Futures margin deposit of \$50, or 50% of S_0, was enough to cover all modeled declines, whereas the Short Futures margin deposit of \$30, or 30% of S_0, was not. Thus there exists the possibility that the Short side defaults and loses the contract.

To compute the probability of a margin call, it is necessary to sum the path probabilities Pr(m) from the model over all the paths m that terminate in a first negative margin account balance. Each such path is found with a depth-first search of the margin account tree. This is a problem well suited to a *recursive programming* solution:

```
1  function Pneg = PrNeg(X,Pr,m,n)
2  % Recursive Octave/MATLAB function to compute the
3  % probability of hitting a negative element in a
4  % non-recombining tree by depth-first search—DFS.
5  % INPUTS:                                    (Example)
6  %    X = data NRT                    ([1  2 -1  3 -4 -5  6])
7  %    Pr = prob. NRT          ([1 .49 .51 .1 .2 .3 .4])
8  %    m = index of the current root              (1)
9  %    N = remaining tree depth, must be >0       (2)
10 % OUTPUTS:
11 %    Pneg = probability of first negatives in DFS
12 % EXAMPLE:
13 %    Pneg = PrNeg([1   2   -1   3   -4   -5   6],...
14 %    [1 .49  .51  .1  .2  .3  .4], 1,2) % Pneg=0.7100
15 %
16     Pneg=0; % default value for several cases
17     if n<0    % leaf termination condition for DFS
18        return % zero contribution from below level N
19     end  % ...so n>=0 hereafter
20     if X(m)<0  % then this nonleaf is negative
21        Pneg=Pr(m); % ...so count its probability
22     else % otherwise this nonleaf is nonnegative
23        P0=PrNeg(X,Pr,2*m,n-1);   % so check path 2m
24        P1=PrNeg(X,Pr,2*m+1,n-1); % ...and path 2m+1
```

```
25      Pneg=P0+P1; % ...and add their probabilities
26    end
27    return % and pass the current Pneg value upwards
28  end
```

PrNeg() descends the binary tree X along every path, in depth-first order, until it encounters a negative element. It then records the probability of that path, discarding any descendent paths, and passes the value back to be added to that of other paths.

For the margin accounts modeled as the output of CRRmargin() above, the probabilities are

```
PrNeg(ML, Pr, 1, N)  % ans = 0,  no risk, as expected
PrNeg(MS, Pr, 1, N)  % ans = 0.072998, somewhat risky
```

Remark. Inputs m=1 and n=N indicate the root $m = 1$ of a tree of depth N.

To estimate the risk of a particular margin requirement, let $\alpha > 0$ be fixed, and suppose that the supervising institution will require αS_0 as an initial margin deposit at the signing of a Futures contract for asset S with spot price S_0. Then the probability of a margin call is computed from the output of CRRmargin() by the following code:

```
[ML,MS,Pr]=CRRmargin(T,S0,alpha*S0,alpha*S0,r,v,N);
PrNeg(ML, Pr, 1, N), PrNeg(MS, Pr, 1, N)
```

Remark. The spot price S_0 plays no role so it may be set to any convenient positive value such as $100.

It is left as an exercise to tabulate or plot these probabilities for various CRR parameters of riskless rate and volatility.

5.3   Exercises

1. Suppose that S and R are modeled with a recombining binomial tree of $N \geq 1$ levels. Prove that

$$G(N - 1, j) = F(N - 1, j) = S(N - 1, j)R(N - 1, j)$$

   for all states j.

2. Suppose that S and R are modeled with a recombining binomial tree of $N = 2$ levels. Prove that if $(R(1, 1) - R(1, 0))(S(1, 1) - S(1, 0)) > 0$, then $G(0, 0) > F(0, 0)$.

3. Suppose that S and R are modeled with a recombining binomial tree of $N > 2$ levels. Prove that if

$$(R(N-1,j+1) - R(N-1,j))(S(N-1,j+1) - S(N-1,j)) > 0$$

for all $j = 0, 1, ..., N - 2$, then $G(0,0) > F(0,0)$.

4. Fix $N = 4$ in the function FwdFut() defined on p.135.

(a) Find inputs S and R such that $F(0) > G(0)$, namely F(1,1)>G(1,1) in the output of [F,G]=FwdFut(S,R,N).

(b) Find S and R such that $F(0) < G(0)$, namely F(1,1)<G(1,1) in the output of [F,G]=FwdFut(S,R,N).

Note: To be valid, solutions S and R in parts (a) and (b) must be positive and must satisfy the no-arbitrage condition

$$S(n+1,j) < R(n,j)S(n,j) < S(n+1,j+1)$$

for all $0 \leq n < N$ and all $0 \leq j \leq n$.

5. Suppose that a commodities exchange wishes to broker Futures contracts on an asset S, expiring in 0.5 years while riskless annual interest rates are expected to remain constant at 0.02%. Under consideration are margin requirements of 20, 30, 50, 80, and 150% of S_0 for each contract.

(a) Compute the probability of a margin call, which will occur if and only if the margin balance falls below 0, for both Long and Short Futures contracts, with these five margin requirements. Use $N = 6$ time steps and volatilities $\sigma \in \{0.10, 0.15, 0.20, 0.25, 0.30, 0.40, 0.50, 0.70\}$. Tabulate the results and compare Long and Short margin requirements.

(b) Profile the computation with $N = 6$ and again with $N = 13$ to compare the run times. Include in the count all times above 1% of the total. Compare the run times to test the $O(2^N)$ order of complexity.

5.4 Further Reading

- Chicago Board of Trade. *The Chicago Board of Trade Handbook of Futures and Options.* McGraw-Hill, New York (2006).

- Kenneth R. French. "A Comparison of Futures and Forward Prices." *Journal of Financial Economics* 3 (1983), pp.311–342.

- Aron Gottesman. *Derivatives Essentials: an Introduction to Forwards, Futures, Options and Swaps.* Wiley, Hoboken (2016).

- John van der Hoek and Robert James Elliot. *Binomial Models in Finance.* Springer, New York (2006).

6

Dividends and Interest

Some assets provide a cash flow to the owner. Bonds with coupons pay periodic interest, some stocks pay dividends, some properties pay rent or royalties, and so on. The prices of derivatives on those assets must account for this cash flow as well as the asset price itself.

6.1 Stocks with Dividends

For starters, suppose that risky asset S is a share of stock in a company that pays a per-share *dividend* d_i at times $\{t_i : i = 1, 2, \dots\}$, with $0 < t_1 < t_2 < \cdots$. In reality, each t_i could be one of four times:

- t_i^d, the *declaration date*, when the company behind S announces what the amount d_i will be,

- t_i^e, the *effective* or *ex-dividend date*, when first the owner of S may sell it but still receive d_i,

- t_i^r, the *record date*, when the company behind S determines the *owner of record* who will receive d_i,

- t_i^p, the *payment date*, at which time d_i is paid to the owner of record.

Exchange-traded shares of stock are tracked by the exchange, to determine ownership at t_i^e-, the day before the ex-dividend date. The owner of record is identified shortly after t_i^e at t_i^r. This allows time to reconcile information from multiple sources. Only afterward is payment made at t_i^p. Thus,

$$t_i^d < t_i^e < t_i^r < t_i^p.$$

For example, Bank of America paid a quarterly cash dividend over the period 2015-2020 with the times and amounts in Table 6.1.[1]

---

[1] This schedule was published on the website

https://www.nasdaq.com/market-activity/stocks/bac/dividend-history

TABLE 6.1

Historical dividends on BAC from 2015–2020.

Time index	Amount (d)	Declaration (t^d)	Ex-dividend (t^e)	Record (t^r)	Payment (t^p)
0	$0.05	2014-10-24	2014-12-03	2014-12-05	2014-12-26
1	$0.05	2015-02-11	2015-03-04	2015-03-06	2015-03-27
2	$0.05	2015-04-17	2015-06-03	2015-06-05	2015-06-26
3	$0.05	2015-07-24	2015-09-02	2015-09-04	2015-09-25
4	$0.05	2015-10-22	2015-12-02	2015-12-04	2015-12-24
5	$0.05	2016-01-21	2016-03-02	2016-03-04	2016-03-25
6	$0.05	2016-04-29	2016-06-01	2016-06-03	2016-06-24
7	$0.075	2016-07-28	2016-08-31	2016-09-02	2016-09-23
8	$0.075	2016-10-28	2016-11-30	2016-12-02	2016-12-30
9	$0.075	2017-01-26	2017-03-01	2017-03-03	2017-03-31
10	$0.075	2017-05-01	2017-05-31	2017-06-02	2017-06-30
11	$0.12	2017-07-28	2017-08-30	2017-09-01	2017-09-29
12	$0.12	2017-10-25	2017-11-30	2017-12-01	2017-12-29
13	$0.12	2018-01-31	2018-03-01	2018-03-02	2018-03-30
14	$0.12	2018-04-25	2018-05-31	2018-06-01	2018-06-29
15	$0.15	2018-07-26	2018-09-06	2018-09-07	2018-09-28
16	$0.15	2018-10-24	2018-12-06	2018-12-07	2018-12-28
17	$0.15	2019-01-30	2019-02-28	2019-03-01	2019-03-29
18	$0.15	2019-04-24	2019-06-06	2019-06-07	2019-06-28
19	$0.18	2019-07-25	2019-09-05	2019-09-06	2019-09-27
20	$0.18	2019-10-22	2019-12-05	2019-12-06	2019-12-27
21	$0.18	2020-01-29	2020-03-05	2020-03-06	2020-03-27
22	$0.18	2020-04-22	2020-06-04	2020-06-05	2020-06-26
23	$0.18	2020-07-22	2020-09-03	2020-09-04	2020-09-25
24	$0.18	2020-10-21	2020-12-03	2020-12-04	2020-12-24
25	$0.18	2021-01-19	2021-03-04	2021-03-05	2021-03-26

The most suitable date for the purpose of evaluating derivatives on stocks is the ex-dividend date t_i^e. The value of the dividend d_i should be discounted by the riskless return over $t_i^p - t_i^e$, but since this is a small and nearly constant discount it may be ignored at the cost of a small error in a known direction.

6.1.1 Effects on Forwards

Known dividends to be paid between now ($t = 0$) and expiry ($t = T$) affect the fair price in Forward contracts. Suppose that S and R are modeled in a recombining binomial tree of N time steps. Let $p(n, j)$ be the risk neutral up probability at time n in state j, as defined in Eq.3.16:

$$p(n, j) = \frac{R(n, j)S(n, j) - S(n + 1, j)}{S(n + 1, j + 1) - S(n + 1, j)}, \quad 0 \le n < N;\ 0 \le j \le n.$$

Let $F_1(n, j)$ be the Forward price, negotiated at time n in state j, at which to buy S at time $n + 1$. Assuming the no-arbitrage-expectation axiom (AE-free, from Definition 3), it must satisfy

$$F_1(n, j) = \mathrm{E}(S(n+1)) = p(n, j)S(n+1, j+1) + (1-p(n, j))S(n+1, j). \quad (6.1)$$

It is also possible to deduce Eq.6.1 from the weaker axiom of no arbitrage opportunities. Let V be the long Forward contract with expiry time $n + 1$, initiated at time n in state j with price $F_1(n, j)$. This is a contingent claim on S. Since its payoff is $S - F_1$ and its initial price is 0, the generalized backward pricing formula in this S, R model uses the risk neutral probability p and gives

$$0 = V(n, j) = \frac{p(n, j)V(n + 1, j + 1) + (1 - p(n, j))V(n + 1, j)}{R(n, j)}$$

Substituting the payoffs

$$\begin{aligned}
V(n + 1, j + 1) &= S(n + 1, j + 1) - F_1(n, j), \\
V(n + 1, j) &= S(n + 1, j) - F_1(n, j),
\end{aligned}$$

and then solving for $F_1(n, j)$ gives Eq.6.1.

The Forward price relationship to S can also be used to derive risk neutral probabilities. Note first that

$$S(n + 1, j) < F_1(n, j) < S(n + 1, j + 1),$$

since otherwise either the long or short Forward is an arbitrage opportunity. Thus it is alway possible to express

$$F_1(n, j) = p_1 S(n + 1, j + 1) + (1 - p_1)S(n + 1, j)$$

for some p_1 with $0 < p_1 < 1$. But in fact p_1 is necessarily the risk neutral probability used to price all contingent claims involving S. To prove this, let W be the Arrow-Debreu portfolio, combining S with a riskless asset B, that pays 1 at $(n + 1, j + 1)$ and 0 in the other state $(n + 1, j)$. It may be hedged with a portfolio containing B and $V = S - F_1$, the long Forward contract on S:

$$\begin{aligned}
W(n + 1, j + 1) = 1 &= H_0 R(n, j) + H_1 V(n + 1, j + 1); \\
W(n + 1, j) = 0 &= H_0 R(n, j) + H_1 V(n + 1, j).
\end{aligned}$$

Substituting for V, then for F_1, then solving for H_0 yields

$$\begin{aligned}
H_0 &= \frac{V(n + 1, j)}{R(n, j)[V(n + 1, j) - V(n + 1, j + 1)]} \\
&= \frac{S(n + 1, j) - F_1(n, j)}{R(n, j)[S(n + 1, j) - S(n + 1, j + 1)]} \\
&= \frac{S(n + 1, j) - p_1 S(n + 1, j + 1) - (1 - p_1)S(n + 1, j)}{R(n, j)[S(n + 1, j) - S(n + 1, j + 1)]} \\
&= \frac{p_1}{R(n, j)},
\end{aligned}$$

after canceling nonzero $S(n + 1, j) - S(n + 1, j + 1)$. But then One Price Theorem 1.2 applied to W and the hedge portfolio implies

$$W(n, j) = H_0 + H_1 V(n, j) = H_0 = \frac{p_1}{R(n, j)},$$

since $V(n, j) = 0$. At the same time, the generalized backward pricing formula applied to W implies

$$W(n, j) = \frac{p(n, j)W(n + 1, j + 1) + (1 - p(n, j))W(n + 1, j)}{R(n, j)} = \frac{p(n, j)}{R(n, j)},$$

from which it follows that $p_1 = p(n, j)$. The two derivations of p may be combined into a technical lemma:

Lemma 6.1 *In the binomial model S, R with no arbitrage opportunities, the Forward price $F_1(n, j)$ at $(n, j,)$ with expiry $n + 1$ satisfies*

$$F_1(n, j) = p_1 S(n+1, j + 1) + (1-p_1)S(n+1, j)$$

if and only if

$$p_1 = p(n, j) = \frac{R(n, j)S(n, j) - S(n + 1, j)}{S(n + 1, j + 1) - S(n + 1, j)},$$

namely p_1 is the risk neutral up probability.   □

Now suppose that S has a per-share dividend d_{n+1} with record date $t^p = n$ and ex-dividend date $t^e = n + 1$, for some $0 \le n < N$. For simplicity, suppose also that the dividend is paid immediately, so $t^p = n + 1$ as well. Then the generalized backward pricing formula Eq.3.18 for S implies that

$$
\begin{aligned}
S(n, j) &= \frac{p(n, j)[S(n+1, j+1) + d_{n+1}] + (1-p(n, j))[S(n+1, j) + d_{n+1}]}{R(n, j)} \\
&= \frac{p(n, j)S(n+1, j+1) + (1-p(n, j))S(n+1, j) + d_{n+1}}{R(n, j)}, \quad (6.2)
\end{aligned}
$$

since the owner of record at (n, j) benefits from d_{n+1} in either state, $(n+1, j)$ or $(n + 1, j + 1)$. Substitution into Eq.6.1 yields

$$F_1(n, j) = S(n, j)R(n, j) - d_{n+1}. \quad (6.3)$$

A similar effect results if S pays a *proportional dividend* $d_{n+1} = \delta_{n+1}S$, but then the payment is state-dependent:

$$d_{n+1}(n, j) = \delta_{n+1}S(n, j).$$

Substitution into Eq.6.3 yields

$$F_1(n, j) = S(n, j)\left[R(n, j) - \delta_{n+1}\right]. \quad (6.4)$$

Dividends as annual rates

The further simplification that the dividend proportion is a constant, $\delta_{n+1} = \delta$ paid at all $0 \le n < N$, allows its effect to be expressed in terms of an annualized rate q satisfying

$$q = \frac{1}{T} \sum_{i=0}^{N-1} \delta_{n+1} = \frac{N\delta}{T}, \qquad \Longrightarrow \delta = qT/N$$

Note the similarity with the annualized riskless return rate r if $R(n,j) = R$ is a constant independent of (n,j):

$$R = \exp(rT/N) \approx 1 + rT/N.$$

Rewriting Eq.6.4 with this exponential approximation gives

$$F_1(n,j) \approx S(n,j)\left[1 + (r-q)T/N\right] \approx S(n,j)\exp((r-q)T/N), \qquad (6.5)$$

so that a proportional dividend behaves like a reduction in the riskless return rate. This is a common approximation used for Forwards and European-style options on dividend paying stocks even if the dividends are fixed amounts per share rather than proportions. The sum $\sum_i d_i$ of the dividends to be paid over $[0, T]$ is divided by T and then by the spot price S_0 to get an annualized proportion rate q using

$$q \approx \frac{\sum_i d_i}{TS_0}. \qquad (6.6)$$

This a good approximation if $\sum_i d_i$ is much less than TS_0, which is true for most stocks. However, it ignores the time value of money and treats S as approximately constant over $[0, T]$. Nevertheless, Eq.6.5 shows the right direction and the approximate amount by which dividends adjust Forward prices. The rate q from Eq.6.6 is a common extra input for pricing options on dividend paying stocks in both continuous and discrete models.

6.1.2 Effects on American Call Options

Dividends influence the decision to exercise American Call options before expiry. Taking ownership of the underlying asset just before the ex-dividend date provides a cash bonus. If the dividend amount is sufficiently high and the Call option is sufficiently in the money, then early exercise is optimal. This follows from Lemma 6.3 below.

But first, to prove that a dividend payment is necessary for optimal early Call exercise, suppose that S and R are modeled by recombining binomial trees with positive interest rates, so $R(n,j) > 1$ for all (n,j). Let $C(n,j)$ be the price of an American-style Call option with strike price K and expiry T

corresponding to time step N. Then

$$C(n,j) \geq \frac{p(n,j)C(n+1,j+1) + (1 - p(n,j))C(n+1,j)}{R(n,j)}$$

$$\geq \frac{p(n,j)\,[S(n+1,j+1) - K]^+ + (1 - p(n,j))\,[S(n+1,j) - K]^+}{R(n,j)}$$

$$\geq \frac{[p(n,j)(S(n+1,j+1) - K) + (1 - p(n,j))(S(n+1,j) - K)]^+}{R(n,j)}$$

$$= \frac{[p(n,j)S(n+1,j+1) + (1 - p(n,j))S(n+1,j) - K]^+}{R(n,j)}$$

$$= \frac{[F_1(n,j) - K]^+}{R(n,j)}, \qquad \text{using Lemma 6.1.}$$

The inequalities of the first two steps follow from the American option property that its price is the larger of its exercise value and its backward pricing value. The third inequality follows from the general fact that $[X]^+ + [Y]^+ \geq [X + Y]^+$ for all X, Y.

Now let d_{n+1} be a dividend for S with ex-dividend date $t^e = n + 1$. This affects the one-step Forward price $F_1(n,j)$, so by Eq.6.4,

$$\frac{[F_1(n,j) - K]^+}{R(n,j)} = \frac{[S(n,j)R(n,j) - d_{n+1} - K]^+}{R(n,j)} = \left[S(n,j) - \frac{K + d_{n+1}}{R(n,j)}\right]^+$$

If there is no dividend, so $d_{n+1} = 0$, then this sequence of inequalities implies

$$C(n,j) \geq \left[S(n,j) - \frac{K}{R(n,j)}\right]^+ \tag{6.7}$$

Considering whether to exercise the option at time n, note that it is worthless unless $S(n,j) > K$. In that case, since $K > 0$ and $R(n,j) > 1$,

$$S(n,j) - \frac{K}{R(n,j)} > S(n,j) - K > 0,$$

and since both differences are positive,

$$C(n,j) \geq \left[S(n,j) - \frac{K}{R(n,j)}\right]^+ > [S(n,j) - K]^+.$$

Therefore, if $d_{n+1} = 0$, then the Call option is worth more than any positive early exercise value at n. The equivalent contrapositive of this statement is:

Lemma 6.2 *If early exercise at time n is optimal, then $d_{n+1} > 0$.* □

The converse to Lemma 6.2 requires additional hypotheses on the S, R model as well as on the dividend amount. For simplicity, suppose that $N = 1$ is both the expiry date and the ex-dividend date. Let $d_1 > 0$ be the dividend, and consider under which conditions it is optimal to exercise an American-style Call option at time 0.

Lemma 6.3 *If $d_1 > (R-1)K$ and $S(1,1) > S(1,0) > K$, then early exercise is optimal at time 0.*

Proof: In this one-step, two-state model, there is only one riskless return R and only one risk neutral up probability p. The generalized backward pricing formula gives the positive binomial value

$$
\begin{aligned}
C_B(0,0) &= \frac{pC(1,1) + (1-p)C(1,0)}{R} \\
&= \frac{p\left[S(1,1) - K\right]^+ + (1-p)\left[S(1,0) - K\right]^+}{R} \\
&= \frac{p[S(1,1) - K] + (1-p)[S(1,0) - K]}{R} \\
&= \frac{F_1(0,0) - K}{R}, \qquad \text{by Lemma 6.1,} \\
&= \frac{S(0,0)R - d_1 - K}{R}, \qquad \text{by Eq.6.3,} \\
&= S(0,0) - \frac{K + d_1}{R}.
\end{aligned}
$$

On the other hand, the early exercise value is

$$
C_X(0,0) = [S(0,0) - K]^+ \geq S(0,0) - K,
$$

since $[Z]^+ \geq Z$ for every Z. Finally, if $d_1 > (R-1)K$, then

$$
K < \frac{K + d_1}{R},
$$

and so $C_X(0,0) > C_B(0,0) > 0$. Conclude that early exercise at time 0 is optimal. $\square$

Remark. The proof of Lemma 6.2, at Eq.6.7, actually leads to the stronger result that if early exercise is optimal, then $d_{n+1} > [R(n,j) - 1]K$. Thus no smaller dividend will suffice in Lemma 6.3.

6.1.3 Dividends as Cash Flows

Another way to model a risky asset S that pays dividends is to decompose its price into an *ex-dividend portion*, S', and a *cash flow portion*, I:

$$
S(t,\omega) = S'(t,\omega) + I(t),
$$

where only S' requires a model of the future, continuous or discrete, depending on states ω.

The cash flow portion I is treated as riskless because dividends are typically constant or change slowly over time and are paid at regular intervals. Eq.1.7 gives the present value, at time t, of the dividend cash flow:

$$I(t) = \sum_{\{i:T_i>t\}} d_i Z(t, T_i),$$

where d_i is the dividend paid at time T_i, and $Z(t,T)$ is the discount factor at time t for a zero-coupon bond maturing at time T. In a discrete model with times $t_0 < t_1 < t_2 < \cdots$ and a riskless return R_n over $[t_n, t_{n+1}]$, the zero-coupon bond discount has a simple form:

$$Z(t_n, t_{n+1}) = \frac{1}{R_n}, \quad Z(t_n, t_{n+k}) = \frac{1}{R_n \cdots R_{n+k-1}} = \prod_{i=0}^{k-1} \frac{1}{R_{n+i}}, \quad k = 1, 2, \ldots$$

Then the present value of the cash flow at time step n is

$$I(n) = \sum_{k=1}^{\infty} \frac{d_{n+k}}{R_n \cdots R_{n+k-1}}, \tag{6.8}$$

(abusing notation by writing $I(n) \overset{\text{def}}{=} I(t_n)$).

Present value of a dividend cash flow

Under a few simplifying assumptions, there is an efficient way to compute $I(n)$ by induction, without infinite series. Suppose that:

- R_n depends only on the time t_n. This avoids path-dependent cash flows.

- $d_{n+1} \geq 0$, it has ex-dividend date $t^e = t_{n+1}$, it is paid immediately at $t^p = t_{n+1}$, and it is credited to the asset owner of record at $t^{e-} = t_n$. Thus $d_{n+1} > 0 \implies I(n) > 0$.

- There is some finite future time step $N > 0$, so $0 < t_N < \infty$, after which dividends are ignored: $d_{n+1} = 0$ for all $n > N$. This makes Eq.6.8 a finite sum and implies that $I(n) = 0$ for all $n > N$.

The proof of the following Lemma is left as an exercise:

Lemma 6.4 *The present value sequence $\{I(n) : n = 0, 1, 2, \ldots\}$ satisfies the equations:*

$$I(N) = \frac{d_{N+1}}{R_N}; \quad I(n) = \frac{I(n+1) + d_{n+1}}{R_n}, \quad n = N-1, \ldots, 1, 0, \tag{6.9}$$

corresponding to the terminal condition $I(N+1) = 0$ from which all previous values $I(N), I(N-1), \ldots, I(0)$ are computed by backward recursion. □

Octave implementation and indexing conventions

The recursion in Lemma 6.4 may be implemented as a single loop in Octave, given N, the dividend sequence $\{d_1, \ldots, d_{N+1}\}$ =d(2:N+2), and the riskless returns sequence $\{R_0, R_1, \ldots, R_N\}$ =R(1:N+1). The output will be the present value cash flow sequence $I(0), I(1), \ldots, I(N)$ =I(1:N+1), part of an infinite sequence that has the conventional values $I(n) = 0$ for all $n > N$.

For consistency, all the time indices in Octave/MATLAB arrays are one more than the time indices in the textbook formulas:

```
I(N+1) = d(N+2)/R(N+1); % ignore later dividends
for n=(N-1):(-1):0 % textbook time index n={N-1,...,1,0} is
   I(n+1)=(I(n+2)+d(n+2))/R(n+1); % ..Octave array index n+1
end
```

Inputs for this calculation may be obtained from several sources. The future dividend sequence may be inferred from historical data such as that in Table 6.1. Future riskless returns must be modeled from bond interest rates, as will be done in Section 6.2 below. Past riskless returns, to use with past dividend sequences as in Exercise 3 below, may be obtained from historical interest rates. Table 6.2 shows an example sequence[2] of three-month US Treasury annualized interest rates which may be used to compute riskless returns over one year, one month, and three months (one quarter). Dates were chosen to align with the ex-dividend dates from Table 6.1.

Modified CRR model with separated dividend cash flow

In the recombining binomial tree model, separating the dividend cash flow from S to get the risky ex-dividend portion S' yields the equation

$$S'(n,j) \stackrel{\text{def}}{=} S(n,j) - I(n). \tag{6.10}$$

Given a tree of riskless returns $R(n,j)$, the up and down factors and risk neutral probabilities are given by Eqs. 3.14–3.17, substituting $S \leftarrow S'$. Thus the ex-dividend portion S' satisfies the generalized backward pricing formula

$$S'(n,j) = \frac{p(n,j)S'(n+1,j+1) + (1-p(n,j))S'(n+1,j)}{R(n,j)}$$

In particular, $S'(n+1, j+1) = u(n,j)S'(n,j)$ and $S'(n+1, j) = d(n,j)S'(n,j)$, where $u(n,j), d(n,j)$ are the time and state dependent up and down factors, respectively, from the discrete model for S'.

---

[2]This data was published on the website

https://fred.stlouisfed.org/series/DGS3MO

TABLE 6.2
Quarterly three-month US Treasury rates from 2015–2020.

Time index	Date (yyyy-mm-dd)	DGS3MO ($r \times 100$)	R/yr (e^r)	R/mo ($e^{r/12}$)	R/qtr ($e^{r/4}$)
0	2014-12-04	0.02	1.000200	1.0000167	1.0000500
1	2015-03-03	0.02	1.000200	1.0000167	1.0000500
2	2015-06-03	0.02	1.000200	1.0000167	1.0000500
3	2015-09-02	0.03	1.000300	1.0000250	1.0000750
4	2015-12-02	0.21	1.002102	1.0001750	1.0005251
5	2016-03-02	0.36	1.003606	1.0003000	1.0009004
6	2016-06-01	0.30	1.003005	1.0002500	1.0007503
7	2016-08-31	0.33	1.003305	1.0002750	1.0008253
8	2016-11-30	0.48	1.004812	1.0004001	1.0012007
9	2017-03-01	0.63	1.006320	1.0005251	1.0015762
10	2017-05-31	0.98	1.009848	1.0008170	1.0024530
11	2017-08-30	1.03	1.010353	1.0008587	1.0025783
12	2017-11-30	1.27	1.012781	1.0010589	1.0031800
13	2018-03-01	1.63	1.016434	1.0013593	1.0040833
14	2018-05-31	1.93	1.019487	1.0016096	1.0048367
15	2018-09-06	2.13	1.021528	1.0017766	1.0053392
16	2018-12-06	2.41	1.024393	1.0020104	1.0060432
17	2019-02-28	2.35	1.023778	1.0019603	1.0058923
18	2019-06-06	2.33	1.023574	1.0019436	1.0058420
19	2019-09-05	1.97	1.019895	1.0016430	1.0049371
20	2019-12-05	1.54	1.015519	1.0012842	1.0038574
21	2020-03-05	0.62	1.006219	1.0005168	1.0015512
22	2020-06-04	0.15	1.001501	1.0001250	1.0003751
23	2020-09-03	0.11	1.001101	1.0000917	1.0002750
24	2020-12-03	0.08	1.000800	1.0000667	1.0002000
25	2021-03-04	0.04	1.000800	1.0000667	1.0002000

Now suppose that the up and down factors u, d are constant in all times and states, and given by the CRR model in terms of volatility σ and the number of time steps N to expiry T:

$$u = e^{\sigma\sqrt{T/N}}, \qquad d = \frac{1}{u} = e^{-\sigma\sqrt{T/N}}.$$

Then there is a familiar formula for S':

$$S'(n,j) = S'(0,0)u^j d^{n-j}. \tag{6.11}$$

Since $S'(0,0) = S(0,0) - I(0)$, Eqs.6.10 and 6.11 may be combined as a formula for the original S:

$$S(n,j) = [S(0,0) - I(0)]u^j d^{n-j} + I(n).$$

This has two applications: to compute the initial ex-dividend price $S'(0,0)$ from the spot price $S_0 = S(0,0)$, and then to obtain the asset price model $S(n,j)$ from $S'(n,j)$ and $I(n)$.

Next, modify the constant riskless return formula $R = e^{rT/N}$ to use a time-varying *yield curve* with annualized rate r_n at time step t_n:

$$R_n = e^{r_n T/N},$$

where it is assumed that $t_n - t_{n-1} = T/N$ for every $n = 1, \ldots, N$. In this modified CRR model, the risk neutral probabilities are time-dependent:

$$p_n \overset{\text{def}}{=} \frac{R_n - d}{u - d}, \qquad \Longrightarrow \quad 1 - p_n = \frac{u - R_n}{u - d}.$$

The usual backward pricing formula for S' is also modified:

$$S'(n,j) = \frac{p_n S'(n+1, j+1) + (1 - p_n) S'(n+1, j)}{R_n}. \tag{6.12}$$

The riskless cash flow portion $I(n)$ satisfies Eq.6.9 with riskless returns R_n, so there is a similar backward pricing formula for I:

$$I(n) = \frac{p_n I(n+1) + (1 - p_n) I(n+1) + d_{n+1}}{R_n}. \tag{6.13}$$

Furthermore, since $S(n,j) = S'(n,j) + I(n)$, the two backward pricing formulas in Eqs.6.12 and 6.13 may be added to get

$$S(n,j) = \frac{p_n S(n+1, j+1) + (1 - p_n) S(n+1, j) + d_{n+1}}{R_n}.$$

This is Eq.6.2 for the modified CRR model.

Octave implementation of modified CRR with dividends

To illustrate the effect of dividends on option prices, here is a function using the modified CRR model with constant volatility and time-dependent riskless return. It builds separate models of the ex-dividend `Sx` and cash flow `Ipv` portions of the asset `S` with a function `CRRmD()`, using the modified Cox-Ross-Rubinstein simplification:

```
1  function [S,Sx,Ipv] = CRRmD(T,S0,Di,ri,v,N)
2  % Octave/MATLAB function to model an asset price
3  % as a risky ex-dividend price Sx plus a riskless
4  % dividend cash flow present value Ipv, using the
5  % modified Cox-Ross-Rubinstein (CRR) binomial
6  % pricing model with time-varying rates ri.
7  % INPUTS:                          (Example)
8  %    T  =  expiration time         (1 year)
9  %    S0 =  spot stock price        (100)
```

```
10 %    Di =   2:N+2 dividends in a vector (see below)
11 %    ri =   1:N+1 annual yield curve    (see below)
12 %    v  =   volatility; must be >0           (0.15)
13 %    N  =   height of the binomial tree        (12)
14 % OUTPUTS:
15 %    S  =   cum-dividend binomial tree at all (n,j).
16 %    Sx =   ex-dividend binomial tree at all (n,j).
17 %    Ipv =  dividend present value at all n.
18 % EXAMPLE:
19 %    Di=[0 1 0 0 1 0 0 2 0 0 2 0 0 2]; % Di(k+1)=D_k
20 %    ri=[1 1 1 1 1 2 2 2 2 2 3 3 3 3]/100; % ri(k+1)=r_k
21 %    [S,Sx,Ipv]=CRRD(1,100,Di, ri, 0.15, 12);
22 %
23    [pui,up,Ri] = CRRparams(T,ri,v,N); % Modified CRR
24    Sx=zeros(N+1,N+1); S=zeros(N+1,N+1);
25    Ipv=zeros(1,N+1); % ... allocate output matrices
26    Ipv(N+1)=Di(N+2)/Ri(N+1); % ignore d_n if n>N+1
27    for n=(N-1):(-1):0  % textbook time index n is..
28      Ipv(n+1)=(Ipv(n+2)+Di(n+2))/Ri(n+1); % at n+1
29    end % ... backward recursion for Ipv.
30    Sx(1,1)=S0-Ipv(1); S(1,1)=S0; % initial values
31    for n=1:N      % textbook time indices
32      for j=0:n   % states j={0,1,...,n} at time n
33        Sx(n+1,j+1)=Sx(1,1)*up^(2*j-n);
34        S(n+1,j+1)=Sx(n+1,j+1)+Ipv(n+1);
35      end
36    end % ... forward recursion for Sx and S.
37    return; % Prices are in S, Sx, and Ipv.
38  end
```

Remark. Given a vector input `ri` corresponding to a time-varying sequence of annualized riskless rates, the function `CRRparams()` returns vector outputs `pui` and `Ri` corresponding to time-varying risk neutral up probabilities and riskless returns.

The model may be used to compare American and European Call option prices. It marks when early exercise is optimal, for an asset with known dividends over the period from now to expiry:

```
1  function [Ca,Ce,EE] = CRRmDaeC(T,S0,K,Di,ri,v,N)
2  % Octave/MATLAB function to price American and
3  % European Call options on an asset decomposed
4  % into S=Sx+Ipv by CRRD(), using the modified
5  % Cox-Ross-Rubinstein (CRR) yield curve model.
6  % INPUTS:                          (Example)
7  %    T =   expiration time          (1 year)
8  %    S0 =  stock  price                (100)
9  %    K =   strike  price               (101)
10 %    Di =  dividend sequence 2:N+2  (see below)
11 %    ri =  risk-free APRs 1:N+1     (see below)
```

```
12  %    v  =   volatility;  must  be  >0            (0.15)
13  %    N  =   height  of  the  binomial  tree      (12)
14  % OUTPUT:
15  %    Ca  =   price  of  American  Call  at  all  (n,j).
16  %    Ce  =   price  of  European  Call  at  all  (n,j).
17  %    EE  =   Is  early  exercise  optimal  at  (n,j)?
18  % EXAMPLE:
19  %    Di=[0  1  0  0  1  0  0  2  0  0  2  0  0  2];  % Di(k+1)=D_k
20  %    ri =[1  1  1  1  1  2  2  2  2  2  3  3  3]/100;  % ri (k+1)=r_k
21  %    [Ca,Ce,EE]=CRRmDaeC(1 ,100 ,101 ,Di,ri ,0.15 ,12);
22  %
23     [pui ,up,Ri]=CRRparams(T,ri ,v,N);  % Modified  CRR
24     [S,Sx,Ipv]=CRRmD(T,S0,Di,ri ,v,N);  % decompose
25     Ca=zeros(N+1,N+1);  Ce=zeros(N+1,N+1);  % output
26     EE=zeros(N,N);  % early  exercise  optimal?  T/F
27     for  j =0:N  % to  set  terminal  values  at  (N, j )
28        xC=S(N+1,j+1)-K;  % Call  payoff  at  expiry  is  the
29        Ca(N+1,j+1)=Ce(N+1,j+1)=max(xC,0);  % plus  part
30     end
31     for  n=(N-1):(-1):0  % textbook  time  indices
32        for  j =0:n  % states  j={0,1,...,n}  at  time  n
33        % Backward  pricing  for  A  and  E:
34        bCe=(pui(n+1)*Ce(n+2,j+2)  +  ...
35             (1-pui(n+1))*Ce(n+2,j+1))/Ri(n+1);
36        bCa=(pui(n+1)*Ca(n+2,j+2)  +  ...
37             (1-pui(n+1))*Ca(n+2,j+1))/Ri(n+1);
38        xC=S(n+1,j+1)-K;  % Call  exercise  value
39        % Set  prices  at  node  (n, j ):
40        Ce(n+1,j+1)=bCe;  % always  binomial  price
41        Ca(n+1,j+1)=max(bCa,xC);  % highest  price
42        % Is  early  exercise  optimal?
43        EE(n+1,j+1)=(xC>bCa);  % Yes,  if  xC>bCa
44        end
45     end  % ...backward  induction  pricing:
46     return;  % Ca,Ce,  and  EE  are  defined.
47  end
```

These two functions may be tested on historical data from Table 6.1 (p.146) and Table 6.2 (p.154). Suppose that it is January 2, 2015, and consider January 2021 Calls, expiring in 6 years, on the dividend-paying stock BAC.

First, define BACdivQ, the sequence of 24 quarterly dividends from Table 6.1, starting at time index 0 (which will not be used), noting that there is a dividend every three months:

```
% Quarterly BAC dividends 0:25 from the table:
BACdivQ=[0.05,...    % textbook time index 0 value
    0.05,0.05,0.05,0.05, 0.05,0.05,0.075,0.075,...
    0.075,0.075,0.12,0.12, 0.12,0.12,0.15,0.15,...
    0.15,0.15,0.18,0.18, 0.18,0.18,0.18,0.18, 0.18];
```

Next, define `ri`, the annualized riskless rates assumed constant for one quarter, using the `DGS3MO` column and starting at time index 0:

```
% Annualized riskless interest rates 0:24 from the table:
ri=[0.02,...    % textbook time index 0 value
   0.02,0.02,0.03,0.21, 0.36,0.30,0.33,0.48,...
   0.63,0.98,1.03,1.27, 1.63,1.93,2.13,2.41,...
   2.35,2.33,1.97,1.54, 0.62,0.15,0.11,0.08]/100;
```

(These will be converted into quarterly returns by `CRRparams()`.)

Finally, apply `CRRmD()` to find the ex-dividend model and the dividend cash flow model, and apply `CRRmDaeC()` to price the American and European style Call options with dividends, find early exercise opportunities, and compare their premiums:

```
T=6; S0=17.90; K=30; N=24; sigma=0.1779;
[S,Sx,Ipv]=CRRmD(T,S0,BACdivQ,ri,sigma,N);
Ipv(1)  % =  2.847, present value of dividend cash flow
Sx(1,1) % = 15.053, ex-dividend spot price.
[Ca,Ce,EE]=CRRmDaeC(T,S0,K,BACdivQ,ri,sigma,N);
Ca(1,1),Ce(1,1) % 0.29852, 0.28688 ==> early exercise premium
EE    % TRUE (=1) shows where early exercise is optimal
Ca>Ce % TRUE (=1) where American-style > European-style
```

Remark. The volatility $\sigma = 17.79\%$ used here is the VIX closing value for January 2, 2015, published at

> `https://finance.yahoo.com/quote/^VIX/history`

Its range that day was 17.05–20.14%.
The spot price $S_0 = \$17.90$ used here is the closing price for BAC on January 2, 2015, published at

> `https://finance.yahoo.com/quote/BAC/history`

There is also an *adjusted closing price* $S_0' = \$15.65$ for BAC on that day on the same web page. That is an approximation to the ex-dividend spot price as it is "adjusted for splits and dividend and/or capital gain distributions."

Interpolation from quarterly to monthly time steps

Modeling monthly samples of the dividend cash flow and ex-dividend prices may be done by interpolating, or rather injecting, the quarterly data.

For the example above, reuse `BACdivQ` at every third position in `BACdivM` to get a sequence of 76 monthly values with two zero dividends between adjacent dividend months:

```
% Monthly BAC dividends:
BACdivM=zeros(1,76); % allocate the array
for m=0:25 % indices of quarters from Dec, 2014
  BACdivM(3*m+1) = BACdivQ(m+1); % every 3rd month
end
```

Then triplicate `ri` into an array `r3i` of riskless one-month returns, which is equivalent to assuming that the quarterly rate holds for three months:

```
% Monthly riskless return factors:
r3i=zeros(1,75);  % textbook time indices 0,1,...,74
for q=0:24        % triplicated quarterly rates
   r3i(3*q+1)=r3i(3*q+2)=r3i(3*q+3)=ri(q+1);
end
```

(These will be converted into monthly returns by `CRRparams()`.)

Repeating `CRRmD()` and `CRRmDaeC()` with these longer arrays and larger N but otherwise identical parameters T, S_0, K, σ gives nearly the same prices:

```
T=6; S0=17.90; K=30; N=72; sigma=0.1779;
[S,Sx,Ipv]=CRRmD(T,S0,BACdivM,r3i,sigma,N);
Ipv(1)  % = 2.677, present value of dividend cash flow
Sx(1,1) % = 15.223, ex-dividend spot price
[Ca,Ce,EE]=CRRmDaeC(T,S0,K,BACdivM,r3i,sigma,N);
Ca(1,1),Ce(1,1) % 0.32185, 0.30802  ==> early exercise premium
EE    % TRUE (=1) shows where early exercise is optimal
Ca>Ce % TRUE (=1) where American-style > European-style
```

Octave implementation of plain CRR with dividends

The unmodified CRR model with constant volatility and riskless return may be implemented by changing just a few lines in `CRRmD.m` and `CRRmDaeC.m`:

```
1   function [S, Sx, Ipv] = CRRD(T, S0, Di, r, v, N)
2   % Octave/MATLAB function to model an asset price
3   % as a risky ex-dividend price Sx plus a riskless
4   % dividend cash flow present value Ipv, using the
5   % Cox-Ross-Rubinstein (CRR) binomial pricing model.
6   % INPUTS:                                 (Example)
7   %   T  =  expiration time                 (1 year)
8   %   S0 =  spot stock price                (100)
9   %   Di =  2:N+2 dividends in a vector (see below)
10  %   r  =  constant risk-free annual yield (0.02)
11  %   v  =  volatility; must be >0          (0.15)
12  %   N  =  height of the binomial tree     (12)
13  % OUTPUTS:
14  %   S  =  cum-dividend binomial tree at all (n,j).
15  %   Sx =  ex-dividend binomial tree at all (n,j).
16  %   Ipv = dividend present value at all n.
17  % EXAMPLE:
18  %   Di=[0 1 0 0 1 0 0 2 0 0 2 0 0 2]; % Di(k+1)=d_k
19  %   [S,Sx,Ipv]=CRRD(1,100,Di, 0.02, 0.15, 12);
20  %
21     [pu,up,R] = CRRparams(T,r,v,N); % Use CRR values
22     Sx=zeros(N+1,N+1); S=zeros(N+1,N+1);
23     Ipv=zeros(1,N+1); % ...allocate output matrices
24     Ipv(N+1)=Di(N+2)/R; % ignore Di(n+1)=d_n if n>N+1
```

```
25    for n=(N-1):(-1):0  % textbook time indices
26      Ipv(n+1)=(Ipv(n+2)+Di(n+2))/R; % Di(n)=d_{n+1}
27    end % ...backward recursion for Ipv.
28    Sx(1,1)=S0-Ipv(1); S(1,1)=S0; % initial values
29    for n=1:N      % textbook time indices
30      for j=0:n    % states j={0,1,...,n} at time n
31        Sx(n+1,j+1)=Sx(1,1)*up^(2*j-n);
32        S(n+1,j+1)=Sx(n+1,j+1)+Ipv(n+1);
33      end
34    end % ...forward recursion for Sx and S.
35    return; % Prices are in S, Sx, and Ipv.
36  end
```

```
1   function [Ca,Ce,EE] = CRRDaeC(T,S0,K,Di, r,v,N)
2   % Octave/MATLAB function to price American and
3   % European Call options on an asset decomposed
4   % into S=Sx+Ipv by CRRD(), using the CRR model.
5   % INPUTS:                               (Example)
6   %    T =  expiration time               (1 year)
7   %    S0 = stock price                     (100)
8   %    K =  strike price                    (101)
9   %    Di = dividend sequence 2:N+2     (see below)
10  %    r =  annualized risk-free yield      (0.02)
11  %    v =  volatility; must be >0          (0.15)
12  %    N =  height of the binomial tree      (12)
13  % OUTPUT:
14  %    Ca =  price of American Call at all (n,j).
15  %    Ce =  price of European Call at all (n,j).
16  %    EE =  Is early exercise optimal at (n,j)?
17  % EXAMPLE:
18  %    Di=[0 1 0 0 1 0 0 2 0 0 2 0 0 2]; % Di(k+1)=D_k
19  %    [Ca,Ce,EE]=CRRDaeC(1,100,101,Di,0.02,0.15,12);
20  %
21    [pu,up,R]=CRRparams(T,r,v,N); % Use CRR values
22    [S,Sx,Ipv]=CRRD(T,S0,Di,r,v,N); % decompose
23    Ca=zeros(N+1,N+1); Ce=zeros(N+1,N+1); % output
24    EE=zeros(N,N); % early exercise T/F
25    for j = 0:N % to set terminal values at (N,j)
26      xC=S(N+1,j+1)-K; % Call payoff at expiry...
27      Ca(N+1,j+1)=Ce(N+1,j+1)=max(xC,0); % plus part
28    end
29    for n = (N-1):(-1):0 % textbook time indices
30      for j = 0:n % states j={0,1,...,n} at time n
31        % Backward pricing for A and E:
32        bCe=(pu*Ce(n+2,j+2)+(1-pu)*Ce(n+2,j+1))/R;
33        bCa=(pu*Ca(n+2,j+2)+(1-pu)*Ca(n+2,j+1))/R;
34        xC=S(n+1,j+1)-K; % Call exercise value
35        % Set prices at node (n,j):
36        Ce(n+1,j+1)=bCe; % always binomial price
```

```
37      Ca(n+1,j+1)=max(bCa,xC); % highest  price
38      % Is  early  exercise  optimal?
39      EE(n+1,j+1)=(xC>bCa); % Yes, if xC>bCa
40    end
41   end % ...backward induction pricing:
42   return; % Ca,Ce, and EE are defined.
43 end
```

The example in the comments is a simple experiment:

```
Di=[0 1 0 0 1 0 0 2 0 0 2 0 0 2]; % Di(2:N+2)=d_{1:N+1}
[Ca,Ce,EE]=CRRDaeC(1,100,Di,101,0.02,0.15,12);
EE     % shows when early exercise is optimal
Ca>Ce  % shows how early exercise adds value
```

It illustrates how early exercise optimality propagates backward to make the American Call option more valuable at predecessor states and times.

6.2   Interest Rates

Historical interest rates are not trustworthy predictors of future interest rates. That job requires a model of the future, typically based on bond prices.

A bond traded on an exchange is actually a risky asset. Its price depends on several factors:

- time to *maturity*, when it must be repaid,

- *coupons*, partial repayments required before maturity,

- *creditworthiness*, the probability of repayment without default,

- money supply, adjusted from time to time by government actions,

- market demand for capital.

Of these factors, the first two are fixed in every contract between lender (buyer of the bond) and borrower (seller). The remaining factors contribute to risk which is managed by other contract terms and derivatives.

In Chapter 1, Eq.1.1, a bond's price was denoted by $B(t)$ and it was treated as a riskless asset because its value at maturity $T > 0$ was known. Then in Chapter 2, Eq.2.3, a simple continuous model using the differential Eq.2.1 gave $B(t) = B_0 e^{rt}$, where B_0 was the spot price at $t = 0$, r was a rate parameter from the model, and maturity T was not mentioned. If that differential equation is solved for a specified terminal, or *face value* $F \stackrel{\text{def}}{=} B(T)$, then the formula becomes

$$B(t) = Fe^{r[t-T]}.$$

But even if both $B(0)$ and $B(T)$ are known, a bond's price at intermediate times t with $0 < t < T$ may not follow the simple exponential formula from continuous compounding at a constant riskless rate. A more accurate financial model must in fact treat it as a risky asset, with market forces resulting in deviations from the exponential function. So, denote the price at time t of a bond by $B(t,T)$, with the goal of estimating this function from data available at time $t = 0$.

Bond prices are sometimes quoted after normalization to a *discount* (or premium, in some cases) through division by the face value $F = B(T,T)$. It is more common, however, to use the *yield to maturity*, a fluctuating interest rate $r = r(t,T)$ implied by the discount (or premium) to face value, plus any remaining coupons.

For example, a zero-coupon bond with maturity date T has discount Z and yield to maturity r that are related at time $t < T$ by:

$$Z = e^{r[t-T]}, \qquad \Longrightarrow \quad r = \frac{\log Z}{t - T}, \tag{6.14}$$

Both are risky and fluctuate with t. Discounts are dimensionless quantities. The units for T and t are usually years, making r an annualized interest rate.

Octave function to convert between discounts and yields

These anonymous functions work with vector inputs `t`,`T` because of the point-wise operators `./` and `.*`. They also work if `t=0` but `T` is a vector, since the expression `0-T` is correctly evaluated as `-T`.

```
ZtT=@(r,t,T)exp(r.*(t-T));   rtT=@(Z,t,T)log(Z)./(t-T);
```

6.2.1 Zero-Coupon Bonds

Spot discount factors $Z(0,T)$ are priced by auction in markets. Their reciprocals provide riskless returns over various periods in financial models. Only a few standard maturity times T are available, however. For other periods, it is necessary to interpolate.

One method is exponential interpolation with the expected forward discount of Eq.1.6. This is estimated from spot discounts at nearby maturities. Namely, suppose that $T_1 < T < T_2$ and that $Z(0,T_1)$ and $Z(0,T_2)$ are known. Determine the expected forward rate r from Eq.1.6:

$$\bar{Z}(T_1,T_2) = e^{r[T_1-T_2]} = \frac{Z(0,T_2)}{Z(0,T_1)},$$

and then interpolate the exponential function with that rate r to get an expected spot discount estimate

$$
\begin{aligned}
Z(0,T) \quad &\approx \quad \bar{Z}(0,T) \quad \stackrel{\text{def}}{=} \quad Z(0,T_1)e^{r[T_1-T]} \tag{6.15}\\
&= \quad Z(0,T_1)\left[\frac{Z(0,T_2)}{Z(0,T_1)}\right]^{\frac{T_1-T}{T_1-T_2}} = \quad Z(0,T_1)^{1-\alpha}Z(0,T_2)^{\alpha},
\end{aligned}
$$

where $\alpha \stackrel{\text{def}}{=} \frac{T_1-T}{T_1-T_2}$, so $1 - \alpha = \frac{T-T_2}{T_1-T_2}$.

Another method is spline interpolation. For discount factors, the interpolation set is $\{(T_n, Z(0,T_n)) : n = 1,\ldots,N\}$ for maturities $0 < T_1 < \cdots T_N$. When $N > 2$, the spline interpolation function is smoother than the piecewise exponential function obtained from Eq.6.15 on adjacent $[T_n, T_n+1]$. Discounts for $0 < T < T_1$ may use $T_0 = 0$ as an endpoint for which $Z(0,T_0) = 1$.

Octave function to interpolate $Z(0,T)$ by two methods.

```
1  function [Ze,Zs] = ZT(Tn, Zn, Ti)
2  % Octave/MATLAB function to compute Zero Coupon
3  % Bond spot discounts by interpolation.
4  % INPUTS:                              (Example)
5  %    Tn = maturities with known discounts   (1:5)
6  %    Zn = spot discount at each Tn    (cos(Tn/10))
7  %    Ti = evaluation points    ([0.5,1.5,3.2,4.7])
8  % OUTPUTS:
9  %    Ze = piecewise exponential values at Ti
10 %    Zs = spline interpolation values at Ti
11 % EXAMPLE:
12 %    Tn=1:5; Zn= cos(Tn/10); Ti=[0.5,1.5,3.2,4.7];
13 %    [Ze,Zs]=ZT(Tn,Zn,Ti)
14 %    plot(Tn,Zn,"k",Ti,Ze,"r",Ti,Zs,"b");
15 %
16    [Tn,nsort]=sort(Tn); Zn=Zn(nsort); % 0<T1<T2<..
17    Ze = zeros(size(Ti)); % initialize Ze
18    for i=1:length(Ti)  % evaluation times
19       T = Ti(i);
20       t1=0; t2=Tn(1); z1=1; z2=Zn(1); % 0<T<T1
21       for j=length(Tn):-1:2 % seek from high end
22          if(T>Tn(j-1))
23             t1=Tn(j-1); t2=Tn(j);
24             z1=Zn(j-1); z2=Zn(j);
25             break;
26          end
27       end
28       alpha=(t1-T)/(t1-t2); % Compute Ze by exp()
29       Ze(i)=z1^(1-alpha)*z2^alpha; % interpolation
30    end                    % Then, compute Zs by spline
31    Zs=spline([0,Tn],[1,Zn],Ti); % interpolation
32    return;
33 end
```

6.2.2   Coupon Bonds

Zero-coupon bond discounts may be computed from coupon bond prices, using the present value formula Eq.1.7.

Price and Yield to Maturity

Traditionally, prices for bonds are computed after an auction where the bids are made in percentages called *yield to maturity (YTM)*. That percentage is the rate r that appears in the present value computation for the bond. Leaving r as a variable, the other parameters are

- F, the face value, often a nominal amount such as $100;

- T, the time to maturity, in years but with a precision of one day;

- I, the interest rate, annualized by adding up all coupons in one year;

- N, the number of coupon payments per year, such as two for semiannual coupons or four for quarterly coupons;

- t, the time to the first coupon payment, in years but with a precision of one day.

Thus the coupon rate is I/N and the coupon amount is FI/N.

At the issuance of the bond, T is an integer such as 2, 3, 5, 7, or 10 for US Treasury notes, and then the price of the bond will be its present value:

$$P = Fe^{-rT} + \sum_{n=1}^{NT} \frac{FI}{N} e^{-rn/N} \approx \frac{F}{(1+r)^T} + \sum_{n=1}^{NT} \frac{FI/N}{(1+r/N)^n} \qquad (6.16)$$

where the second formula uses the Taylor approximation $e^{\theta} \approx 1 + \theta$ which is quite accurate for small $|\theta|$.

Suppose that a company offers a series of coupon bonds with maturities $0 < T_1 < T_2 < \cdots T_M$, each with a coupon paying an amount c_i per dollar of face value, $i = 1, \ldots, M$ at times Δ, 2Δ, $3\Delta \ldots$. Suppose for simplicity that each $T_i = N_i\Delta$ is an integer multiple of this interval, and that each bond is repaid at T_i together with the last coupon payment. Denote the spot prices of these bonds, per unit face value, by $B_1, \ldots, B_M$. If the company is as creditworthy as the US government, then these spot prices should be the present value of the cash flow from the coupons plus the repayment at maturity. Then Eq.1.7 produces the linear system of equations

$$B_i = Z(0, T_i) + c_i \sum_{n=1}^{N_i} Z(0, n\Delta), \qquad i = 1, \ldots, M; \quad T_i = N_i\Delta.$$

Here $Z(0, T)$ is the US Treasury zero-coupon bond discount with maturity T. The matrix A for this linear system written as $A\mathbf{z} = \mathbf{b}$ is

$$A \stackrel{\text{def}}{=} \begin{pmatrix} c_1 & \cdots & c_1 & 1+c_1 & 0 & \cdots & \cdots & \cdots & \cdots & 0 \\ c_2 & \cdots & & & \cdots & c_2 & 1+c_2 & 0 & \cdots & 0 \\ \vdots & & & & & & & \ddots & & \vdots \\ c_M & \cdots & & & & & & \cdots & c_M & 1+c_M \end{pmatrix}.$$

The unknowns $\mathbf{z}$ and right-hand side $\mathbf{b}$ are

$$
\mathbf{z} \stackrel{\text{def}}{=} \begin{pmatrix} Z(0,T_1) \\ \vdots \\ Z(0,T_M) \end{pmatrix} = \begin{pmatrix} Z(0,\Delta) \\ \vdots \\ Z(0,N_M\Delta) \end{pmatrix} ; \qquad \mathbf{b} \stackrel{\text{def}}{=} \begin{pmatrix} B_1 \\ \vdots \\ B_M \end{pmatrix} .
$$

There are M rows and N_M columns. There is a unique solution if $N_M = M$, for then A is a nonsingular matrix: lower triangular with nonzeros on the main diagonal, so $A\mathbf{z} = \mathbf{b}$ can be solved by forward substitution. This will be the case for a series of bonds with maturities $T = 1, 2, \ldots, M$ and annual coupons, so $\Delta = 1$ and $N_M = M$:

$$
\begin{aligned}
Z(0,1) &= B_1/(1+c_1) \\
Z(0,2) &= (B_2 - c_2 Z(0,1))/(1+c_2) \\
&\vdots \\
Z(0,M) &= (B_M - c_M[Z(0,1) + \cdots + Z(0,M-1)])/(1+c_M)
\end{aligned} \tag{6.17}
$$

Interpolation with noninvertible A

When matrix A is not invertible, it is still possible to estimate $Z(0,T)$ by using interpolation. Given M coupon bonds, use a model with M parameters:

$$
Z(0,T) = a_M T^M + \cdots + a_1 T + 1 = 1 + \sum_{m=1}^{M} a_m T^m. \tag{6.18}
$$

The last coefficient is 1 since $Z(0,0) = 1$ in all cases. The system of equations derived from the present value identity becomes a linear system of equations for the remaining coefficients $\mathbf{a} = (a_1, \ldots, a_M)$:

$$
\begin{aligned}
B_i &= 1 + \sum_{m=1}^{M} (N_i\Delta)^m a_m + c_i \sum_{n=1}^{N_i} \left(1 + \sum_{m=1}^{M} (n\Delta)^m a_m \right) \\
&= 1 + N_i c_i + \sum_{m=1}^{M} \left((N_i\Delta)^m + c_i \sum_{n=1}^{N_i} (n\Delta)^m \right) a_m,
\end{aligned}
$$

for $i = 1, \ldots, M$. The right-hand side vector will now be

$$
\hat{\mathbf{b}} \stackrel{\text{def}}{=} \begin{pmatrix} B_1 - 1 - N_1 c_1 \\ \vdots \\ B_M - 1 - N_M c_M \end{pmatrix},
$$

where $N_i c_i$ is the sum of the coupons over the N_i payments. The matrix $\hat{A}$ for the system will be $M \times M$ and its i, j coefficient will be

$$
\hat{A}(i,j) \stackrel{\text{def}}{=} (N_i\Delta)^j + c_i \sum_{n=1}^{N_i} (n\Delta)^j.
$$

Solving the system $\hat{A}\mathbf{a} = \hat{\mathbf{b}}$ for $\mathbf{a}$ will provide the coefficients from which the zero-coupon bond discounts may be calculated.

Estimating creditworthiness

One application is to estimate the creditworthiness of a company issuing coupon bonds but not zero-coupon bonds. Using the company's zero-coupon debt discount $\hat{Z}(0,T)$ in the present value equation and solving for it gives a market-based estimate for default risk. The ratio $\hat{Z}(0,T)/Z(0,T)$ of the company's zero-coupon discount is a measure of creditworthiness since US government debt is considered maximally creditworthy.

6.2.3 Cash Flow Swaps

Suppose that the owner of an asset that pays a risky, time-varying interest rate $\{r(t) : 0 \leq t < T\}$ wishes to *swap*, or exchange it for a steady cash flow at a fixed interest rate κ. The fair rate κ must be computed with a model of the future, used to estimate the risky rates as they change. Zero-coupon bond discounts $Z(0,T)$ for standard choices of T provide the input data when first estimating κ. Conversely, market rates for κ with various terms provide input data to estimate $Z(0,t)$ at all values of t.

Fair swap rates

In practice, estimates are needed only at discrete equispaced times

$$t_n = nT/N \overset{\text{def}}{=} n\Delta t, \qquad n = 1, 2, \ldots, N,$$

when the variable interest payments are made. Assume that the interest rate remains constant over these small time intervals:

$$r(t) = r_n, \qquad t_n \leq t < t_{n+1}.$$

The expected values $\bar{r}_n$ of these future piecewise constant rates satisfy

$$e^{-\bar{r}_n[t_{n+1}-t_n]} = e^{-\bar{r}_n\Delta t} = \bar{Z}(t_n, t_{n+1}) = \frac{Z(0,t_{n+1})}{Z(0,t_n)}, \qquad (6.19)$$

for $n = 0, 1, \ldots, N-1$, by Eqs.1.4 and 1.6 on p.3. Thus

$$\bar{r}_n\Delta t = \log\frac{Z(0,t_n)}{Z(0,t_{n+1})}, \qquad n = 0, 1, \ldots, N-1. \qquad (6.20)$$

But $\bar{r}_n\Delta t$ is very small for typical interest rates and payment intervals so that the approximation $e^{\bar{r}_n\Delta t} = 1 + \bar{r}_n\Delta t$ is accurate. Use it when exponentiating both sides of Eq.6.20 to get

$$1 + \bar{r}_n\Delta t = \frac{Z(0,t_n)}{Z(0,t_{n+1})}, \qquad n = 0, 1, \ldots, N-1. \qquad (6.21)$$

Now consider the expected cash flow to the party that pays variable interest and receives constant interest. At t_{n+1}, the net flow per dollar of principal is $\kappa\Delta t - \bar{r}_n\Delta t$, which is the same as $[1 + \kappa\Delta t] - [1 + \bar{r}_n\Delta t]$, so the present value of the expected net flow may be computed by Eq.1.7:

$$
\begin{aligned}
PV &= \sum_{n=0}^{N-1}([1 + \kappa\Delta t] - [1 + \bar{r}_n\Delta t])Z(0, t_{n+1}) \\
&= [1 + \kappa\Delta t]\sum_{n=0}^{N-1} Z(0, t_{n+1}) - \sum_{n=0}^{N-1}[1 + \bar{r}_n\Delta t]Z(0, t_{n+1}) \\
&= [1 + \kappa\Delta t]\sum_{n=1}^{N} Z(0, t_n) - \sum_{n=0}^{N-1} Z(0, t_n)
\end{aligned}
$$

after reindexing the first sum and using Eq.6.21 in the second. But $PV = 0$ with a fair fixed rate κ, implying

$$
\kappa\Delta t = \frac{\sum_{n=0}^{N-1} Z(0, t_n)}{\sum_{n=1}^{N} Z(0, t_n)} - 1 = \frac{1 - Z(0, T)}{\sum_{n=1}^{N} Z(0, t_n)},
$$

using the fact that $Z(0, t_0) = Z(0, 0) = 1$ and $Z(0, t_N) = Z(0, T)$. Thus

$$
\kappa = \frac{1}{\Delta t}\left(\frac{1 - Z(0, T)}{\sum_{n=1}^{N} Z(0, t_n)}\right), \tag{6.22}
$$

which is commonly used in practice.

For example, consider the published rates for US T-Bills, which are zero-coupon bonds with maturities of 4, 8, 13, 26, and 52 weeks. These were collected at the website:

https://www.treasury.gov/resource-center/data-chart-center/
interest-rates/Pages/TextView.aspx?data=billrates

A selection from December 15, 2021 is in Table 6.3. The fair swap rate for quarterly exchanges requires three of the zero-coupon bond discounts in

TABLE 6.3
Spot US T-bill rates and discounts on December 15, 2021.

Maturity (weeks)	T (years)	Yield (APR)	Discount $Z(0, T)$
4	4/52	0.03%	0.99998 = ZtT(.03/100,0,4/52)
8	8/52	0.05%	0.99992 = ZtT(.05/100,0,8/52)
13	1/4	0.06%	0.99985 = ZtT(.06/100,0,1/4)
26	1/2	0.13%	0.99935 = ZtT(.13/100,0,1/2)
52	1	0.27%	0.99730 = ZtT(.27/100,0,1)

TABLE 6.4
Quarterly US T-bill rates and discounts, with interpolation.

Maturity (weeks)	T (years)	Discount $Z(0,T)$	Source (or formula)	Expected $\bar{r}$ (APR)
13	1/4	0.99985	$Z(0,1/4)$ from Table 6.3	0.060005%
26	2/4	0.99935	$Z(0,1/2)$ from Table 6.3	0.20008%
39	3/4	0.99832	$Z(0,1/2)^{\frac{1}{2}}Z(0,1)^{\frac{1}{2}}$	0.41069%
52	4/4	0.99730	$Z(0,1)$ from Table 6.3	0.41069%

Table 6.3, plus a fourth that must be interpolated. This is done in Table 6.4, where $\alpha = (3/4 - 1/2)/(1 - 1/2) = 1/2$ is used in Eq.6.15 to find $Z(0,3/4)$.

Apply Eq.6.22 to the data in Table 6.4 to compute

$$\kappa = \frac{1}{1/4}\left(\frac{1 - Z(0,1)}{\sum_{n=1}^{4} Z(0,n/4)}\right) = 0.0027035 = 0.27035\%,$$

which may be done with the Octave commands

```
Z=[0.99985 0.99935 0 0.99730]; Z(3)=sqrt(Z(2)*Z(4))
kappa=4*(1-Z(4))/sum(Z), kappapct=kappa*100
r1bar=-log(Z(1))*400, r2bar=-log(Z(2)/Z(1))*400
r3bar=-log(Z(3)/Z(2))*400, r4bar=-log(Z(4)/Z(3))*400
```

Remark. Notice how increasing riskless yields, over $[0,T]$ with increasing maturity $T \in \{\frac{1}{4}, \frac{1}{2}, \frac{3}{4}\}$, are amplified into increasing expected forward yields over $[t, t+1/4]$ with increasing $t \in \{0, \frac{1}{4}, \frac{1}{2}, \frac{3}{4}\}$. The last expected forward yield over $[\frac{3}{4}, 1]$ is the same as for $[\frac{1}{2}, \frac{3}{4}]$ since the intermediate discount $Z(0, \frac{3}{4})$ is exponentially interpolated.

Relating swap rates

Denote by $\kappa(T, \Delta t)$ the market swap rate for the period $[0,T]$ with cash exchanges taking place at intervals Δt. In practice, with units of years, T will be a positive integer and Δt will be either $1/2$ for semiannual swaps, $1/4$ for quarterly swaps, or $1/12$ for monthly swaps. The rates are always annualized in the published quotes.

These market rates are not available for all combinations of T and Δt. In particular, semiannual swap rates are published for large T but only quarterly swap rates are published for $T = 1$. An approximation for $\kappa(1, 0.5)$ may be obtained by equating compounded returns over one year, regardless of T:

$$\left(1 + \frac{\kappa(T, 0.5)}{2}\right)^2 = \left(1 + \frac{\kappa(T, 0.25)}{4}\right)^4$$

Solving gives

$$\kappa(T, 0.5) = 2 \left[\left(1 + \frac{\kappa(T, 0.25)}{4} \right)^2 - 1 \right]. \tag{6.23}$$

Put $T = 1$ to estimate an unavailable $\kappa(1, 0.5)$ from a published $\kappa(1, 0.25)$. For example, the quarterly swap rate $\kappa = \kappa(1, 0.25)$ computed from Table 6.4 gives an estimate

$$\hat{\kappa}(1, 0.5) = 2 \left[\left(1 + \frac{\kappa(1, 0.25)}{4} \right)^2 - 1 \right] = 0.0027044 = 0.27044\%.$$

If both $\kappa(T, 0.5)$ and $\kappa(T, 0.25)$ are published, the error in the approximation may be estimated by comparing the published $\kappa(T, 0.5)$ with the approximation from Eq.6.23. For example, suppose that the market rate $\kappa(1, 0.5)$ agrees exactly with the value given by Eq.6.22 with Table 6.3 data. Then

$$\kappa(1, 0.5) = \frac{1}{1/2} \left(\frac{1 - Z(0, 1)}{Z(0, 1/2) + Z(0, 1)} \right) = 0.0027045 = 0.27045\%,$$

giving a difference $\kappa(1, 0.5) - \hat{\kappa}(1, 0.5) = 0.0000001$.

More generally, with two exchange intervals Δ_1 and Δ_2, the swap rate relation is

$$(1 + \kappa(T, \Delta_1)\Delta_1)^{1/\Delta_1} = (1 + \kappa(T, \Delta_2)\Delta_2)^{1/\Delta_2},$$
$$\implies \kappa(T, \Delta_1) = \frac{1}{\Delta_1} \left[(1 + \kappa(T, \Delta_2)\Delta_2)^{\Delta_1/\Delta_2} - 1 \right]. \tag{6.24}$$

6.2.4 Benchmarks

Riskless returns may be calculated from various published zero-coupon bond discounts, or coupon bond prices, or swap rates. There are many so-called benchmarks, available for specific currencies, that are compiled and published by both private and public institutions.

Benchmark interest rates are typically calculated from market rates using published formulas. The market and formula are chosen to make the result robust against manipulation. In stochastic process language, the benchmark should be a martingale. This is impossible to guarantee mathematically in practice, but there are some accepted principles that boost faith:

- large numbers of constituent rates, averaged together,

- volume weighting of the averages,

- consistency among related benchmarks.

Overnight lending rates among many large banks satisfy the first two conditions for very short terms. Longer-term rates may be computed from government bond discounts set at auctions and may be checked for consistency with market-derived swap rates.

AUD benchmarks. Short-term riskless rates for Australian dollars (AUD) may be obtained directly from the benchmark *BBSW*, or Bank Bill SWaps rate. A full description of this security and its market may be found at the following URLs:

https://www.rba.gov.au/mkt-operations/resources/interest-rate-
benchmark-reform.html
https://www2.asx.com.au/content/dam/asx/benchmarks/asx-bbsw-
conventions.pdf

The Australian Securities Exchange (ASX) publishes the values at the URL

https://www.asx.com.au/data/benchmarks/bbsw-10-day-rolling-
history.pdf

A 10-day sample of these rates, collected in Table 6.5, provides a good illustration of interest rate variability. This data consists of annualized percentage rates for *tenors* of $k \in \{1, 2, 3, 4, 5, 6\}$ months.[3]

TABLE 6.5
ASX Australian dollar Bank Bill Swaps rates by tenor.

BBSW date (dd/mm/yyyy)	1 mo (%)	2 mo (%)	3 mo (%)	4 mo (%)	5 mo (%)	6 mo (%)
11/03/2022	0.0150	0.0661	0.1450	0.2550	0.3859	0.5100
10/03/2022	0.0186	0.0700	0.1450	0.2531	0.3700	0.5000
09/03/2022	0.0150	0.0700	0.1458	0.2550	0.3650	0.5086
08/03/2022	0.0150	0.0650	0.1385	0.2254	0.3250	0.4450
07/03/2022	0.0150	0.0650	0.1350	0.2050	0.2950	0.4043
04/03/2022	0.0156	0.0500	0.1183	0.1800	0.2700	0.3798
03/03/2022	0.0150	0.0450	0.0983	0.1583	0.2300	0.3385
02/03/2022	0.0161	0.0350	0.0850	0.1400	0.2050	0.2810
01/03/2022	0.0175	0.0430	0.0850	0.1250	0.1850	0.2550
28/02/2022	0.0166	0.0406	0.0798	0.1250	0.1767	0.2500

The correspondence with zero-coupon bond discounts is

$$Z(0, k/12) = \exp(-\text{BBSW}(k)/12) \qquad k = 1, 2, \ldots, \qquad (6.25)$$

using the 30/360 convention that treats one year as 12 equal 30-day months. To get the expected riskless rates $\bar{r}_0, \ldots, \bar{r}_5$, put $\Delta t = 1/12$ and combine Eqs.6.25 and 6.20:

$$\bar{r}_k \Delta t = \log \frac{Z(0, k\Delta t)}{Z(0, [k+1]\Delta t)} = \text{BBSW}(k+1) - \text{BBSW}(k), \quad k = 0, \ldots, 5, \quad (6.26)$$

---

[3]By convention, "tenor" is used instead of "maturity" for calculated benchmark rates.

with the convention that $\mathrm{BBSW}(0) = 0$ since $Z(0,0\Delta t) = 1$. Note that $\bar{r}_k$ is annualized: the one-month expected riskless return is $\bar{R} = \exp(\bar{r}\Delta t) = \exp(\bar{r}/12)$. For example, get monthly expected riskless rates from the 1 March 2022 row (01/03/2022) with these Octave commands:

```
bbsw=[0.0175,0.0430,0.0850,0.1250,0.1850,0.2550]; % 1/3/2022
rbar=12*diff([0,bbsw]) % 0.210 0.306 0.504 0.480 0.720 0.840
```

Here `diff(x)=[x(2)-x(1),x(3)-x(2),...]` is the discrete difference function. It computes Eq.6.26 correctly, with leading term `x(1)`, if given the zero-padded input `x=[0,bbsw]`.

EUR rates. The European Central Bank (ECB) currently has two short-term benchmarks, EONIA and €STR, and an intermediate-term benchmark called EURIBOR. Their values are published by the ECB at the URLs

https://www.ecb.europa.eu/stats/financial_markets_and_interest_
rates/euro_short-term_rate/html/index.en.html
https://www.ecb.europa.eu/paym/interest_rate_benchmarks/WG_euro_
risk-free_rates/html/index.en.html

The €STR was negative, at -0.577%, computed on 11 March 2022 from 565 transactions among 32 banks totaling more than €55 billion.

There are also informal *indicators*, published by brokerages and similar institutions on URLs such as

https://www.fxempire.com/macro/euro-area/interest-rate

A best *lending rate*, which was around 0.25% in January 2022, is typically adjusted from the ECB rates to reflect creditworthiness and monetary policy. As a result, different European Union member states and other nations that trade in euros have different lending rates.

USD benchmarks. Riskless rates for US dollars (USD) can be derived from several benchmarks, listed here with their range of tenors or maturities:

- Fed funds effective rate, overnight;

- Secured Overnight Financing Rate (SOFR), overnight;

- USD London Interbank Offered Rate (LIBOR), 1–12 months;

- US Treasury bills, notes, and bonds, 1–30 years;

- 1-month Term SOFR swap rates, 1–30 years;

- Semiannual coupon bond swap rates, 1–30 years;

- Prime lending rate for consumer credit and real estate loans, 1 month – 30 years.

Their market values are published online, for example at the URL

https://www.chathamfinancial.com/technology/us-market-rates

See Tables 6.6, 6.7, and 6.8 below for some examples.

TABLE 6.6
Historical benchmark USD rates (%): Prime lending rate; Fed Funds overnight rate; Secured Overnight Financing Rate (SOFR) with 30- and 90-day moving averages (MA).

Date (d/m/y)	Prime	Fed Funds	SOFR	30-Day MA	90-Day MA
11/03/22	3.250	0.080	0.05000	0.04967	0.04934
10/03/22	3.250	0.080	0.05000	0.04967	0.04934
11/02/22	3.250	0.080	0.05000	0.04833	0.04934
11/03/21	3.250	0.070	0.01000	0.03133	0.05812

TABLE 6.7
USD LIBOR benchmark interest rates at closing on 11 March 2022 and three earlier dates.

USD LIBOR	1 mo	3 mo	6 mo	12 mo
11 Mar 2022	0.39657	0.82600	1.13057	1.59600
10 Mar 2022	0.38700	0.80286	1.10286	1.53486
11 Feb 2022	0.19114	0.50643	0.84043	1.39229
11 Mar 2021	0.10600	0.18388	0.19275	0.27725

6.3   Exercises

1. Suppose that S is a dividend paying stock with a declared dividend amount D at ex-dividend date t^e. Let $S(t^e-)$ denote the closing price of S on the day before the ex-dividend date, and $S(t^e+)$ denote the opening price of S on the ex-dividend date. Show that

$$S(t^e+) = S(t^e-) - D$$

by constructing an arbitrage otherwise.

2. Prove Eq.6.9 from Eq.6.8.

3. Use the data in Table 6.1 (p.146) and Table 6.2 (p.154) to compute the present value sequence of BAC dividends from December 2014 through March 2021, namely time indices 0–25. Disregard any dividends outside of that time period.

   (a) Find the values quarterly.

   (b) Find the values monthly.

   (c) Plot the two dividend present value sequences, and the dividends themselves, on the same graph.

TABLE 6.8

Intermediate and long-term benchmark USD interest rates at closing on 11 March 2022 and two earlier dates.

BENCH-MARK	Date (d/m/y)	1 yr (%)	2 yr (%)	3 yr (%)	5 yr (%)	7 yr (%)	10 yr (%)	30 yr (%)
U.S.A.	11/03/22	1.155	1.734	1.906	1.941	1.996	1.990	2.355
Treasury	14/02/22	1.086	1.581	1.800	1.911	1.989	1.992	2.296
bonds	12/03/21	0.074	0.150	0.338	0.839	1.283	1.624	2.385
1-month	11/03/22	1.168	1.630	1.736	1.729	1.738	1.786	1.790
SOFR	14/02/22	1.108	1.557	1.692	1.740	1.764	1.803	1.801
swaps	15/03/21	0.060	0.107	0.282	0.693	1.022	1.359	1.768
Swaps—	11/03/22	1.356	1.796	1.875	1.868	1.875	1.911	1.914
monthly	14/02/22	1.229	1.682	1.814	1.865	1.889	1.925	1.924
money	15/03/21	0.141	0.196	0.369	0.803	1.154	1.474	1.886
Swaps—	11/03/22	1.506	1.949	2.035	2.030	2.040	2.078	2.084
semi-	14/02/22	1.293	1.785	1.941	2.011	2.045	2.086	2.096
bond	15/03/21	0.205	0.256	0.440	0.911	1.281	1.618	2.057

4. For the experiments below, use `CRRDaeC()` with the parameters $T = 1$, $S_0 = 100$, $K = 101$, $r = 0.02$, $v = 0.20$, and $N = 12$.

   (a) Find a nonzero dividend sequence for which early Call exercise is sometimes optimal.

   (b) Find another nonzero dividend sequence for which early Call exercise is never optimal.

5. Implement CRR pricing for American and European Put options on an asset with dividends, using decomposition into risky ex-dividend and riskless cash flow portions. (Hint: Make a few changes to `CRRDaeC()`.)

   For the experiments below, set $T = 1$, $S_0 = 100$, $K = 101$, $r = 0.02$, $v = 0.20$, and $N = 12$.

   (a) Find a nonzero dividend sequence for which early Put exercise is sometimes optimal.

   (b) Find a nonzero dividend sequence for which early Put exercise is never optimal.

6. Suppose that government bonds are available with maturities of 1, 2, 3, and 4 years, with annual coupons for 0.5, 0.6, 0.8, and 1.1% of face value, respectively. Their spot prices are respectively 0.9994, 0.9992, 0.9989, and 0.9985 times face value.

   (a) Compute the zero-coupon bond discounts $Z(0,1)$, $Z(0,2)$, $Z(0,3)$, and $Z(0,4)$ implied by these inputs.

   (b) Plot the yield curve implied by $\{Z(0,T) : T = 1, 2, 3, 4\}$.

7. Suppose that a company offers coupon bonds with maturities of 1, 2, 3, and 4 years, with semiannual coupons at 0.3, 0.4, 0.5, and 0.6% of face value, respectively. Their spot prices are respectively 0.9994, 0.9992, 0.9989, and 0.9985 times face value.

   (a) Compute the zero-coupon bond discounts $\hat{Z}(0,1)$, $\hat{Z}(0,2)$, $\hat{Z}(0,3)$, and $\hat{Z}(0,4)$ implied by these inputs.

   (b) Plot the yield curve implied by $\{\hat{Z}(0,T) : 0 < T \leq 4\}$.

8. Compute the price at issuance of a US Treasury Note with the following parameters:

   - Maturity in 7 years.
   - Semiannual coupon at 1.500% annual interest.
   - Face value $1000.
   - Yield to maturity 1.414%.

   Use both e^x and its approximations $1 + x$ and $1 - x$ in the present value calculation and compare the results.

9. Compute the expected monthly riskless rates and returns over 6 months for two currencies using tabulated benchmarks:

   (a) Use the data in Table 6.5 for 11 March 2022 to compute the Australian dollar values.

   (b) Use the LIBOR data in Table 6.7 for 11 March 2022 to compute the US dollar values. Justify your interpolation method.

6.4 Further Reading

- Serena Alim and Ellis Connolly. "Interest Rate Benchmarks for the Australian Dollar." *Bulletin*, Reserve Bank of Australia, September 2018. https://www.rba.gov.au/publications/bulletin/2018/sep/interest-rate-benchmarks-for-the-australian-dollar.html

- Vipul K. Bansal, M.E. Ellis, and John Marshall. "The Pricing of Short-Dated and Forward Interest Rate Swaps." *Financial Analyst's Journal* (1993), pp.82–87.

- Kathryn Barraclough and Robert E. Whaley. "Early Exercise of Put Options on Stocks." *Journal of Finance* 67:4 (August 2012), pp.1423–1456.

- H.D. Grove and J. D. Bazley. "Interest Rate Swaps." *Journal of Accounting Education* 15:4 (1997), pp.591–614.

7

Implied Volatility

Both the Black-Scholes pricing formula and its CRR approximation assume constant volatility σ for the duration of the options contract. Since the Call and Put premiums vary with volatility, the actual market prices for C and P at various expiries and strike prices can be used to estimate σ. Existence of this *implied volatility* for the model can be shown using Calculus, and there is an efficient numerical method for actually calculating it.

7.1   The Inverse Problem for Volatility

For any model of the future that gives a price function with volatility σ as an input parameter, there is an inverse mapping that fits volatility to actual market prices. This inverse mapping will be a function, giving a single σ for each actual price, if the model's dependence on σ is one-to-one. That is the case for Black-Scholes Calls and Puts because of the *Inverse Function Theorem*:

Theorem 7.1 (Inverse Function) *If $f(\sigma)$ is a continuous function that is differentiable at a point σ_0 where $f(\sigma_0) = y_0$, and $f'(\sigma_0) > 0$, then*
    (a) f is one-to-one, in fact increasing, near σ_0,
    (b) $f^{-1}(y)$ exists and is continuous near y_0, and
    (c) $f^{-1}(y)$ is differentiable at y_0 with $(f^{-1})'(y_0) = 1/f'(\sigma_0)$.   □

A proof of this theorem[1] may be found in most Calculus textbooks. Apply the theorem with $f(\sigma_0) = y_0$ being the spot Call price from the Black-Scholes model with volatility σ_0. The derivative exists at σ_0: it is the Vega of the option, one of the Greeks, which by Eq.2.35 is positive, the product of three positive factors:

$$f'(\sigma_0) = \kappa_C = S_0\phi(d_1)\sqrt{T} > 0, \qquad \text{where } d_1 = \frac{\log \frac{S_0}{K} + \left(r + \frac{\sigma_0^2}{2}\right)T}{\sigma_0\sqrt{T}}.$$

The same holds for the Put price as a function of σ, since $\kappa_P = \kappa_C$ by Eq.2.36. Hence there is a unique implied volatility σ for the Black-Scholes model, for any Call (or Put) spot price in the range of Eq.2.25 (or Eq.2.26).

---

[1] This version also uses the Calculus fact that $f' > 0$ implies f is an increasing function.

To determine this range, consider d_1, d_2 in Eq.2.24 (with $v \leftarrow \sigma$ as in the current notation). As $\sigma \to \infty$, $d_1 \to \infty$ and $d_2 \to \infty$, so since Φ is a c.d.f., so $\Phi(d) \to 0$ as $d \to -\infty$ and $\Phi(d) \to 1$ as $d \to +\infty$, it follows that $C(0) < S_0$ and $P(0) < K$. But in fact these inequalities must hold in any arbitrage-free market: no-arbitrage premiums will always be in the range of Black-Scholes prices for some volatility. The Cox-Ross-Rubinstein model has the same range, by a similar analysis. The proof is left as an exercise.

Octave/MATLAB bisection search

By the *Intermediate Value Theorem*, or IVT: if f is continuous on an interval $[a, b]$, and $f(a) \leq y \leq f(b)$, then there exists some $x \in [a, b]$ such that $f(x) = y$.

The *bisection search method* exploits this theorem iteratively: Evaluate $f(m)$ at the midpoint $m = (a + b)/2$. Either $f(a) \leq y \leq f(m)$, in which case $x \in [a, m]$, or else $f(m) \leq y \leq f(b)$, and thus $x \in [m, b]$. In either case, the interval containing x is halved. Repeat k times to find x within a tolerance interval of length $2^{-k}|b - a|$, which can be as small as desired.

```
1   function [a,b]=bisection(f,y,a,b,tol)
2   % Octave/MATLAB function to perform a bisection
3   % search to find an interval [a,b] of width less
4   % than tolerance tol containing a point p such
5   % that f(p)=y.  Assume f(a)<y and f(b)>y.
6   % INPUTS:                            (Example)
7   %    f =    function handle          (@sin)
8   %    y =    target value for f(p)    (0.5)
9   %    a =    initial left endpoint    (0.0)
10  %    b =    initial right endpoint   (1.0)
11  %    tol  stop when abs(b-a)<tol     (1e-5)
12  % OUTPUTS:
13  %    [a,b] =   final interval endpoints
14  % EXAMPLE:
15  %    [a,b] = bisection(@sin,0.5,0.0,1.0,1e-5)
16  %
17      do
18         m=(a+b)/2;          % midpoint
19         if(f(m)<y) a=m;     % then use [m,b]
20         else b=m;           % else use [a,m]
21         end
22      until(abs(b-a)<tol) % close enough!
23      return
24  end
```

To find the implied volatility $x = \sigma$ for a Call (or Put) premium y, use bisection with f being a function that returns the price of a European-style Call (or Put) option computed from some model. Fixed parameters are spot price S_0, strike price K, time T to expiry, and riskless rate r. Volatility σ is the only variable x to solve for. The target value y is the market price of the option. The initial interval $0 < a < b < \infty$ must satisfy $f(a) \leq y \leq f(b)$, but this is assured by choosing a close to 0 and b sufficiently large. A solution to $f(x) = y$ is therefore the implied volatility parameter for the model that uniquely fits the market conditions.

Volatility from the Black-Scholes formula

Implied volatility may be found using BS() defined in Chapter 2 on p.34:

```
1  % EXAMPLE: Implied volatility using Black-Scholes:
2  T=1; S0=100; K=S0; r=0.03;
3  f = @(sig) BS(T,S0,K,r,sig)   % Call option price
4  a=0.05; b=0.95;  % initial range of volatilities
5  y = 10.00;         % make sure f(a)<y and f(b)>y !
6  bisection(f,y,a,b,0.001)
```

Volatility from Cox-Ross-Rubinstein models

Implied volatility may also be computed using a discrete model for pricing European-style Call premiums, such as CRR() defined in Chapter 3 on p.88:

```
1  % EXAMPLE:  Implied volatility using CRR:
2  T=1; S0=100; K=S0; r=0.03; N=100;
3  g = @(sig) CRR(T,S0,K,r,sig,N)(1,1) % returns C(0)
4  a=0.05; b=0.95;  % initial range of volatilities
5  y = 10.00;         % make sure g(a)<y and g(b)>y !
6  bisection(g,y,a,b,0.001)
```

7.2   Implied Volatility Surfaces

An *options chain* is a table of market prices for Call and Put options on a single underlying asset S, for example a share of stock in Bank of America. These tables are published by exchanges and other financial institutions, for example at this URL:

https://www.nasdaq.com/market-activity/stocks/bac/option-chain

Assuming that the riskless return rate is known and constant over the period $[0, T]$, and using market values for the spot price S_0 and option premiums $C(0)$ and $P(0)$ in the options chain, it is possible to estimate the implied volatility $\sigma(T_i, K_j)$ for various expiries $0 < T_i < T$ and strike prices $K_j \approx S_0$. The plot of $\sigma(T_i, K_j)$ over the grid of values (T_i, K_j) may be interpreted as points on a *volatility surface*.

Values of $\sigma(T_i, K_j)$ are found by solving the inverse problem

$$C(0, T_i, K_j) = f(T_i, S_0, K_j, r, \sigma(T_i, K_j))$$

with target market values for $C(0, T_i, K_j)$ at the available expiries and strike prices in the options chain. Here f is a European-style Call option premium pricing function such as CRR or Black-Scholes. Each value costs one call to `bisection()` and a bounded number of f evaluations.

Some options chain publishers compute these implied volatilities and include them in the table, for example:

https://finance.yahoo.com/quote/BAC/options

Table 7.1 shows American-style Call option premiums for BAC on December 14, 2021 ($t = 0$), when the closing price was \$44.13. The most recent ex-dividend date was December 2, 2021, and the one before that was September 2, 2021. Both were \$0.21. BAC has a history of quarterly dividends with ex-dividend dates near the beginning of March, June, September, and December, so no dividend is expected between $t = 0$ and the latest expiry, February 18, 2022. By Lemma 6.2, early exercise will not be optimal without a dividend before expiry, so the American-style and European-style options will have the same premium. Hence the implied volatilities may be computed using the Black-Scholes formula for pricing Call options.

For the riskless rate, use recently published US Treasury Bill APRs for maturities similar to the option expiries. On December 14, 2021, the value from https://www.treasury.gov/, found by searching online for "T-bill rates," was 0.05% APR for 56 days. For time to expiry, use fractions of a year computed by (days to expiry)/365.

TABLE 7.1
Implied volatilities from December 14, 2021 closing prices for BAC American-style Call options at $r = 0.05\%$ and $S_0 = 44.13$.

	Dec17 (3)		Dec23 (9)		Jan7 (21)		Jan21 (35)		Feb18 (63)	
K	C	σ	C	σ	C	σ	C	σ	C	σ
40	4.14	.50	4.35	.54	4.85	.55	4.60	.35	5.15	.37
41	3.20	.56	3.42	.49	3.50	.35	4.00	.39	4.35	.35
42	2.24	.47	2.72	.52	2.89	.39	3.15	.36	3.65	.34
43	1.30	.36	1.51	.30	1.96	.32	2.44	.34	2.87	.31
44	0.63	.35	0.86	.29	1.34	.30	1.71	.30	2.39	.32
45	0.23	.35	0.45	.29	0.84	.29	1.24	.30	1.76	.29
46	0.07	.36	0.22	.30	0.52	.29	0.88	.30	1.39	.29

Once the Call market prices C have been read from Table 7.1, a bisection search for the implied volatility produces the values σ_{imp} in the table.

Octave commands to graph an implied volatility surface.

The following Octave code uses a double loop to put these volatilities into a matrix `Vimp`. They may then be plotted with `mesh(x,y,z)`. This produces a perspective plot, displayed in Figure 7.1, of a polygonal surface approximating the function whose value at `z(row,col)` is determined by `x(col)` and `y(row)`.

```
Ks=40:46; Ts=[3,9,21,35,63]; % strikes and days to expiry
C=[4.14 4.35 4.85 4.60 5.15;  3.20 3.42 3.50 4.00 4.35; ...
   2.24 2.72 2.89 3.15 3.65;  1.30 1.51 1.96 2.44 2.87; ...
   0.63 0.86 1.34 1.71 2.39;  0.23 0.45 0.84 1.24 1.76; ...
   0.07 0.22 0.52 0.88 1.39]; % Call premiums on 2021-12-14
S0=44.13; r=0.05; % spot price and riskless APR on 2021-12-14
Vimp=zeros(length(Ks),length(Ts)); % implied volatilities
minv=0.01; maxv=0.99; tol=0.00001; % bisection parameters
for col=1:length(Ts)
  T=Ts(col)/365; % time to expiry in years
  for row=1:length(Ks)
    f=@(v) BS(T,S0,Ks(row),r/100,v); % Black-Scholes Call
    Vimp(row,col)=bisection(f,C(row,col),minv,maxv,tol);
  end
end
mesh(Ts,Ks,Vimp); % note the transposed order (T,K,V(K,T))
title("Implied Volatility Surface");
xlabel("T (days)"); ylabel("K");
```

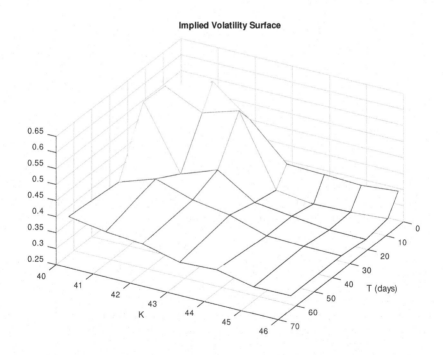

FIGURE 7.1
Implied volatility from BAC Call options as a function of strike price K and days to expiry T.

7.3   Implied Binomial Trees

Consider the problem of fitting a recombining binomial tree model of risky asset prices $\{S(n,j) : 0 \leq n \leq N, \ 0 \leq j \leq n\}$ to this market data:

- known spot price $S(0)$,

- constant riskless rate r,

- options premiums $\{C_i(0)\}$ for European-style Calls on S with fixed expiry $T > 0$ at various strike prices $\{K_i : i = 1, \ldots, m\}$.

It may be assumed that depth N of the tree is sufficiently large to provide enough free parameters to fit the target spot price and Call option premiums.

One idea is to fix N large enough, model $S(N,j) = S(0)u^{2j-N}$ using CRR volatility parameter v in $u = \exp(v\sqrt{T/N})$, and then find a probability mass function $Q(N,j)$ on the terminal states $0 \leq j \leq N$ at depth N such that

$$
1 = \sum_{j=0}^{N} Q(N,j), \quad \text{with } Q(N,j) > 0, \text{ all } j;
$$

$$
C_i(0) = e^{-rT} \sum_{j=0}^{N} Q(N,j) \left[S(N,j) - K_i\right]^{+}, \qquad i = 1, \ldots, m.
$$

$$
S(0) = e^{-rT} \sum_{j=0}^{N} Q(N,j)S(N,j), \tag{7.1}
$$

These are evidently Arrow-Debreu expansions. The discounted probability $e^{-rT}Q(N,j)$ plays the role of $\lambda(N,j)$, the spot price of the unit security that, at time N, pays 1 in state j and 0 in all other states. The system of equations may be written in matrix form using the payoff matrix

$$
A(i,j) = \begin{cases} e^{rT}, & i = 0; \ j = 0, 1, \ldots, N, \\ \left[S(N,j) - K_i\right]^{+}, & i = 1, \ldots, m; \ j = 0, 1, \ldots, N, \\ S(N,j), & i = m+1; \ j = 0, 1, \ldots, N. \end{cases} \tag{7.2}
$$

Its rows are the values at expiry time T of a riskless bond, the various Call options, and the underlying asset S. The vectors it relates are

$$
\mathbf{q} = \begin{pmatrix} 1 \\ C_1(0) \\ \vdots \\ C_m(0) \\ S(0) \end{pmatrix}, \qquad \mathbf{k} = \begin{pmatrix} Q(N,0) \\ Q(N,1) \\ \vdots \\ Q(N,N) \end{pmatrix}, \tag{7.3}
$$

which are the spot prices of the assets and the probabilities of the future states, respectively. The system in Eqs.7.1 then becomes

$$e^{rT}\mathbf{q} = A\mathbf{k}, \tag{7.4}$$

meaning that the spot prices are the present values of the expected payoffs.

The existence of a solution to Eq.7.4 is guaranteed by Theorem 8.4 in Chapter 8, under the hypothesis that the combination of spot prices $\mathbf{q}$ and payoffs A offers no arbitrage opportunities. To avoid arbitrage, N and v must satisfy

$$d < R < u, \qquad \Longleftrightarrow \qquad e^{-v\sqrt{T/N}} < e^{rT/N} < e^{v\sqrt{T/N}}.$$

The left inequality is immediate since $r > 0$ and $v > 0$. The right inequality holds if and only if

$$r\frac{T}{N} < v\sqrt{\frac{T}{N}}, \qquad \Longleftrightarrow \quad r\sqrt{T} < v\sqrt{N}.$$

Additionally, the modeled range of prices $\{S(N,j); j = 0, 1, \ldots, N\}$ must include all the strike prices, so

$$(\forall i)\ S(0)u^{-N} < K_i < S(0)u^N, \qquad \Longleftrightarrow \quad (\forall i)\ e^{-v\sqrt{TN}} < \frac{K_i}{S(0)} < e^{v\sqrt{TN}}.$$

These inequalities will hold if and only if

$$\frac{1}{\sqrt{T}} \log \max \left\{ \frac{S(0)}{\min_i K_i}, \frac{\max_i K_i}{S(0)} \right\} < v\sqrt{N}.$$

In both cases, the inequalities hold for all sufficiently large N.

Remark. These conditions are necessary but not sufficient to preclude arbitrage since some of the option premiums in $\mathbf{q}$ may be mispriced.

7.3.1 Path Independent Probabilities

Suppose that a positive solution $\{Q(N,j) : j = 0, \ldots, N\}$ is found for Eq.7.4. It may be used to define a probability $Q(n,j)$ for each state (n,j) in a binomial tree of S prices. Then Arrow-Debreu prices and risk neutral probabilities may be computed from $\{Q(n,j) : n = 0, \ldots, N;\ j = 0, \ldots, n\}$.

The first construction assumes that $Q(n,j)$ is the sum of the equal probabilities of all paths from $(0,0)$ to (n,j). The number of such paths is the number of ways to move up j times in n steps, which is

$$\binom{n}{j} = \frac{n!}{j!(n-j)!}.$$

Thus each individual path $(0,0) \to (n,j)$ has probability $Q(n,j)/\binom{n}{j}$.

Now let $p(n, j)$ be the risk neutral up probability at state (n, j). It satisfies

$$Q(n+1, j+1)/\binom{n+1}{j+1} \;=\; p(n,j)Q(n,j)/\binom{n}{j}$$

$$Q(n+1, j)/\binom{n+1}{j} \;=\; (1 - p(n,j))Q(n,j)\binom{n}{j}$$

since one of the equiprobable paths $(0,0) \to (n+1, j+1)$ first gets to (n, j) and then goes up to $(n+1, j+1)$ with probability $p(n, j)$. Likewise, one path $(0,0) \to (n+1, j)$ first gets to (n, j) and then goes down with probability $1 - p(n, j)$. Solving for p and $1 - p$ gives

$$p(n,j) \;=\; \left(\frac{j+1}{n+1}\right)\frac{Q(n+1, j+1)}{Q(n,j)} \qquad (7.5)$$

$$1 - p(n,j) \;=\; \left(1 - \frac{j}{n+1}\right)\frac{Q(n+1, j)}{Q(n,j)}$$

Multiplying by $Q(n, j)$ and then adding gives the backward induction formula for this construction:

$$Q(n,j) = \left(\frac{j+1}{n+1}\right)Q(n+1, j+1) + \left(1 - \frac{j}{n+1}\right)Q(n+1, j). \qquad (7.6)$$

Finally, the Arrow-Debreu spot prices may be computed by discounting the probabilities:

$$\lambda(n,j) = \frac{Q(n,j)}{R^n}, \qquad n = 0, 1, \ldots, N; \; j = 0, 1, \ldots, n. \qquad (7.7)$$

7.3.2   Jackwerth's Generalization

Risk neutral probabilities to be used in the implied binomial tree are not uniquely defined by the probability function on states at expiry. A weight function w may be introduced to change the backward induction:

$$Q(n,j) = w\left(\frac{j+1}{n+1}\right)Q(n+1, j+1) + \left(1 - w\left(\frac{j}{n+1}\right)\right)Q(n+1, j), \qquad (7.8)$$

for $n = N-1, \ldots, 1, 0$ and $j = 0, \ldots, n$, with the solution to Eqs.7.1 giving the terminal values $\{Q(N, j) : j = 0, \ldots, N\}$. Equation 7.6 is of this form with $w(\theta) = \theta$, but w may be any fixed function defined on $[0, 1]$ that satisfies

$$0 < \theta < 1 \implies 0 < w(\theta) < 1; \qquad w(0) = 0; \qquad w(1) = 1. \qquad (7.9)$$

For any such weight, $Q(n, j)$ remains a positive probability function on the states $j = 0, 1, \ldots, n$ at each time step n:

Lemma 7.2 *Suppose that $Q(N, j) > 0$ for all j and $\sum_{j=0}^{N} Q(N, j) = 1$. Then $Q(n, j)$ defined by Jackwerth's formula (Eq. 7.8) satisfies*

$$(\forall j)\, Q(n, j) > 0 \qquad and \qquad \sum_{j=0}^{n} Q(n, j) = 1,$$

for every $0 \le n \le N$.

Proof: Proceed by induction on n, starting with $n = N$ where the conclusion holds by hypothesis.

Now suppose that the conclusion holds for $n + 1$. Fix j with $0 \le j \le n$ and note that $Q(n, j) > 0$ since all terms on the right-hand side of Eq. 7.8 are positive. Then, to show that $\sum_{j=0}^{n} Q(n, j) = 1$, use Eq. 7.8 to separate it into two sums, extract the $j = 0$ and $j = n$ terms where $w(0/(n+1)) = 0$ and $w((n+1)/(n+1)) = 1$, and recombine the two sums after substituting j for $j + 1$ in the first to get the same index range:

$$
\begin{aligned}
\sum_{j=0}^{n} Q(n, j) \;=\;& \sum_{j=0}^{n} w\left(\frac{j+1}{n+1}\right) Q(n+1, j+1) \\
&+ \sum_{j=0}^{n} \left(1 - w\left(\frac{j}{n+1}\right)\right) Q(n+1, j) \\
=\;& Q(n+1, n+1) + \sum_{j=0}^{n-1} w\left(\frac{j+1}{n+1}\right) Q(n+1, j+1) \\
&+ Q(n+1, 0) + \sum_{j=1}^{n} \left(1 - w\left(\frac{j}{n+1}\right)\right) Q(n+1, j) \\
=\;& Q(n+1, n+1) + \sum_{j=1}^{n} w\left(\frac{j}{n+1}\right) Q(n+1, j) \\
&+ Q(n+1, 0) + \sum_{j=1}^{n} \left(1 - w\left(\frac{j}{n+1}\right)\right) Q(n+1, j) \\
=\;& \sum_{j=0}^{n+1} Q(n+1, j) \;=\; 1,
\end{aligned}
$$

by the inductive hypothesis. Conclude that the result holds for n as well, and thus for all $0 \le n \le N$. □

Note that Lemma 7.2 implies that $Q(0, 0) = 1$.

$Q(n, j)$ may be interpreted as the probability of state (n, j) in a recombining binomial model of the future. Apply a constant riskless return R per time step as a discount to get the spot prices for Arrow-Debreu unit securities in the model:

$$\lambda(n, j) \overset{\text{def}}{=} \frac{Q(n, j)}{R^n}. \tag{7.10}$$

From Q and λ so defined, the risk neutral up probability in state (n, j) should be

$$p(n, j) \overset{\text{def}}{=} w \left(\frac{j+1}{n+1} \right) \frac{Q(n+1, j+1)}{Q(n, j)} > 0, \tag{7.11}$$

which then implies

$$1 - p(n, j) = \frac{Q(n, j) - w \left(\frac{j+1}{n+1} \right) Q(n+1, j+1)}{Q(n, j)}$$

$$= \left(1 - w \left(\frac{j}{n+1} \right) \right) \frac{Q(n+1, j)}{Q(n, j)} > 0. \tag{7.12}$$

It is left as an exercise to show that the Eq.7.11 and Eq.7.12 probabilities produce the Arrow-Debreu spot prices $\lambda(n, j)$ in Eq.7.10 using Jamshidian's forward induction, Eq.3.21 on p.69. They may therefore be used to price any asset that satisfies a backward induction with p, $1 - p$, and R:

Theorem 7.3 *Suppose that V is an asset that satisfies*

$$V(n, j) = \frac{p(n, j)V(n+1, j+1) + (1 - p(n, j))V(n+1, j)}{R},$$

for $n = 0, 1, \ldots, N - 1$ and $j = 0, 1, \ldots, n$. Then for any n,

$$V(0, 0) = \sum_{j=0}^{n} V(n, j)\lambda(n, j)$$

namely the spot price of V has an Arrow-Debreu expansion as a weighted sum of its future values.

Proof: Proceed by induction on n. The result is evidently true at $n = 0$, where the sum contains a single term:

$$V(0, 0)\lambda(0, 0) = V(0, 0)\frac{Q(0, 0)}{R^0} = V(0, 0).$$

Now suppose that $0 \le n < N$. Put

$$Z(n, j) \overset{\text{def}}{=} \frac{V(n, j)}{R^n}, \qquad \implies V(n, j)\lambda(n, j) = Q(n, j)Z(n, j),$$

noting that therefore

$$Z(n, j) = p(n, j)Z(n+1, j+1) + (1 - p(n, j))Z(n+1, j).$$

Apply this formula, substitute for $p(n,j)$ and $1 - p(n,j)$, cancel $Q(n,j)$, and separate the sums as in the proof of Lemma 7.2 to get

$$
\begin{aligned}
\sum_{j=0}^{n} V(n,j)\lambda(n,j) &= \sum_{j=0}^{n} Q(n,j)Z(n,j) \\
&= \sum_{j=0}^{n} Q(n,j)\Big[p(n,j)Z(n+1,j+1) + (1-p(n,j))Z(n+1,j)\Big] \\
&= \sum_{j=0}^{n} w\left(\frac{j+1}{n+1}\right) Q(n+1,j+1)Z(n+1,j+1) \\
&\quad + \sum_{j=0}^{n} \left(1 - w\left(\frac{j}{n+1}\right)\right) Q(n+1,j)Z(n+1,j) \\
&= \sum_{j=0}^{n+1} Q(n+1,j)Z(n+1,j) = \sum_{j=0}^{n+1} V(n+1,j)\lambda(n+1,j),
\end{aligned}
$$

after combining the telescoping sums and canceling terms. Conclude that the expansion is independent of n, so it equals $V(0,0)$ for all $0 \le n \le N$. $\qquad \square$

7.3.3 Rubinstein's One-Two-Three Algorithm

There is an explicit fit to the spot price and options chain due to Mark Rubinstein. The inputs are as follows:

- The spot price S_0 of the stock.

- A sequence of equispaced, ascending strike prices:

$$ K_1 < K_2 < \cdots K_m; \qquad \Delta \overset{\text{def}}{=} K_{i+1} - K_i > 0, \text{ all } i = 1, \ldots, m-1, $$

  These should be *near the money*, namely close to the spot price.

- Market prices for the corresponding European-style Call options with common expiry T:
$$ C_1, \ldots, C_m, $$

  where C_i has strike price K, $i = 1, \ldots, m$. By convention, the ask price is used. If there is no dividend in the interval $[0, T]$, then American-style options may be used as they will have the same premiums.

- The riskless return factor for the period $[0, T]$:

$$ R = \exp(rT) = 1/Z(0,T), $$

  where r is the riskless annual interest rate determined from zero-coupon government bonds.

The algorithm builds a minimal tree of depth $N = m + 1$ that fits some of the market data. The model for $S(T)$ is the bottom row $\{S(N, j) : j = 0, \ldots, N\}$ of a recombining binomial tree, mostly consisting of the strike prices:

$$
\begin{aligned}
S(N, j) &= K_j, \qquad j = 1, 2, \ldots, N - 1 (= m); \\
S(N, 0) &= \frac{S_0 - C_1 + K_1(C_2 - C_1)/\Delta}{1/R + (C_2 - C_1)/\Delta}, \\
S(N, N) &= \frac{\Delta C_m}{C_{m-1} - C_m} + K_m
\end{aligned}
\tag{7.13}
$$

Rather than solve a linear system of equations, the time-T probabilities are mostly chosen to be butterfly spread premiums:

$$
\begin{aligned}
Q(N, j) &= \frac{R}{\Delta}[C_{j-1} - 2C_j + C_{j+1}], \qquad j = 2, \ldots, N - 2 (= m - 1); \\
Q(N, 1) &= 0; \qquad Q(N, N - 1) = 0; \\
Q(N, N) &= \frac{R}{\Delta}[C_{m-1} - C_m]; \qquad Q(N, 0) = 1 - \sum_{j=1}^{N} Q(N, j).
\end{aligned}
\tag{7.14}
$$

If there is no arbitrage, then these will all be nonnegative, with $Q(N, 0)$ chosen so that they sum to one.

The S and Q trees are then pruned to remove the bottom row elements corresponding to negative and zero values $Q(N, j)$. At least two will be removed, since $Q(N, 1) = Q(N, N - 1) = 0$ by construction. Negative values should not occur, but when they do they indicate mispriced options, or arbitrage. If N_1 is the largest index in the pruned and re-indexed bottom row, then $N_1 \leq N - 2$.

Octave functions to find an implied binomial tree.

Rubinstein's 1-2-3 algorithm is readily implemented with Jackwerth's weight function. The original version uses the weight $w(x) = x$, defined by an anonymous function as in the example in the comments.

```
1  function [S,Q,up,down,pu,N1]=IBT123J(S0,Ks,Cs,rho,w)
2  % Octave/MATLAB function to compute an implied binomial
3  % tree from spot prices and market European-style Call
4  % option ask prices, by Rubinstein's "1,2,3" method
5  % with Jackwerth's generalization.
6  % INPUTS:                                    (Example)
7  %    S0= spot stock price       (BAC 2021/12/11:    44.52)
8  %    Ks= equispaced, increasing strike prices    (40:46)
9  %    Cs= Call asks     ([5.6,4.8,4.1,3.45,2.87,2.29,1.81])
10 %    rho= riskless return (to 2022/3/20:(exp(r*109/365))
11 %    w= weight function                      (w=@(x)x)
12 % OUTPUT:
13 %    S = binomial tree of stock prices at all (n,j)
```

```
14 %    Q = tree of probabilities of all states (n,j)
15 %    up = upfactors at (n,j)
16 %    down = downfactors at (n,j)
17 %    pu = risk neutral up probabilities at (n,j)
18 %    N1 = depth of the pruned tree
19 % NOTE: since Octave array indices start at 1, output
20 %    array value for (n,j) is stored at index (n+1,j+1).
21 % EXAMPLE:
22 %    S0=44.22; Ks=40:46; % BAC spot price and strikes
23 %    Cs=[5.6,4.8,4.1,3.45,2.87,2.29,1.81]; % Call asks
24 %    r=0.06/100; rho=exp(r*109/365); % 13-wk T-bill APR
25 %    [S,Q,up,down,pu,N1] = IBT123J(S0,Ks,Cs,rho,w)
26 %
27 m = length(Ks);          % expect valid Ks(1),...,Ks(m)
28 % Check that Ks and Cs have the same length
29 if(length(Cs) != m ) % expect valid Cs(1),...,Cs(m)
30   error ("length(Cs)!=length(Ks)");
31 end
32 N=m+1;                   % initial tree depth, before pruning
33 Del=Ks(2)-Ks(1); % expect Del=Ks(j)-Ks(j-1), all j
34 if(!(Del>0))             % expect Ks(1)<Ks(2)<...
35   error("must have Ks(1)<Ks(2)")
36 end
37 for j=3:m                % expect equispaced Ks
38   if( Ks(j)!= Ks(j-1)+Del )
39     error("must have equispaced Ks")
40   end
41 end
42 %% Generate time-T risk neutral probabilities Q(N,j):
43 QN=ones(1,N+1); % so QN(j+1)=Q(N,j) for j=0,...,N=m+1
44 for j=2:m-1     % normalized butterfly spread premiums:
45   QN(j+1)=rho*(Cs(j-1)-2*Cs(j)+Cs(j+1))/Del;
46 end
47 QN(2)=0; QN(m+1)=0;     % Q(N,1) and Q(N,m) are special
48 QN(N+1)=rho*(Cs(m-1)-Cs(m))/Del; % Q(N,N) is special
49 QN(1)=1-sum(QN(2:(N+1)));   % so sum(Q(N,j))=1, as prob.
50 %% Generate the initial bottom row of stock prices
51 SN=zeros(1,N+1); % so SN(j+1)=S(N,j) for the initial tree
52 b=(Cs(2)-Cs(1))/Del;    % temporary "slope" variable
53 SN(1)=rho*(S0-Cs(1)+Ks(1)*b)/(1+rho*b); % S(N,0) is special
54 SN(2:m+1)=Ks(1:m);     % S(N,j)=Ks(j) for j=1:m
55 SN(N+1)=Ks(m)+Del*Cs(m)/(Cs(m-1)-Cs(m)); % S(N,N) special
56 %% Prune the bottom row of S to use only positive Qs:
57 Q1=QN(QN>0);           % subvector of the positive Q(N,j)
58 S1=SN(QN>0);           % ...and the corresponding S(N,j)
59 N1=length(Q1)-1; % depth of the pruned tree
60 R=rho^(1/N1);          % new one-step riskless return
61 %% Initialize and assign the output binomial trees
62 S = zeros(N1+1,N1+1);   % array of stock prices.
```

```
63  Q = zeros(N1+1,N1+1);     % array of path probabilities.
64  pu = zeros(N1+1,N1+1);    % array of one-step up probs.
65  up = zeros(N1+1,N1+1);    % up factors
66  down = zeros(N1+1,N1+1);  % down factors
67  S(N1+1,:)=S1; % Last row of pruned stock price tree
68  Q(N1+1,:)=Q1; % Last row of state probabilities tree
69  % Backward induction to fill the trees
70  for n=N1-1:-1:0
71    for j=0:n
72      Qup=w((j+1)/(n+1))*Q(n+2,j+2);
73      Qdown=(1-w(j/(n+1)))*Q(n+2,j+1);
74      Q(n+1,j+1)=Qup+Qdown;
75      pu(n+1,j+1) = Qup/Q(n+1,j+1); % Eq.7.11
76      S(n+1,j+1) = (pu(n+1,j+1)*S(n+2,j+2)
77                  + (1-pu(n+1,j+1))*S(n+2,j+1))/R;
78      up(n+1,j+1) = S(n+2,j+2)/S(n+1,j+1);   % upfactor
79      down(n+1,j+1) = S(n+2,j+1)/S(n+1,j+1); % downfactor
80    end
81  end
82  return;
83  end
```

It is left as an exercise to compute the implied binomial tree with different Jackwerth weights and to compare the results.

Experiments and simulations may be performed with implied binomial trees, as they yield a model of the future that incorporates market information. In particular, other contingent claims on S, including novel exotic ones, may be priced using the implied recombining binomial tree built from Call option premiums. This is left as a project for the ambitious reader.

Remark. Two alternatives to Rubinstein's construction are the Derman and Kani algorithm and the Barle and Cakici algorithm, described in the Further Reading section below.

7.4 Exercises

1. Suppose that C and P are European-style Call and Put options, respectively, at strike price K and expiry T, for a risky underlying asset S with spot price S_0. Show that $0 < C(0) \leq S_0$ and $0 < P(0) \leq K$. (Hint: construct an arbitrage otherwise.)

2. Show that the Eq.7.11 and Eq.7.12 probabilities produce the Arrow-Debreu spot prices $\lambda(n,j)$ in Eq.7.10 using Jamshidian's forward induction, Eq.3.21 on p.69.

3. Compute implied volatility for the data in Table 7.1 using both Black-Scholes and CRR with $N = 20$. Tabulate and compare the results.

4. The table below gives part of the options chain for American-style Calls on Bank of America common stock (BAC) as of closing on March 17, 2022, when the spot price was $43.03:

Strike price K	T= 1 d (3/18)	8 d (3/25)	15 d (4/01)	21 d (4/08)	27 d (4/14)
42.00	1.10	1.44	1.76	1.96	2.18
43.00	0.34	0.86	1.10	1.33	1.58
44.00	0.06	0.44	0.69	0.88	1.06
45.00	0.02	0.18	0.35	0.53	0.71
46.00	0.01	0.07	0.16	0.31	0.44
47.00	0.01	0.02	0.09	0.17	0.27

Also, the US T-bill rates for various maturities were

Date	4 wk	8 wk	13 wk	26 wk	52 wk
03/14/2022	0.22	0.30	0.45	0.84	1.20
03/15/2022	0.21	0.29	0.46	0.84	1.19
03/16/2022	0.23	0.28	0.43	0.84	1.26
03/17/2022	0.20	0.30	0.40	0.79	1.20

Use this data to compute and plot the volatility surface for BAC.

5. Suppose that a share of XYZ has a spot price of $47.12, that riskless interest rates for the next month are expected to be a constant 0.66% APR, and that the premiums for European-style Call options expiring in 4 weeks $(T = 4/52)$ are as follows:

Strike price:	45.00	46.00	47.00	48.00	49.00
Call premium:	3.52	2.78	2.10	1.44	1.37

(a) Construct an implied binomial tree for these inputs using Rubinstein's 1-2-3 algorithm. Display it along with the implied risk neutral up probabilities.

(b) Plot the three weight functions $w_1(x) = \sqrt{x}$, $w_2(x) = x^2$, and $w_3(x) = (1 - \cos(x/\pi))/2$, for $0 \leq x \leq 1$, on the same graph.

(c) Apply Rubinstein's 1-2-3 algorithm with Jackwerth's generalization to the data, using weights w_1, w_2, w_3 from part (b). Compare S, p, and Q for the three weights.

7.5 Further Reading

- Stanko Barle and Nusret Cakici, "How to Grow a Smiling Tree." *Journal of Financial Engineering* 7 (1998), pp.127–146.

- Emanuel Derman and Iraj Kani. "Riding on a Smile." *RISK* 7:2 (1994), pp.139–145.

- Wolfgang K. Härdle and Alena Myšičková. *Numerics of Implied Binomial Trees.* SFB 649 Discussion Paper 2008-044, ISSN 1860-5664, Humboldt-Universität zu Berlin (2008).

- Jens Carsten Jackwerth and Mark Rubinstein. "Recovering Probability Distributions from Option Prices." *Journal of Finance* 51 (1996), pp.1611–1631.

- Jens Carsten Jackwerth, "Generalized Binomial Trees." *Journal of Derivatives* 5:2 (1997), pp.7–17.

- Mark Rubinstein. "Implied Binomial Trees." *Journal of Finance* 49:3 (1994), pp.771–818.

8

Fundamental Theorems

In Chapter 7, market-negotiated premiums for derivatives were used to build a model of the future for the underlying risky asset. As defined in Chapter 1, such a model is a set of future states, representing possible future prices, with a probability function on the states.

The goal in this chapter is to prove that such a model exists, or more precisely that there exists a probability function for future states in any market without arbitrage. This can be done rigorously for discrete financial models using only Linear Algebra, Geometry, and Calculus in $\mathbf{R}^n$. The result is called the "Fundamental Theorem on Asset Pricing." There are two versions, depending on the type of no-arbitrage hypothesis. The two versions have subtly different proofs. They are presented here to illuminate some of the abstract ideas underlying financial models and calculations.

8.1   Finite Financial Models

Recall that in general, an *asset* $a : T \times \Omega \to \mathbf{R}$ is a stochastic process, a time-varying random variable on a set of times T, over probability space Ω. Then $a(t, \omega)$ is the price of the asset at time $t \in T$ in state $\omega \in \Omega$.

T contains time $t = 0$, the *present*, and at least one future time $t > 0$, since contingent claims have future payoffs. Ω likewise contains at least two states, to model uncertainty about the future.

A *finite financial model* simplifies the choices for T and Ω to the finite sets $T = \{0, 1\}$ and $\Omega = \{1, 2, \ldots, n\}$, with fixed $n \geq 2$. Then calculations are performed using just the prices at one of two times in one of n states:

- The *spot price* $a(0)$, of asset a, assumed to be the same in all states at time $t = 0$.

- The *payoff* $a(1, j)$, of asset a, at future time $t = 1$, in state $\omega = j$.

The *payoff vector* $\mathbf{a} = (a(1, 1), \ldots, a(1, j), \ldots, a(1, n))$ lists all the modeled future prices for the asset. A market will have a finite number $m \geq 1$ of possibly risky assets, indexed as $a_i : T \times \Omega \to \mathbf{R}$, $1 \leq i \leq m$, which will each have spot price $a_i(0)$, payoff $a_i(1, j)$, and payoff vector $\mathbf{a}_i$, respectively.

It will be assumed throughout that there is a *numeraire*, also called *cash*, that determines the units in which all other assets are priced. This implies the existence of at least one riskless asset, denoted here by a_0 with spot price $a_0(0) = 1$ and constant payoff vector $\mathbf{a}_0 = (R, \ldots, R)$ of riskless returns $R \geq 1$.

These notions correspond to some familiar abstractions from linear algebra:

Definition 4 (Market Matrices) *Using $T = \{0, 1\}$ and $\Omega = \{1, 2, \ldots, n\}$, a finite market with m assets and cash is modeled by a matrix A and a vector $\mathbf{q}$, namely:*

- *Payoff matrix*

$$
A \overset{\text{def}}{=} \begin{pmatrix} 1 \\ \mathbf{a}_1 \\ \vdots \\ \mathbf{a}_m \end{pmatrix} = \begin{pmatrix} 1 & \cdots & 1 \\ a_1(1,1) & \cdots & a_1(1,n) \\ \vdots & \ddots & \vdots \\ a_m(1,1) & \cdots & a_m(1,n) \end{pmatrix},
$$

  where $a_i(1, j)$ is the payoff of asset i in state j, and

- *Spot price vector*

$$
\mathbf{q} \overset{\text{def}}{=} \begin{pmatrix} 1 \\ a_1(0) \\ \vdots \\ a_m(0) \end{pmatrix}.
$$

The top row of A is the numeraire, or cash. It may be replaced by a row of riskless returns $R \geq 1$, the same in all states, without loss of generality.

A *portfolio* is a linear combination of assets from $\{a_i : i = 0, 1, \ldots, m\}$, each present in amount x_i. It may be written as the column vector

$$
\mathbf{x} = (x_i) = \begin{pmatrix} x_0 \\ x_1 \\ \vdots \\ x_m \end{pmatrix}.
$$

In the finite market modeled by $A, \mathbf{q}$, any portfolio $\sum_i x_i a_i(t, \omega)$ represented by the (column) vector of weights $\mathbf{x}$ has

- spot price $\mathbf{x}^T \mathbf{q}$, and

- payoff vector $\mathbf{x}^T A$.

Remark.   For consistency in linear algebra computations, payoffs will be row vectors while spot price vectors, portfolio weight vectors, and probability mass functions will be column vectors. Unfortunately, this is only one of several conventions in use.

The spot price of a portfolio is a number, so it is obvious what it means for it to be positive or nonnegative. For the payoff vector, those concepts should be defined componentwise:

Definition 5 (Vector Ordering) *For* $\mathbf{v} = (v_1, \ldots, v_n) \in \mathbf{R}^n$,

- *Say that* $\mathbf{v}$ *is* positive, *and write* $\mathbf{v} > \mathbf{0}$, *iff* $(\forall j)\, v_j > 0$.

- *Say that* $\mathbf{v}$ *is* nonnegative, *written* $\mathbf{v} \geq \mathbf{0}$, *iff* $(\forall j)\, v_j \geq 0$.

Here $\mathbf{0} = (0, \ldots, 0) \in \mathbf{R}^n$ is the zero vector. Additionally, for another vector $\mathbf{w} \in \mathbf{R}^n$, write $\mathbf{v} > \mathbf{w}$ to mean $\mathbf{v} - \mathbf{w} > \mathbf{0}$ and write $\mathbf{v} \geq \mathbf{w}$ to mean $\mathbf{v} - \mathbf{w} \geq \mathbf{0}$.

Such positivity or nonnegativity is a property of *orthants*, which are subsets of $\mathbf{R}^n$ that generalize the positive real line and the first quadrant of the plane:

Definition 6 (Orthants) *Using the vector ordering notation,*

- *The* closed orthant *of vectors with nonnegative coordinates is*

$$K \overset{\text{def}}{=} \{\mathbf{v} \in \mathbf{R}^n : \mathbf{v} \geq \mathbf{0}\}.$$

- *Remove the point* $\mathbf{0} = (0, \ldots, 0)$ *to get the* pointless orthant

$$K \setminus \mathbf{0} = \{\mathbf{v} \in \mathbf{R}^n : \mathbf{v} \geq \mathbf{0},\ (\exists j)\, v_j > 0\}.$$

- *The interior of* K *is the* open orthant:

$$K^o \overset{\text{def}}{=} \{\mathbf{v} \in \mathbf{R}^n : (\forall j)\, v_j > 0\} = \{\mathbf{v} \in \mathbf{R}^n : \mathbf{v} > \mathbf{0}\}.$$

Definition 7 (Profitable Portfolios) *Suppose that* A *is a market matrix.*

- *Say that* $\mathbf{p}$ *is a* profitable portfolio *iff it has a nonnegative payoff vector:* $\mathbf{p}^T A \geq \mathbf{0}$. *Equivalently,* $\mathbf{p}^T A \in K$.

- *Say that* $\mathbf{s}$ *is a* strictly profitable portfolio *iff it is profitable and also has positive payoff in some state:* $(\exists j)\, \mathbf{s}^T A(j) > 0$. *Equivalently,* $\mathbf{s}^T A \in K \setminus \mathbf{0}$.

Remark. The trivial portfolio $\mathbf{p} = \mathbf{0}$ is profitable for every A, but a market without a numeraire might have no strictly profitable portfolios. For example, let $A = \begin{pmatrix} -1 & 1 \end{pmatrix}$ be the market with two states for just one asset. A portfolio here will consist of a one-component vector $\mathbf{p} = (x)$, so

$$\mathbf{p}^T A = \begin{pmatrix} -x & x \end{pmatrix}.$$

If either component of $\mathbf{p}^T A$ is positive, the other must be negative, so it is not possible to have $\mathbf{p}^T A \in K \setminus \mathbf{0}$.

A market with a numeraire, or cash, as its zero row allows the riskless, all-cash portfolio $\mathbf{x} = (1, 0, \ldots, 0)$. Since this satisfies $\mathbf{x}^T A(j) = 1$ for all j, it is both profitable and strictly profitable. Likewise, a market with any riskless asset contains a riskless porfolio that is both profitable and strictly profitable.

8.1.1   Arbitrage and Positivity

As before, an *arbitrage* is a portfolio **x** that yields profit without risk. The definitions of the two deterministic types from Chapter 1 may be translated into the notation of finite markets $A, \mathbf{q}$ and orthants K and $K \setminus \mathbf{0}$.

First note that in a finite financial model, it may be assumed that every state has positive probability or else it would be omitted as superfluous.

Definition 8 *An immediate or type one arbitrage (IA, as in Definition 1 on p.11), is a portfolio* **x** *that satisfies the following:*

IA1: $\mathbf{x}^T \mathbf{q} < 0$, *so it leaves a surplus at assembly,*

IA2: $\mathbf{x}^T A \geq \mathbf{0}$, *so its price in all future states is nonnegative. This is equivalent to* $\mathbf{x}^T A \in K$.

The absence of such an arbitrage in a finite market $A, \mathbf{q}$ may be described as an implication that one of the necessary properties negates the other:

Lemma 8.1 (IA-Free) *Finite market A with spot prices $\mathbf{q}$ is immediate-arbitrage-free iff any profitable portfolio must have a nonnegative price:*

$$\mathbf{x}^T A \in K \implies \mathbf{x}^T \mathbf{q} \geq 0.$$

Proof: Read this as "If IA2, then not IA1." This implication holds iff IA1 and IA2 cannot both be true, iff there cannot exist an immediate arbitrage.   □

Definition 9 *A type two arbitrage opportunity (AO, as in Definition 2 on p.11) is a portfolio satisfying:*

AO1: $\mathbf{x}^T \mathbf{q} \leq 0$, *so it costs nothing to assemble,*

AO2: $\mathbf{x}^T A \geq \mathbf{0}$, *so it does not lose money in any state, and*

AO3: $(\exists j)\, \mathbf{x}^T A(j) > 0$, *so there is a state in which it makes a profit.*

Note that AO2 and AO3 together are equivalent to $\mathbf{x}^T A \in K \setminus \mathbf{0}$.

The absence of this type of arbitrage opportunity may likewise be written as an implication:

Lemma 8.2 (AO-Free) *Finite market A with spot prices $\mathbf{q}$ is arbitrage-opportunity-free iff any strictly profitable portfolio must have a positive price:*

$$\mathbf{x}^T A \in K \setminus \mathbf{0} \implies \mathbf{x}^T \mathbf{q} > 0.$$

Proof: Read this implication as "If AO2 and AO3, then not AO1," which prevents all three conditions from holding simultaneously for any portfolio **x**. Thus there cannot exist an arbitrage opportunity.   □

8.1.2 Fundamental Theorems of Asset Pricing

In any arbitrage-free market, the price vector $\mathbf{q}$ is a weighted average of the payoffs in the states of Ω. This is a consequence of the previous lemmas and some geometric facts to be proved in Section 8.3 below. For even the weakest hypothesis of no immediate arbitrages, the weights must be nonnegative:

Theorem 8.3 (FT from IA-Free) *Market A with spot prices $\mathbf{q}$ is immediate-arbitrage-free if and only if there is a nonnegative vector $\mathbf{k}$ such that*

$$\mathbf{q} = A\mathbf{k}.$$

Proof: ($\Longleftarrow$): Suppose that there exists $\mathbf{k} \geq \mathbf{0}$ that solves $\mathbf{q} = A\mathbf{k}$, and let $\mathbf{x}$ be a profitable portfolio. Then $\mathbf{x}^T A \geq \mathbf{0}$, so

$$\mathbf{x}^T \mathbf{q} = \mathbf{x}^T (A\mathbf{k}) = (\mathbf{x}^T A)\mathbf{k} \geq 0,$$

Thus, by definition, $A, \mathbf{q}$ is IA-free.

($\Longrightarrow$): Suppose that $A, \mathbf{q}$ is IA-free. Then $\mathbf{q} \in ((AK)')' = AK$ by Corollary 8.20 and Theorem 8.21. Conclude that there is some $\mathbf{k} \in K$, namely some $\mathbf{k} \geq \mathbf{0}$, such that $\mathbf{q} = A\mathbf{k}$. $\quad\square$

Under the stronger hypothesis that there are no arbitrage opportunities, the weight vector must be strictly positive:

Theorem 8.4 (FT from AO-Free) *Market A with spot prices $\mathbf{q}$ is AO-free if and only if there is a positive vector $\mathbf{k}$ such that*

$$\mathbf{q} = A\mathbf{k}.$$

Proof: ($\Longleftarrow$): Suppose that $\mathbf{k} > \mathbf{0}$ solves $\mathbf{q} = A\mathbf{k}$ and let $\mathbf{x}$ be a strictly profitable portfolio. Then

$$\mathbf{x}^T \mathbf{q} = \mathbf{x}^T (A\mathbf{k}) = (\mathbf{x}^T A)\mathbf{k} > 0,$$

since $\mathbf{x}^T A \in K \setminus \mathbf{0}$. Thus, by definition, $A, \mathbf{q}$ is AO-free.

($\Longrightarrow$): Suppose that $A, \mathbf{q}$ is AO-free. Then $\mathbf{q} \in ((AK^o)^*)^* = AK^o$ by Corollary 8.20 and Theorem 8.21 Conclude that there is some $\mathbf{k} > \mathbf{0}$, namely $\mathbf{k} \in K^o$, such that $\mathbf{q} = A\mathbf{k}$. $\quad\square$

8.2 Applications of the Fundamental Theorems

Suppose that payoff matrix A with spot price vector $\mathbf{q}$ corresponds to an arbitrage-free market. The Fundamental Theorem implies that there is a vector $\mathbf{k} > \mathbf{0}$ such that $\mathbf{q} = A\mathbf{k}$. This $\mathbf{k}$ is a discounted risk neutral probability function, expanding the spot prices of assets as the present value of a weighted average of their modeled future values:

$$\mathbf{q}(i) = A\mathbf{k}(i) = \mathrm{PV}(\mathrm{E}(\mathbf{a}_i)), \qquad i = 0, 1, \ldots, m,$$

where $\mathbf{a}_i$ is the payoff vector of one of the assets in the market. The expectation is taken with respect to the risk neutral probabilities, $\mathbf{k}$ without its discount.

The same $\mathbf{k}$ may be used to find the spot price of any derivative with a known payoff $\mathbf{d}$ at the same future time as A. This $\mathbf{d}$ is a row vector, like the asset payoffs $\{\mathbf{a}_i\}$ making up the rows of A. Its spot price is therefore

$$d_0 \overset{\text{def}}{=} \mathbf{d}\mathbf{k} = \mathrm{PV}(\mathrm{E}(\mathbf{d})),$$

where again the expectation uses the risk neutral probabilities implied by $A, \mathbf{q}$.

8.2.1   Hedges

Sellers of a contingent claim incur a liability, the payoff vector $\mathbf{d}$ of the derivative. To protect against loss, they seek to *hedge*, or replicate it, with a portfolio of other assets, often including the underlying as well as some riskless assets. To cover the liability exactly at minimal cost to the seller, a hedge portfolio $\mathbf{h}_s$ for $\mathbf{d}$ in a market $A, \mathbf{q}$ must satisfy

$$\text{Minimize } \mathbf{h}_s^T \mathbf{q}, \text{ subject to the constraint } \mathbf{h}_s^T A = \mathbf{d}. \qquad (8.1)$$

At spot prices $\mathbf{q}$, the cost of the hedge portfolio is $\mathbf{h}_s^T \mathbf{q}$, so the seller of the derivative must charge at least this much in order to hedge the sale. Hence the "ask price" is the minimum cost found by solving the problem in Eq.8.1.

The potential buyer of a derivative contingent claim has the opposite perspective on hedges. They provide the same payoff, so if one were available at a lower spot price than the derivative, there would be an immediate arbitrage. Assuming no arbitrages, the prospective buyer has a maximum "bid price" for any hedge portfolio $\mathbf{h}_b$ that has a payoff equal to the derivative's. That is the maximum cost found by solving

$$\text{Maximize } \mathbf{h}_b^T \mathbf{q}, \text{ subject to the constraint that } \mathbf{h}_b^T A = \mathbf{d}. \qquad (8.2)$$

Theorem 8.5 *In an arbitrage-free market $A, \mathbf{q}$, the bid and ask prices for any derivative with an exact hedge must be equal.*

Proof: First note that if there exists an exact hedge, then Eqs.8.1 and 8.2 both have solutions. These may be denoted $\mathbf{h}_s$ and $\mathbf{h}_b$, respectively, and they satisfy

$$\mathbf{h}_s^T A = \mathbf{d} = \mathbf{h}_b^T A.$$

Since $A, \mathbf{q}$ is arbitrage-free, there exists $\mathbf{k} \in K^o$ such that $\mathbf{q} = A\mathbf{k}$. But then

$$\mathbf{h}_b^T \mathbf{q} = \mathbf{h}_b^T (A\mathbf{k}) = (\mathbf{h}_b^T A)\mathbf{k} = (\mathbf{h}_s^T A)\mathbf{k} = \mathbf{h}_s^T (A\mathbf{k}) = \mathbf{h}_s^T \mathbf{q}.$$

Conclude that the bid price must equal the ask price. □

Remark. Two exact hedge portfolios $\mathbf{h}_b$ and $\mathbf{h}_s$ need not be the same, only their payoffs (and thus their spot prices) must be equal.

8.2.2 Complete Markets

Arbitrage-free market A with n states is *complete* iff every contingent claim can be hedged exactly. This is true iff the row space of A is all of $\mathbf{R}^n$, iff A has full rank n. The row space is dependent on the discrete financial model, and this full rank condition does not hold in general. For example, it cannot hold if A has fewer rows (assets) than columns (states). However, it can be guaranteed under additional assumptions. In particular, two-state market models with a riskless asset and at least one risky asset are always complete.

Example: invertible market matrices

If A is invertible, then it has full rank and models a complete market. Thus there exists a hedge $\mathbf{h}$ for any payoff $\mathbf{d}$, such as any contingent claim on the other modeled asset payoffs. Furthermore, this hedge portfolio is uniquely determined:

$$\mathbf{h}^T A = \mathbf{d} \implies \mathbf{h} = \left(\mathbf{d} A^{-1}\right)^T.$$

By Theorem 8.4, this market will be arbitrage-opportunity-free (AO-Free) if and only if there exists $\mathbf{k} \in K^o$ such that $\mathbf{q} = A\mathbf{k}$. But since A is invertible, there is a unique solution:

$$A\mathbf{k} = \mathbf{q} \implies \mathbf{k} = A^{-1}\mathbf{q}.$$

This leads to a simple test for arbitrages:

Corollary 8.6 *Finite market model $A, \mathbf{q}$ with invertible A is arbitrage-free if and only if $A^{-1}\mathbf{q} > 0$.* □

It also provides an explicit formula for risk neutral probabilities. For example, the two-asset, two-state market considered in Chapter 3 has finite model

$$A = \begin{pmatrix} R & R \\ S_\downarrow & S_\uparrow \end{pmatrix}, \qquad \mathbf{q} = \begin{pmatrix} 1 \\ S_0 \end{pmatrix}.$$

Here 1 is the spot price of one unit of the riskless asset that returns $R > 1$, and S_0 is the spot price for the risky asset with payoffs $S_\downarrow \neq S_\uparrow$. It may be assumed, by renaming the two states if necessary, that $S_\downarrow < S_\uparrow$. Then solution $\mathbf{k}$ has an explicit formula:

$$A^{-1} = \frac{1}{R(S_\uparrow - S_\downarrow)} \begin{pmatrix} S_\uparrow & -R \\ -S_\downarrow & R \end{pmatrix}, \implies \mathbf{k} = A^{-1}\mathbf{q} = \begin{pmatrix} \dfrac{S_\uparrow - RS_0}{R(S_\uparrow - S_\downarrow)} \\ \dfrac{RS_0 - S_\downarrow}{R(S_\uparrow - S_\downarrow)} \end{pmatrix}.$$

The AO-Free condition holds, by Corollary 8.6, iff $\mathbf{k} > 0$, iff

$$S_\downarrow < RS_0 < S_\uparrow.$$

The positive components of $\mathbf{k}$ are the arbitrage-free weights to use when pricing contingent claims on S that pay off when the riskless asset returns R.

Since the top row of A has R in every column, the top row of vector $A\mathbf{k}$ is R times the sum of the coefficients $\mathbf{k}$. But that equals the top row coefficient of $\mathbf{q}$, which is 1. Thus $p \overset{\text{def}}{=} R\mathbf{k}$ is a probability mass function on the states $\Omega = \{\downarrow, \uparrow\}$:

$$p(\downarrow) = \frac{S_\uparrow - RS_0}{S_\uparrow - S_\downarrow}, \qquad p(\uparrow) = \frac{RS_0 - S_\downarrow}{S_\uparrow - S_\downarrow}.$$

These are risk neutral probabilities since $A, \mathbf{q}$ is arbitrage-free, and they are uniquely defined because A is invertible.

Example: noninvertible complete markets

In the general case of complete markets with n states but more than n assets, A will have rank n but will not be invertible. Even so, if $A, \mathbf{q}$ is arbitrage-free, there exists a unique $\mathbf{k} \in K^o$ such that $\mathbf{q} = A\mathbf{k}$. That is because the Fundamental Theorem guarantees that such $\mathbf{k}$ exists, and $A^T A$ is invertible because A has full row space rank, so $\mathbf{k}$ is uniquely determined by

$$A^T \mathbf{q} = A^T A\mathbf{k} \implies \mathbf{k} = (A^T A)^{-1} A^T \mathbf{q}.$$

For example, let

$$A = \begin{pmatrix} R & R \\ a_\downarrow & a_\uparrow \\ b_\downarrow & b_\uparrow \end{pmatrix}; \qquad \mathbf{q} = \begin{pmatrix} 1 \\ a_0 \\ b_0 \end{pmatrix}$$

be a three-asset, two-state finite market model, where $R > 1$ represents riskless return, and $a_\downarrow < a_\uparrow$. By the Fundamental Theorem, $A, \mathbf{q}$ is arbitrage-free iff $(\exists \mathbf{k} \in K^o)\ \mathbf{q} = A\mathbf{k}$. Writing $\mathbf{k} = (k_1, k_2)$, this is three simultaneous linear equations:

$$\begin{aligned} Rk_1 + Rk_2 &= 1, \\ a_\downarrow k_1 + a_\uparrow k_2 &= a_0, \\ b_\downarrow k_1 + b_\uparrow k_2 &= b_0. \end{aligned}$$

There exists a solution $\mathbf{k} > 0$ only if $a_\downarrow < Ra_0 < a_\uparrow$, which together with the first equation forces the unique solution

$$k_1 = \frac{a_\uparrow - Ra_0}{a_\uparrow - a_\downarrow}, \qquad k_2 = \frac{Ra_0 - a_\downarrow}{a_\uparrow - a_\downarrow},$$

which determines b_0 in terms of $b_\downarrow$ and $b_\uparrow$, from the third equation. Conclude that this arbitrage-free complete market must have a consistent risk neutral weighting on future states, so that

$$a_0 = \text{PV}(\text{E}(\mathbf{a})), \qquad b_0 = \text{PV}(\text{E}(\mathbf{b})),$$

and both expectations use the same probability function. In particular, a portfolio containing just a and the riskless asset can be hedged by one containing just b and riskless. It follows that there are infinitely many hedge portfolios, combining a and b, for any derivative in this market.

Remark. This example illustrates how new assets added to an arbitrage-free complete market have unique no-arbitrage spot prices determined by their modeled future payoffs. Derivatives or contingent claims on existing assets fall into this category.

8.2.3 Incomplete Markets

In an *incomplete market*, matrix A has smaller rank than the number n of modeled states. Therefore, not every payoff can be hedged. There may still be derivatives with hedges, however, as an exercise below will show. There may even be multiple hedges for some contingent claims, but there will be model-dependent constraints on those derivatives.

Nevertheless, the seller of an arbitrary contingent claim in an incomplete market can still protect against loss by *superreplicating* the payoff $\mathbf{d}$. The seller constructs a portfolio $\mathbf{h}_s$ by solving

Minimize $\mathbf{h}_s^T \mathbf{q}$ subject to $\mathbf{h}_s^T A \geq \mathbf{d}$.

This is *superreplication*, creating the cheapest portfolio with a payoff that is at least as much as the liability incurred by selling the derivative asset.

Conversely, the buyer of the contingent claim $\mathbf{d}$ compares its price to the alternative portfolio $\mathbf{h}_b$ solving

Maximize $\mathbf{h}_b^T \mathbf{q}$ subject to $\mathbf{h}_b^T A \leq \mathbf{d}$.

This is *subreplication*, pricing the most expensive of all portfolios worth no more than the offered derivative.

Both superreplication and subreplication are *convex optimization problems*, solved by a minimum or maximum of a linear function, $\mathbf{h} \mapsto \mathbf{h}^T \mathbf{q}$, on the closed set $\{\mathbf{h} : \mathbf{h}^T A \geq \mathbf{d}\}$ or $\{\mathbf{h} : \mathbf{h}^T A \leq \mathbf{d}\}$, respectively. The search set for solutions $\mathbf{h}$ is the intersection of n half-spaces, one for each column of A. Each half-space is a convex set, and the intersection of convex sets is convex. Such problems are efficiently solvable by *linear programming*, as will be described below.

If market A with prices $\mathbf{q}$ is arbitrage-free, then any profitable portfolio $\mathbf{p}$ must have a nonnegative price:

$$\mathbf{p}^T A \geq \mathbf{0} \implies \mathbf{p}^T \mathbf{q} \geq 0.$$

Let $\mathbf{p} = \mathbf{h}_s - \mathbf{h}_b$ be the difference of the portfolios solving the hedge optimization problems. Then

$$\mathbf{p}^T A = \mathbf{h}_s^T A - \mathbf{h}_b^T A \geq \mathbf{d} - \mathbf{d} = \mathbf{0},$$

---

so $\mathbf{p}$ is a profitable portfolio, so $\mathbf{p}^T\mathbf{q} \geq 0$ and thus $\mathbf{h}_s^T\mathbf{q} \geq \mathbf{h}_b^T\mathbf{q}$. The nonempty interval

$$[\mathbf{h}_b^T\mathbf{q}, \mathbf{h}_s^T\mathbf{q}]$$

is the *no-arbitrage bid-ask spread* for the contingent claim $\mathbf{d}$.

Remark. This interval is a single point if the market is complete and there is an exact hedge portfolio for every contingent claim.

The calculations above may be stated as follows:

Theorem 8.7 *If $A, \mathbf{q}$ is arbitrage-free, then any portfolio has a nonempty bid-ask interval of no-arbitrage prices. If the market is complete, then this interval consists of a single point.* □

Octave computations with GLPK

Reliable open-source software for linear programming, applicable to these sub-replication and superreplication problems of financial mathematics, may be found in the GLPK toolkit from

https://www.gnu.org/software/glpk/

Download, compile, and install this package on your computer. It is usable with many software systems. Documentation on how to use it within Octave is available online:

https://octave.org/doc/v4.4.1/Linear-Programming.html

This software has its own syntax for defining the abstract optimization problem, stated in terms of given (column) vectors $\mathbf{b}, \mathbf{c}$ and matrix A:

"Find a (column) vector $\mathbf{x}$ such that $\mathbf{c}^T\mathbf{x}$ is optimal given constraints on $A\mathbf{x}$ versus $\mathbf{b}$."

In the notation conventions of this chapter, the target $\mathbf{x} = \mathbf{h}$ will be the hedge portfolio column vector, A will be the transpose of the market payoff matrix (so asset payoffs will be column vectors), $\mathbf{c} = \mathbf{q}$ will be the spot price vector, and $\mathbf{b} = \mathbf{d}^T$ will be the column vector transpose of the contingent claim payoff row vector.

Example: Three-state model with one risky asset

Suppose that a market model contains one risky asset S and one riskless asset B with the following attributes:

- Spot price $S(0) = 90$,
- Unit riskless asset spot price $B(0) = 1$,
- Modeled prices $S(T) \in \{85, 90, 95\}$ at future time $T > 0$,

- Riskless return $B(T) = R = 1.02$ to time T.

There are more states than assets so the market is incomplete.

Begin the analysis in Octave by determining whether $A\mathbf{q}$ is arbitrage-free. But augmented row echelon form shows that $A\mathbf{k} = \mathbf{q}$ is consistent and has a one parameter family of solutions:

```
R=1.02; S=[85,90,95]; S0=90; q=[1;S0]; A=[R R R; S];
rref([A R*q])
%  1.00000   0.00000  -1.00000  -0.36000
%  0.00000   1.00000   2.00000   1.36000
```

(This is in the notation of Definition 4.) Use R*q in the augmented matrix to get the risk neutral probabilities $R\mathbf{k} \overset{\text{def}}{=} \mathbf{p} = (p_1, p_2, p_3)$, which sum to 1, instead of the discounted vector $\mathbf{k}$ which will sum to $\frac{1}{R}$. The echelon form implies that p_3 is the unique free variable and that the general solution is

$$
\begin{aligned}
p_1 &= p_3 - 0.36 \\
p_2 &= 1.36 - 2p_3
\end{aligned}
$$

(Notice that $\sum_i p_i = 1$.) To find a positive solution $\mathbf{p} > \mathbf{0}$ requires solving the simultaneous inequalities

$$
p_3 > 0, \quad p_3 - 0.36 = p_1 > 0, \quad 1.36 - 2p_3 = p_2 > 0,
$$

which imply that $0.36 < p_3 < 0.72$. Taking the midpoint gives one of the infinitely many positive solutions:

```
p3=(0.36+0.72)/2; p1=p3-0.36; p2=1.36-2*p3;
p=[p1 p2 p3] % p =   0.18000    0.28000    0.54000
k=p/R        % k =   0.17647    0.27451    0.52941
```

Conclude by Theorem 8.4 that $A, \mathbf{q}$ is arbitrage-free.

Remark. Octave computes the unique solution closest to 0 for such an underdetermined system with

```
A\q*R % p = 0.15333    0.33333    0.51333
A\q   % k = 0.15033    0.32680    0.50327
```

For general $A, \mathbf{q}$, a minimal norm solution $\mathbf{k}$ to $\mathbf{q} = A\mathbf{k}$ may not be positive even if there exists a positive solution. ($A = 0$ with $\mathbf{q} = \mathbf{0}$ gives a trivial example.) However, if A contains a riskless asset with payoff $R > 0$, then every solution satisfies $\sum_i k_i = 1/R > 0$. The intersection of this hyperplane with the orthant K contains all the positive solutions, all of which are closer to $\mathbf{0}$ than any other points on the hyperplane outside K. This gives a simple computational test for arbitrages in an incomplete market:

Corollary 8.8 *Finite incomplete market model $A, \mathbf{q}$ is arbitrage-free if and only if $A\mathbf{k} = \mathbf{q}$ has a minimal norm solution $\mathbf{k} > \mathbf{0}$.* □

The market in this example is incomplete, so there exist derivatives without exact hedges. However, some derivatives may have exact hedges simply because their payoffs happen to lie in the row space of A. There is a simple way to detect this, using Theorem 8.7:

Corollary 8.9 *Suppose that* $A, \mathbf{q}$ *is an arbitrage-free market model. Then the following are equivalent for a derivative with payoff* $\mathbf{d}$:

   (a) $\mathbf{d}$ *is in the row space of* A.
   (b) $\mathbf{d}$ *has an exact hedge.*
   (c) $\mathbf{d}$ *has a unique no-arbitrage spot price.*   □

In particular, there is no exact hedge for an in-the-money European-style Call option at strike price $K = 88$ in this market, though by Theorem 8.7, such a derivative may be both superreplicated and subreplicated. Its payoff $[S - K]^+$ may be computed with an anonymous function of the underlying price:

```
K=88; payoff=@(s)max(s-K,0); % European-style Call at strike K
```

To begin, put the market matrix and spot prices into the format used by GLPK:

```
S0=90; % Spot price of the risky asset
c=[1;S0]; % q: spot prices of all market assets
R=1.02; % B(T): riskless return to expiry.
S1=[85;90;95]; % S(T): modeled payoffs for S
b=payoff(S1); % C(T): modeled payoffs for Call
A=[R*ones(size(S1)),S1]; % A: market transposed into GLPK format
```

Next, set parameters for the linear programming function `glpk()`:

```
vartype="CC";    % ==> x(j) is Continuous, j=1,2
param.msglev=1;  % ==> use a low verbosity level
param.itlim=1000; % ==> huge iteration limit
ub=100*[S0;1]; % huge Upper bounds on x: buy =< 1000*S0
lb= -ub;         % huge Lower bounds on x: sell =< 1000*S0
```

Then, find a superreplication, taking the derivative seller's perspective:

```
sellctype="LLL"; % Lower constraint type A*x(j)>=b(j), j=1,2,3
sellsense=1; % Optimization direction for c'*x: "1" ==> "min"
[hs,ask]=glpk(c,A,b,lb,ub,sellctype,vartype,sellsense,param);
hs % = [ -58.33333; 0.70000]; Cost-minimizing hedge portfolio
ask % = 4.6667; minimum cost to superreplicate the Call
```

Finally, find a subreplication, taking the derivative buyer's perspective:

```
buyctype="UUU"; % Upper constraint type A*x(j)=<b(j), j=1,2,3
buysense=-1; % Optimization direction for c'*x: "-1" ==> "max"
[hb,bid]=glpk(c,A,b,lb,ub,buyctype,vartype,buysense,param);
hb % = [ -86.2745; 1.0000]; Cost-maximizing hedge portfolio
bid % = 3.7255; maximum cost to subreplicate the Call
```

Since bid is strictly less than ask, by Corollary 8.9 there is no exact hedge for this derivative in this market.

8.3 Cones, Convexity, and Duals

The proofs of Fundamental Theorems 8.3 and 8.4, as well as Farkas's Lemma 8.26, follow from the properties of *convex cones*. These are intuitive concepts about shapes that are made precise using analytic geometry:

Definition 10 (Convex Set) *A set $S \subset \mathbf{R}^n$ is said to be* convex *iff*

$$\mathbf{x}, \mathbf{y} \in S \implies (\forall \lambda \in [0,1])\ \lambda\mathbf{x} + (1-\lambda)\mathbf{y} \in S.$$

Thus, convexity means that if two points are in the set, then the line segment connecting them lies entirely within the set as well.

Definition 11 (Cone) *Set $S \subset \mathbf{R}^n$ is said to be a* cone *iff*

$$\mathbf{x} \in S \implies (\forall \lambda > 0)\lambda\mathbf{x} \in S.$$

Here the intuition is that every *ray* from $\mathbf{0}$ through any point in the cone is also in the cone. To define a ray precisely, let $\mathbf{v} \in \mathbf{R}^m$ be a nonzero vector and let

$$V \stackrel{\text{def}}{=} \{\lambda\mathbf{v} : \lambda > 0\} \quad \text{and} \quad \bar{V} \stackrel{\text{def}}{=} \{\lambda\mathbf{v} : \lambda \geq 0\}. \tag{8.3}$$

These are both called *rays in the $\mathbf{v}$ direction*, though only $\bar{V}$ contains $\mathbf{0}$. A check of Definitions 10 and 11 shows that each is a convex cone. Likewise, K, $K \setminus \mathbf{0}$, and K^o are all convex cones. However, not all cones are convex: the union of the positive x and y axes in $\mathbf{R}^2$ is a nonconvex cone.

For another example, recall that a *subspace* $V \subset \mathbf{R}^n$ is a set of vectors satisfying three conditions:

Definition 12 (Subspace) *Subset $V \subset \mathbf{R}^n$ is a subspace iff $\mathbf{0} \in V$, $\mathbf{x}, \mathbf{y} \in V \implies \mathbf{x} + \mathbf{y} \in V$, and $\mathbf{x} \in V$, $c \in \mathbf{R} \implies c\mathbf{x} \in V$.*

It is an easy exercise to prove that any subspace $V \subset \mathbf{R}^n$ is a convex cone. But of course, not all convex cones are subspaces.

The smallest subspace of $\mathbf{R}^n$ is just $\{\mathbf{0}\}$. The largest is all of $\mathbf{R}^n$. The *column space* $\mathrm{Col}\, A$ of matrix $A = \{a(i,j)\} \in \mathbf{R}^{m \times n}$ is a subspace of $\mathbf{R}^m$:

$$\mathrm{Col}\, A \stackrel{\text{def}}{=} \left\{\mathbf{x} = \begin{pmatrix} x(1) \\ \vdots \\ x(m) \end{pmatrix} : x(i) = \sum_{j=1}^{n} c_j a(i,j),\ i = 1, \ldots, m\right\}.$$

This subspace is the set of all linear combinations, also called the *span*, of the column vectors in A. Likewise, the *row space* Row A is the subspace of $\mathbf{R}^n$ spanned by the row vectors.

Now suppose that $A \in \mathbf{R}^{m \times n}$ is an $m \times n$ matrix, and that K and K^o are the nonnegative and positive orthants in $\mathbf{R}^n$, respectively. Their images under A are the sets

$$AK \stackrel{\text{def}}{=} \{A\mathbf{k} : \mathbf{k} \in K\}; \qquad AK^o \stackrel{\text{def}}{=} \{A\mathbf{k} : \mathbf{k} \in K^o\}. \tag{8.4}$$

AK and AK^o are both subsets of the column space of A, but with restrictions on the coefficients $\{c_j\}$. Let $\mathbf{v}_i \in \mathbf{R}^m$ denote the ith column of A, for $i = 1, \ldots, n$, and put

$$V_i \stackrel{\text{def}}{=} \{\lambda \mathbf{v}_i : \lambda > 0\}; \qquad \bar{V}_i \stackrel{\text{def}}{=} \{\lambda \mathbf{v}_i : \lambda \geq 0\}. \tag{8.5}$$

These are the rays generated by the column vectors $\mathbf{v}_i$, with or without $\mathbf{0}$. As in the case of Eq.8.3, each V_i or $\bar{V}_i$ is a convex cone. This suggests a decomposition into sums of simpler sets:

Definition 13 (Sums of Sets) *For any finite collection $\{S_i : i = 1, \ldots, n\}$ of subsets of $\mathbf{R}^m$, their sum*

$$\sum_{i=1}^{n} S_i \stackrel{\text{def}}{=} \left\{ \mathbf{s}_1 + \cdots + \mathbf{s}_n : (\forall i)\, \mathbf{s}_i \in S_i \right\} \subset \mathbf{R}^m$$

is the set of all sums of elements, one from each subset.

Apply this definition to write AK and AK^o as sums of rays:

$$AK = \sum_{i=1}^{n} \bar{V}_i; \qquad AK^o = \sum_{i=1}^{n} V_i. \tag{8.6}$$

But sums of convex cones are convex cones:

Lemma 8.10 *Suppose that $C_1, \ldots, C_n$ is a finite set of convex cones in $\mathbf{R}^m$. Then $\sum_{i=1}^{n} C_i$ is a convex cone.*

Proof: Check the properties:

$$\mathbf{x} = \mathbf{x}_1 + \cdots + \mathbf{x}_n \in \sum_{i=1}^{n} C_i \implies \lambda \mathbf{x} = \lambda \mathbf{x}_1 + \cdots + \lambda \mathbf{x}_n \in \sum_{i=1}^{n} C_i,$$

for any $\lambda > 0$, since $\mathbf{x}_i \in C_i \implies \lambda \mathbf{x}_i \in C_i$. Also, for any $\mathbf{x}, \mathbf{y} \in \sum_{i=1}^{n} C_i$, and any $0 \leq \mu \leq 1$,

$$\mu \mathbf{x} + (1 - \mu)\mathbf{y} = [\mu \mathbf{x}_1 + (1 - \mu)\mathbf{y}_1] + \cdots + [\mu \mathbf{x}_n + (1 - \mu)\mathbf{y}_n] \in \sum_{i=1}^{n} C_i,$$

since $\mathbf{x}_i, \mathbf{y}_i \in C_i \implies \mu\mathbf{x}_i + (1-\mu)\mathbf{y}_i \in C_i$ for all $i = 1, \ldots, n.$ □

Lemma 8.11 *Both AK and AK^o are convex cones.*

Proof: By Eqs.8.3 and 8.6, both AK and AK^o are sums of convex cones. By Lemma 8.10, they are therefore both convex cones. □

8.3.1 Open and Closed Sets

Every subspace $V \subset \mathbf{R}^n$ has a *basis*, a minimal set of vectors $\{\mathbf{b}_1, \ldots, \mathbf{b}_k\}$ such that every $\mathbf{v} \in V$ can be written in exactly one way as

$$\mathbf{v} = \sum_{i=1}^{k} c_i \mathbf{b}_i.$$

The *expansion coefficients* $c_i \in \mathbf{R}$ of $\mathbf{v}$ in a fixed basis are uniquely determined by $\mathbf{v}$. Any two bases for V must have the same number of elements k which is called the *dimension* of V, or $\dim V$. For any subspace $V \subset \mathbf{R}^n$,

$$\dim\{\mathbf{0}\} = 0 \le \dim V \le n = \dim \mathbf{R}^n.$$

The *norm* of a vector $\mathbf{x} = (x(1), \ldots, x(n)) \in \mathbf{R}^n$ is defined by

$$\|\mathbf{x}\| \stackrel{\text{def}}{=} \sqrt{x(1)^2 + \cdots + x(n)^2}.$$

It satisfies $\|\mathbf{x}\| \ge 0$, with $\|\mathbf{x}\| = 0$ if and only if $\mathbf{x} = \mathbf{0}$. It gives a distance function $\|\mathbf{x} - \mathbf{y}\|$ between any two vectors $\mathbf{x}, \mathbf{y} \in \mathbf{R}^n$ that satisfies

$$\|\mathbf{x} - \mathbf{y}\| = 0 \iff \mathbf{x} = \mathbf{y}.$$

There is some positive distance between a subspace and any point outside it:

Lemma 8.12 *If $V \subset \mathbf{R}^n$ is a subspace and $\mathbf{x} \in \mathbf{R}^n$ is a point not in V, then there is some $\epsilon > 0$ such that $\|\mathbf{x} - \mathbf{y}\| \ge \epsilon$ for all $\mathbf{y} \in V$.*

Proof: If $V = \mathbf{R}^n$, then there is nothing to prove.

Otherwise, let $k = \dim V$, let $\{\mathbf{b}_1, \ldots, \mathbf{b}_k\}$ be a basis for V, let $B = (b(i,j))$ be the $n \times k$ matrix whose j^{th} column is the basis vector $\mathbf{b}_j$, and define the function

$$s(v_1, \ldots, v_k) \stackrel{\text{def}}{=} \sum_{i=1}^{n} \left(x(i) - \sum_{j=1}^{k} v_j b(i,j) \right)^2 = \|\mathbf{x} - B\mathbf{v}\|^2,$$

which gives the squared distance between $\mathbf{x} = (x(1), \ldots, x(n))$ and the point $B\mathbf{v} = \sum_{j=1}^{k} v_j \mathbf{b}_j \in V$. This s is a quadratic polynomial in the expansion coefficients $v_1, \ldots, v_k$, and $s \to \infty$ as the coefficients grow without bound, so there is a minimum for s at the unique critical point $B\mathbf{v}^*$ given by expansion coefficients $\mathbf{v}^* = (v_1^*, \ldots, v_k^*)$ with $\nabla s(\mathbf{v}^*) = \mathbf{0}$, namely for all $l = 1, \ldots, k$,

$$0 = \frac{\partial s}{\partial v_l}(\mathbf{v}^*) = \sum_{i=1}^{n} \left[-2b(i,l) \left(x(i) - \sum_{j=1}^{k} v_j^* b(i,j) \right) \right],$$

This equation may be written as

$$\sum_{i=1}^{n} b(i,l)x(i) = \sum_{i=1}^{n}\sum_{j=1}^{k} b(i,l)b(i,j)v_j^* \quad \text{which is} \quad B^T\mathbf{x} = B^T B\mathbf{v}^*. \tag{8.7}$$

But $B^T B$ is an invertible $k \times k$ matrix, since B's columns form a basis, so

$$\mathbf{v}^* = (B^T B)^{-1} B^T \mathbf{x} \in \mathbf{R}^k, \qquad \Longrightarrow \quad B\mathbf{v}^* = B(B^T B)^{-1} B^T \mathbf{x} \in V \subset \mathbf{R}^n$$

uniquely defines the closest point $B\mathbf{v}^* \in V \subset \mathbf{R}^n$ to $\mathbf{x}$.

Since $\mathbf{x} \notin V$ while $B\mathbf{v}^* \in V$, it must be that $\epsilon \overset{\text{def}}{=} \|\mathbf{x} - B\mathbf{v}^*\| > 0$. Since $B\mathbf{v}^*$ is the point of minimal squared distance, it is the point of minimal distance, so for all $\mathbf{y} \in V$, $\|\mathbf{x} - \mathbf{y}\| \geq \|\mathbf{x} - B\mathbf{v}^*\| = \epsilon > 0$ as claimed.   □

Remark. For any subspace V with basis matrix B, this closest point in V to $\mathbf{x}$ given by $B(B^T B)^{-1} B^T \mathbf{x}$ is called the *orthogonal projection* of $\mathbf{x}$ onto V. Eq.8.7 is just the normal equation, Eq.2.45 on p.40. The nearest point is a least squares approximation, the best fit to $\mathbf{x}$ by vectors of the form $B\mathbf{v}$.

Two other important properties of sets in $\mathbf{R}^n$ are openness and closedness:

Definition 14 (Open and Closed) *Suppose that $S \subset \mathbf{R}^n$ is a subset.*

- *S is said to be* open *iff* $(\forall \mathbf{x} \in S)(\exists \epsilon > 0)$ $\|\mathbf{y} - \mathbf{x}\| < \epsilon \implies \mathbf{y} \in S$. *Namely, every $\mathbf{x} \in S$ is completely surrounded by points of S.*

- *S is said to be* closed *iff its complement $\mathbf{R}^n \setminus S$ is open.*

The empty set $\emptyset$ is open since it has no points that need to be checked. Thus $\mathbf{R}^n$ is closed, since its complement $\mathbf{R}^n \setminus \mathbf{R}^n = \emptyset$ is open. From these examples it is clear that sets may be both open and closed: $\mathbf{R}^n$ is clearly open, so it is both open and closed. Likewise, $\emptyset$ is both open and closed.

The following are basic facts about open and closed sets in $\mathbf{R}^n$:

- The union of any collection of open sets is open. Apply the definition: any point in the union belongs to one of the open sets, so it is contained in an open ball entirely within that set, which is therefore entirely within the union.

- The intersection of any collection of closed sets is closed. Note that the complement of an intersection of closed sets is the union of the complements of the sets. The complements are open by definition, so this union is open. Conclude that the intersection is closed.

- Any subspace is closed. Use Lemma 8.12 to show that the complement of a subspace is open. The details are left as an exercise.

Remark. The union of finitely many closed sets is closed, and the intersection of finitely many open sets is open. This does not extend to infinite collections:

$$\bigcup_{n=1}^{\infty} \left[\frac{1}{n}, 1\right] = (0, 1] \text{ is a nonclosed union of closed sets;}$$

$$\bigcap_{n=1}^{\infty} \left(-\frac{1}{n}, \frac{1}{n}\right) = \{0\} \text{ is a nonopen intersection of open sets.}$$

Sets may also be neither open nor closed. It is an easy exercise to check these properties for the three orthants defined above:

Lemma 8.13 *K is a closed convex cone; K° is an open convex cone; $K \setminus 0$ is a convex cone that is neither open nor closed.* ☐

8.3.2 Dual Cones and Double Duals

Sets of profitable and strictly profitable portfolios for a finite market $A, \mathbf{q}$ are in fact convex cones. This will be shown below by writing them in terms of A and the orthants K, K°, using the abstract notion of *duality*, defined here:

Definition 15 (Dual and Strict Dual) *Let $S \subset \mathbf{R}^n$ be any set.*

- *The* dual cone *of S is:* $S' \stackrel{\text{def}}{=} \{\mathbf{x} \in \mathbf{R}^n : (\forall \mathbf{y} \in S) \, \mathbf{x}^T \mathbf{y} \geq 0\}$.

- *The* strict dual cone *of S is:* $S^* \stackrel{\text{def}}{=} \{\mathbf{x} \in \mathbf{R}^n : (\forall \mathbf{y} \in S) \, \mathbf{x}^T \mathbf{y} > 0\}$.

Such duals are related to the *orthogonal complement* $S^\perp$ of S, which is defined by zero rather than nonnegative or positive inner products:

$$S^\perp \stackrel{\text{def}}{=} \{\mathbf{x} \in \mathbf{R}^n : (\forall \mathbf{y} \in S) \, \mathbf{x}^T \mathbf{y} = 0\}$$

Remark. By convention, $\emptyset' = \emptyset^* = \emptyset^\perp = \mathbf{R}^n$ since for $S = \emptyset$ there is nothing to check. Also, $\{\mathbf{0}\}' = \{\mathbf{0}\}^\perp = \mathbf{R}^n$, but $\{\mathbf{0}\}^* = \emptyset$.

It is left as an exercise to show that for any subset $S \in \mathbf{R}^n$, the duals S', S^*, and $S^\perp$ are convex cones. In addition, the orthogonal complement $S^\perp$ is a subspace.

The dual of the dual of a subset $S \in \mathbf{R}^n$, also called the *double dual*, will again be S under certain conditions. In particular, the double dual is a convex cone, so one condition is that S is a convex cone. Likewise, to have $(S^\perp)^\perp = S$, it must be that S is a subspace. But in fact, that is a sufficient condition as well: if $S \in \mathbf{R}^n$ is a subspace, then $(S^\perp)^\perp = S$.

Finding sufficient conditions to get $(S^*)^* = S$ and $(S')' = S$ is central to the proofs of the Fundamental Theorems. A method of proving the equality of two sets is showing that each contains the other. One of the containments follows directly from the definitions:

Lemma 8.14 *Let $S \subset \mathbf{R}^n$ be any set. Then $S \subset (S')'$ and $S \subset (S^*)^*$ and $S \subset (S^\perp)^\perp$.*

Proof: First check the trivial cases:

- If $S' = \emptyset$ or $S' = \{0\}$, then $S \subset (S')' = \mathbf{R}^n$.
- If $S^* = \emptyset$ then $S \subset (S^*)^* = \mathbf{R}^n$.
- If $S^\perp = \emptyset$ or $S^\perp = \{0\}$, then $S \subset (S^\perp)^\perp = \mathbf{R}^n$.

Otherwise, suppose S is a nonempty set containing a point other than $\mathbf{0}$. Let $\mathbf{x} \in S$ be given.

- For every $\mathbf{y} \in S'$, $\mathbf{y}^T \mathbf{x} = \mathbf{x}^T \mathbf{y} \geq 0$. Thus $\mathbf{x} \in (S')'$.
- For every $\mathbf{y} \in S^*$, $\mathbf{y}^T \mathbf{x} = \mathbf{x}^T \mathbf{y} > 0$. Thus $\mathbf{x} \in (S^*)^*$.
- For every $\mathbf{y} \in S^\perp$, $\mathbf{y}^T \mathbf{x} = \mathbf{x}^T \mathbf{y} = 0$. Thus $\mathbf{x} \in (S^\perp)^\perp$.

Conclude that $S \subset (S')'$, $S \subset (S^*)^*$, and $S \subset (S^\perp)^\perp$, as claimed.   □

Rays and orthants are specific examples of convex cones for which duals and double duals are easily found:

Let $V, \bar{V}$ be the rays generated by nonzero $\mathbf{v} \in \mathbf{R}^n$ as in Eq.8.3. Then

$$V^* = \{\mathbf{x} \in \mathbf{R}^n : \mathbf{x}^T \mathbf{v} > 0\} \quad \text{and} \quad V' = \{\mathbf{x} \in \mathbf{R}^n : \mathbf{x}^T \mathbf{v} \geq 0\}$$

are, respectively, the open and closed *half-spaces*, with normal vector $\mathbf{v}$. If $\mathbf{v} = \mathbf{0}$, these sets are $\emptyset$ and $\mathbf{R}^n$, respectively, instead of half-spaces.

Lemma 8.15 *For the rays V and $\bar{V}$ defined in Eq.8.3, the double duals are $(V^*)^* = V$ and $(\bar{V}')' = \bar{V}$, respectively.*

Proof: By Lemma 8.14, $V \subset (V^*)^*$. To get $(V^*)^* = V$, it remains to prove that $(V^*)^* \subset V$. So, suppose that $\mathbf{b} \in (V^*)^*$.

Since $\mathbf{v}^T \mathbf{v} > 0$, so that $\mathbf{v} \in V^*$, it follows that $\mathbf{b}^T \mathbf{v} > 0$.

The subspace $\{c\mathbf{v} : c \in \mathbf{R}\}$ is closed and convex, so by Lemma 8.27 it contains a unique nearest point $c_0 \mathbf{v}$ to $\mathbf{b}$, which satisfies $\mathbf{v}^T(\mathbf{b} - c_0\mathbf{v}) = 0$, so

$$c_0 = \frac{\mathbf{v}^T \mathbf{b}}{\mathbf{v}^T \mathbf{v}} > 0, \quad \implies c_0 \mathbf{v} \in V.$$

Put $\mathbf{z} \stackrel{\text{def}}{=} \mathbf{b} - c_0\mathbf{v}$. Note that $\mathbf{z} \in V^\perp$.

If $\mathbf{z} = \mathbf{0}$, then $\mathbf{b} = c_0\mathbf{v} \in V$.

Otherwise, suppose $\mathbf{z} \neq \mathbf{0}$. Since $\mathbf{z} \in V^{\perp}$ and $\mathbf{v} \in V^*$, for every $r \in \mathbf{R}$ the point $\mathbf{v} + r\mathbf{z}$ belongs to V^*, so since $\mathbf{b} \in (V^*)^*$,

$$(\forall r \in \mathbf{R}) \; \mathbf{b}^T(\mathbf{v} + r\mathbf{z}) = \mathbf{b}^T\mathbf{v} + r\mathbf{b}^T\mathbf{z} > 0.$$

If $\mathbf{b}^T\mathbf{z} \neq 0$, then this leads to a contradiction for sufficiently large r of the appropriate sign. Thus $\mathbf{b}^T\mathbf{z} = 0$, which means

$$0 = \mathbf{b}^T(\mathbf{b} - c_0\mathbf{v}) \implies c_0 = \frac{\mathbf{b}^T\mathbf{b}}{\mathbf{b}^T\mathbf{v}} = \frac{\mathbf{v}^T\mathbf{b}}{\mathbf{v}^T\mathbf{v}},$$

from which it follows that $|\mathbf{b}^T\mathbf{v}| = \|\mathbf{b}\|\|\mathbf{v}\|$. This means that $\mathbf{b}$ and $\mathbf{v}$ are parallel, namely $\mathbf{b} = c\mathbf{v}$ for some real c, and then $\mathbf{b}^T\mathbf{v} > 0$ forces $c > 0$, so $\mathbf{b} \in V$. Thus in all cases, $\mathbf{b} \in (V^*)^* \implies \mathbf{b} \in V$, namely $(V^*)^* \subset V$. Conclude that $(V^*)^* = V$.

The very similar proof that $(\bar{V}')' = \bar{V}$ is left as an exercise. $\quad\square$

Theorem 8.16 *For the orthants K, $K \setminus \mathbf{0}$, and K^o defined above,*

(a) $K' = K$, *that is, the nonnegative orthant is a self-dual cone.*

(b) $(K^o)' = K$ *and* $(K^o)^* = K \setminus \mathbf{0}$.

(c) $(K \setminus \mathbf{0})' = K$ *and* $(K \setminus \mathbf{0})^* = K^o$.

(d) $((K^o)^*)^* = K^o$, *that is, the open positive orthant is its own strict double dual cone.*

Proof: This is left as an exercise. $\quad\square$

From this, it follows that profitable and strictly profitable portfolios in a finite market model $A, \mathbf{q}$ must be convex cones, because they are duals:

Corollary 8.17 *The set P of profitable portfolios for a market matrix A is a dual cone: $P = (AK)'$.*

Proof: Since $K' = K$ by Theorem 8.16(a),

$$
\begin{aligned}
\mathbf{p} \in P &\iff \mathbf{p}^T A \in K \\
&\iff (\forall \mathbf{k} \in K')(\mathbf{p}^T A)\mathbf{k} \geq 0 \\
&\iff (\forall \mathbf{k} \in K)(\mathbf{p}^T A)\mathbf{k} \geq 0 \\
&\iff (\forall \mathbf{k} \in K)\mathbf{p}^T(A\mathbf{k}) \geq 0 \\
&\iff (\forall \mathbf{v} \in AK)\mathbf{p}^T\mathbf{v} \geq 0 \\
&\iff \mathbf{p} \in (AK)'.
\end{aligned}
$$

The next to last condition is just the definition of membership (by $\mathbf{p}$) in the dual cone of AK. $\quad\square$

Corollary 8.18 *The set S of strictly profitable portfolios for a market matrix A is a strict dual cone: $S = (AK^o)^*$*

Proof: This is left as an exercise. $\quad\square$

8.3.3   Proofs of the Fundamental Theorems

The absence of arbitrages for $A, \mathbf{q}$ imposes contraints. These constraints may be stated as membership by $\mathbf{q}$ in a convex cone determined by A, at first using profitable or strictly profitable portfolios:

Lemma 8.19 *Suppose that A is a market matrix with spot price vector $\mathbf{q}$. Let P and S be the profitable and strictly profitable portfolios of A, respectively. Then*

(a) $A, \mathbf{q}$ is IA-free iff $\mathbf{q} \in P'$, the dual cone of P.

(b) $A, \mathbf{q}$ is AO-free iff $\mathbf{q} \in S^*$, the strict dual cone of S.

Proof: (a) Observe that $\mathbf{x} \in P \iff \mathbf{x}^T A \geq \mathbf{0}$. Thus, by Lemma 8.1, $A, \mathbf{q}$ is IA-free iff

$$(\forall \mathbf{x} \in P)\, \mathbf{x}^T \mathbf{q} \geq 0$$

which is true (by the definition of dual cone) iff $\mathbf{q} \in P'$.

(b) Observe that $\mathbf{x} \in S \iff \mathbf{x}^T A > \mathbf{0}$. Thus, by Lemma 8.2, $A, \mathbf{q}$ is AO-free iff

$$(\forall \mathbf{x} \in S)\, \mathbf{x}^T \mathbf{q} > 0$$

which is true (by the definition of strict dual cone) iff $\mathbf{q} \in S'$.   □

Both P and S are nonempty and nontrivial, since A has a numeraire, but it is not necessary to mention them when characterizing $\mathbf{q}$ such that $A, \mathbf{q}$ is arbitrage-free. Simply combine Lemma 8.19 with Corollary 8.17, $P = (AK)'$, or Corollary 8.18, $S = (AK^o)^*$:

Corollary 8.20 *Suppose that A is a market matrix with spot price vector $\mathbf{q}$. Then*
(a) $A, \mathbf{q}$ is IA-free if and only if $\mathbf{q} \in ((AK)')'$, the double dual cone.
(b) $A, \mathbf{q}$ is AO-free iff $\mathbf{q} \in ((AK^o)^*)^*$, the strict double dual cone.   □

Thus, $A, \mathbf{q}$ being arbitrage-free is just the geometric property of $\mathbf{q}$ belonging to the double dual of AK or strict double dual of AK^o, respectively.

To complete the proofs of Theorems 8.3 and 8.4, it remains to prove that AK and AK^o are self double duals:

Theorem 8.21 *Let $K, K^o \subset \mathbf{R}^n$ be the nonnegative and positive orthants, respectively. Let $A \in \mathbf{R}^{m \times n}$ be any matrix such that $(AK)'$ and $(AK^o)^*$ are both nonempty. Then $((AK)')' = AK$ and $((AK^o)^*)^* = AK^o$.*

(By Corollaries 8.17 and 8.18 and the remark on p.193, the hypothesis is satisfied by any market matrix with a numeraire.) This will be done in several steps, with an analysis of the convex cones AK and AK^o, which are sums of rays as in Eq.8.6. Let $\bar{V}_i$ and V_i be the rays generated by column $\mathbf{v}_i$ of matrix A, for $i \in \{1, \ldots, n\}$, as in Eq.8.5. Let $\bar{V}_i'$ and V_i^* be their dual cones and strict dual cones, respectively.

Lemma 8.22 $\displaystyle\bigcap_{i=1}^{n} V_i^* \subset (AK^o)^*$, *and* $\displaystyle\bigcap_{i=1}^{n} \bar{V}_i' \subset (AK)'$.

Proof: If $(\forall i)\, \mathbf{y} \in V_i^*$, then $\mathbf{y}^T A = (\mathbf{y}^T \mathbf{v}_1, \dots, \mathbf{y}^T \mathbf{v}_n) > \mathbf{0}$, so for $\mathbf{k} \in K^o$, $\mathbf{y}^T A\mathbf{k} > 0$. Thus $\mathbf{y} \in (AK^o)^*$.

Similarly, if $(\forall i)\, \mathbf{y} \in \bar{V}_i'$, then $\mathbf{y}^T A = (\mathbf{y}^T \mathbf{v}_1, \dots, \mathbf{y}^T \mathbf{v}_n) \geq \mathbf{0}$, so for $\mathbf{k} \in K$, $\mathbf{y}^T A\mathbf{k} \geq 0$. Thus $\mathbf{y} \in (AK)'$. $\qquad\square$

Next, observe that duals and strict duals, as well as orthogonal complements, reverse inclusions:

Lemma 8.23 *If* $A \subset B \subset \mathbf{R}^n$, *then* $B' \subset A'$, $B^* \subset A^*$, *and* $B^\perp \subset A^\perp$.

Proof: Suppose $\mathbf{x} \in B'$. Then

$$(\forall \mathbf{b} \in B)\, \mathbf{b}^T\mathbf{x} \geq 0 \quad \Longrightarrow \quad (\forall \mathbf{a} \in A)\, \mathbf{a}^T\mathbf{x} \geq 0,$$

since $A \subset B$. Thus $\mathbf{x} \in A'$. Conclude that $B' \subset A'$.

The same argument with "$\geq$" replaced with "$>$" shows that $B^* \subset A^*$. Likewise, replace "$\geq$" with "$=$" to show that $B^\perp \subset A^\perp$. $\qquad\square$

Apply Lemma 8.23 to get inclusion relations for double duals:

Corollary 8.24 $\displaystyle ((AK^o)^*)^* \subset \left(\bigcap_{i=1}^{n} V_i^* \right)^*$, *and* $\displaystyle ((AK)')' \subset \left(\bigcap_{i=1}^{n} \bar{V}_i' \right)'$. $\qquad\square$

Finally, compute the dual of the nonempty intersection of finitely many half-spaces as follows:

Lemma 8.25 *Suppose that* $\bigcap_{i=1}^{n} V_i^*$, *and thus also* $\bigcap_{i=1}^{n} \bar{V}_i'$, *are nonempty. Then*

$$\left(\bigcap_{i=1}^{n} V_i^* \right)^* = \sum_{i=1}^{n} V_i \qquad \text{and} \qquad \left(\bigcap_{i=1}^{n} \bar{V}_i' \right)' = \sum_{i=1}^{n} \bar{V}_i$$

Proof: This is left as a challenging exercise. $\qquad\square$

8.3.4 Farkas's Lemma

The use of double duals illustrates the similarity between the two Fundamental Theorems. There is, however, an alternative proof of Theorem 8.3 that uses Farkas's Lemma, a result from 1902. It is presented here since it has applications in many convex optimization algorithms.

Theorem 8.26 (Farkas's Lemma) *Suppose that* $A \in \mathbf{R}^{m \times n}$ *is a matrix and* $\mathbf{q} \in \mathbf{R}^m$ *is a vector. Then exactly one of the following must be true:*

X: *There exists* $\mathbf{x} \in \mathbf{R}^m$ *such that* $\mathbf{x}^T A \geq \mathbf{0}$ *and* $\mathbf{x}^T \mathbf{q} < 0$.

Y: *There exists* $\mathbf{y} \in \mathbf{R}^n$ *such that* $A\mathbf{y} = \mathbf{q}$ *and* $\mathbf{y} \geq \mathbf{0}$.

Proof: First observe that X and Y cannot both hold as that would lead to a contradiction:

$$\mathbf{x}^T A\mathbf{y} = \mathbf{x}^T (A\mathbf{y}) = \mathbf{x}^T \mathbf{q} < 0,$$

while also $\mathbf{x}^T A\mathbf{y} = (\mathbf{x}^T A)\mathbf{y} \geq 0$, since both $(\mathbf{x}^T A) \geq 0$ and $\mathbf{y} \geq \mathbf{0}$.

Evidently, Condition Y holds if and only if

$$\mathbf{q} \in AK = \{A\mathbf{k} : \mathbf{k} \geq \mathbf{0}\},$$

(using the closed orthant K of Definition 6), so if Y fails to hold it must be that $\mathbf{q} \notin AK$. It remains to show that in this case, Condition X must hold.

But AK is a nonempty closed convex cone. By Theorem 8.28, there exists a nonzero vector $\mathbf{x} \in \mathbf{R}^m$ and a constant $\gamma \in \mathbf{R}$ defining a separating hyperplane function

$$f : \mathbf{R}^m \to \mathbf{R}, \quad f(\mathbf{y}) \overset{\text{def}}{=} \mathbf{x}^T \mathbf{y} - \gamma,$$

such that $f(\mathbf{q}) < 0$ but $f(\mathbf{z}) > 0$ for every $\mathbf{z} \in AK$.

Now $\mathbf{0} \in AK$, since $\mathbf{0} \in K$, so $f(\mathbf{0}) = \mathbf{x}^T \mathbf{0} - \gamma = -\gamma > 0$, and therefore $\gamma < 0$. But then

$$\mathbf{x}^T \mathbf{q} - \gamma = f(\mathbf{q}) < 0 \implies \mathbf{x}^T \mathbf{q} < \gamma < 0.$$

On the other hand, since AK is a cone, any $\mathbf{z} \in AK$ and any $\lambda > 0$ result in $\lambda \mathbf{z} \in AK$, so

$$(\forall \lambda > 0)\, \lambda \mathbf{x}^T \mathbf{z} - \gamma = f(\lambda \mathbf{z}) > 0 \implies (\forall \lambda > 0)\, \mathbf{x}^T \mathbf{z} > \gamma/\lambda \implies \mathbf{x}^T \mathbf{z} \geq 0.$$

Thus $\mathbf{x}^T \mathbf{z} \geq 0$ for all $\mathbf{z} \in AK$. Writing $\mathbf{z} = A\mathbf{k}$ gives

$$(\forall \mathbf{k} \in K)\, \mathbf{x}^T A\mathbf{k} \geq 0,$$

so $\mathbf{x}^T A$ is in the dual cone of K. But $K' = K$ by Theorem 8.16(a), so $\mathbf{x}^T A \geq \mathbf{0}$. Conclude that Condition X holds. $\square$

To prove Theorem 8.3 from this lemma, let $A, \mathbf{q}$ be a finite financial model. Then $A, \mathbf{q}$ is immediate-arbitrage-free iff Condition X fails to be true, iff Condition Y holds, iff there is a vector $\mathbf{y} \geq \mathbf{0}$ such that $\mathbf{q} = A\mathbf{y}$.

8.3.5 Hyperplane Separation

Farkas's Lemma, and thus one of the Fundamental Theorems on Asset Pricing, use an important geometrical fact about closed convex sets. Notice that its proof does not use the cone property:

Lemma 8.27 (Closed Convex Minimum) *If Q is a nonempty closed convex set in $\mathbf{R}^m$, and $\mathbf{b} \in \mathbf{R}^m$ is any point, then there is a unique point $\mathbf{q}_0 \in Q$ that is closest to $\mathbf{b}$.*

Proof: It may be assumed without loss of generality that $\mathbf{b} = \mathbf{0}$, since the set $Q + \mathbf{b} \stackrel{\text{def}}{=} \{\mathbf{x} + \mathbf{b} : \mathbf{x} \in Q\}$ is closed, convex, and nonempty just like Q, and $\|(\mathbf{x} + \mathbf{b}) - \mathbf{b}\| = \|\mathbf{x}\|$ implies that points in $Q + \mathbf{b}$ are the same distance from $\mathbf{b}$ as points in Q are from $\mathbf{0}$.

Next, note that if $\mathbf{0} \in Q$, then the unique closest point is $\mathbf{q}_0 = \mathbf{0}$ itself, at distance 0.

More generally, let $\delta = \inf\{\|\mathbf{x}\| : \mathbf{x} \in Q\}$ be the greatest lower bound of all distances from points in Q to $\mathbf{0}$. Then there is a sequence $\{\mathbf{x}_n\} \subset Q$ with $\|\mathbf{x}_n\| \to \delta$ as $n \to \infty$. It remains to show that $\mathbf{x}_n$ converges to a point in Q.

But $(\mathbf{x}_i + \mathbf{x}_j)/2 \in Q$ for all i, j, since Q is convex, and so

$$\|\mathbf{x}_i + \mathbf{x}_j\|^2 \geq 4\delta^2$$

for all i, j. But then

$$\|\mathbf{x}_i - \mathbf{x}_j\|^2 = 2\|\mathbf{x}_i\|^2 + 2\|\mathbf{x}_j\|^2 - \|\mathbf{x}_i + \mathbf{x}_j\|^2 \leq 2\|\mathbf{x}_i\|^2 + 2\|\mathbf{x}_j\|^2 - 4\delta^2 \to 0,$$

as $i, j \to \infty$. Thus $\{\mathbf{x}_n\}$ is a Cauchy sequence which, since Q is closed, converges to a limit point $\mathbf{q}_0 \in Q$ satisfying $\|\mathbf{q}_0\| = \delta$. This limit is unique: if $\mathbf{x} \in Q$ also satisfies $\|\mathbf{x}\| = \delta$, then

$$\|\mathbf{x} - \mathbf{q}_0\|^2 \leq 2\|\mathbf{x}\|^2 + 2\|\mathbf{q}_0\|^2 - 4\delta^2 = 0,$$

from which it follows that $\mathbf{x} = \mathbf{q}_0$. $\qquad\square$

Not only is there positive distance between outside points and a closed convex set, there is a whole hyperplane that lies strictly between them. To prove this, use Lemma 8.27 in combination with some tools from Calculus:

Theorem 8.28 (Hyperplane Separation) *Suppose that $Q \subset \mathbf{R}^m$ is a nonempty closed and convex set, and that $\mathbf{b} \in \mathbf{R}^m$ is a point not in Q. Then there exist a nonzero vector $\mathbf{x} \in \mathbf{R}^m$ and a constant $\gamma \in \mathbf{R}$ defining a hyperplane as the zeros of the function*

$$f(\mathbf{y}) \stackrel{\text{def}}{=} \mathbf{x}^T \mathbf{y} - \gamma,$$

such that $f(\mathbf{b}) < 0$ but $f(\mathbf{q}) > 0$ for every $\mathbf{q} \in Q$.

Proof: **Part I:** Find f to construct the hyperplane.

Define $s : \mathbf{R}^m \to \mathbf{R}$ by $s(\mathbf{y}) \stackrel{\text{def}}{=} \|\mathbf{y} - \mathbf{b}\|^2$, continuous and differentiable with gradient

$$\nabla s(\mathbf{y}) = 2(\mathbf{y} - \mathbf{b}) \in \mathbf{R}^m.$$

It achieves its minimum at the nearest point $\mathbf{q}_0 \in Q$ to $\mathbf{b}$, whose existence and uniqueness follow from Lemma 8.27. So, put $f(\mathbf{y}) \stackrel{\text{def}}{=} \mathbf{x}^T \mathbf{y} - \gamma$ for

$$\mathbf{x} = \mathbf{q}_0 - \mathbf{b}, \quad \gamma = \frac{\|\mathbf{q}_0\|^2 - \|\mathbf{b}\|^2}{2}.$$

Hyperplane $\{\mathbf{y} : f(\mathbf{y}) = 0\}$ is normal to $\mathbf{q}_0 - \mathbf{b}$ and passes through the midpoint between $\mathbf{b}$ and $\mathbf{q}_0$.

It remains to show that f separates $\mathbf{b}$ from Q.

Part II: Show that $f(\mathbf{b}) < 0$.

Compute $f(\mathbf{b}) = \mathbf{q}_0^T \mathbf{b} - \frac{\|\mathbf{q}_0\|^2 + \|\mathbf{b}\|^2}{2}$. The Cauchy-Schwartz inequality and the arithmetic-geometric mean inequality together imply

$$\mathbf{q}_0^T \mathbf{b} \leq \|\mathbf{q}_0\| \|\mathbf{b}\| \leq \frac{\|\mathbf{q}_0\|^2 + \|\mathbf{b}\|^2}{2},$$

with equality only if $\mathbf{q}_0 = \mathbf{b}$. Since $\mathbf{b} \neq \mathbf{q}_0$, conclude that $f(\mathbf{b}) < 0$.

Part III: Show that $f(\mathbf{q}) > 0$.

Take any $\mathbf{q} \in Q$ and suppose toward contradiction that $f(\mathbf{q}) \leq 0$. Then

$$(\mathbf{q}_0 - \mathbf{b})^T \mathbf{q} \leq \frac{\|\mathbf{q}_0\|^2 - \|\mathbf{b}\|^2}{2},$$

so $\nabla s(\mathbf{q}_0)^T (\mathbf{q} - \mathbf{q}_0) \leq -\|\mathbf{q}_0 - \mathbf{b}\|^2 < 0$. Hence there is some small $\lambda \in (0, 1)$ for which

$$s(\mathbf{q}_0 + \lambda[\mathbf{q} - \mathbf{q}_0]) < s(\mathbf{q}_0).$$

But Q is convex, so $\mathbf{q}_0 + \lambda[\mathbf{q} - \mathbf{q}_0] = (1 - \lambda)\mathbf{q}_0 + \lambda\mathbf{q} \in Q$, and this contradicts the extremal property of $\mathbf{q}_0$. Conclude that $f(\mathbf{q}) > 0$.   □

8.4   Exercises

1. Prove that any subspace $V \subset \mathbf{R}^n$ is a closed convex cone.

2. Prove that the closed orthant $K \in \mathbf{R}^n$ of vectors with nonnegative coordinates is a closed convex cone.

3. Prove that the pointless orthant $K \setminus \mathbf{0}$ is a convex cone but is neither open nor closed.

4. Prove that K^o is an open convex cone.

5. Prove that the intersection of any collection of convex sets is convex.

6. Prove Theorem 8.16 on p.209:

   (a) $K' = K$, that is, the nonnegative orthant is a self-dual cone.
   (b) $(K^o)' = K$ and $(K^o)^* = K \setminus \mathbf{0}$.
   (c) $(K \setminus \mathbf{0})' = K$ and $(K \setminus \mathbf{0})^* = K^o$.
   (d) $((K^o)^*)^* = K^o$, that is, the open positive orthant is its own strict double dual cone.

7. Prove Eq.8.6:

$$AK = \sum_{i=1}^{n} \bar{V}_i; \qquad AK^o = \sum_{i=1}^{n} V_i,$$

where $A \in \mathbf{R}^{m \times n}$, and K, K^o are the orthants of Definition 6.

8. Prove Corollary 8.18 on p.209: The set S of strictly profitable portfolios is a strict dual cone: $S = (AK^o)^*$

9. Suppose $S \subset \mathbf{R}^n$ is any set. Prove the following:

   (a) $S^\perp$ is a subspace.

   (b) $S^* \subset S'$ and thus $S^* \cap S' = S^*$.

   (c) $S^\perp \subset S'$ and thus $S^\perp \cap S' = S^\perp$.

   (d) $S^\perp \cap S^* = \emptyset$.

   (e) $S^\perp$, S', and S^* are all convex cones.

   (f) If $\mathbf{0} \in S$, then $S^* = \emptyset$. Thus if S is a subspace, then $S^* = \emptyset$.

10. Suppose that $n > 2$ and market model $A, \mathbf{q}$ has

$$A = \begin{pmatrix} R & \cdots & R \\ a_1 & \cdots & a_n \end{pmatrix},$$

where $R > 1$ is the riskless return and $\mathbf{a} = (a_1, \ldots, a_n)$ is a nonconstant payoff vector for the sole risky asset.

(a) Find necessary and sufficient conditions on $\mathbf{q}$ such that $A, \mathbf{q}$ is arbitrage-free. (Hint: use the Fundamental Theorem.)

(b) Exhibit a derivative payoff $\mathbf{d}$ for which no exact hedge exists. (This shows that A is not a complete market.)

(c) Exhibit a derivative $\mathbf{d}$ for which an exact hedge does exist.

11. Suppose that a market model has five states, a riskless asset returning $R = 1.02$, and two risky assets a, b with spot prices $a_0 = 20$ and $b_0 = 12$ and payoffs $\mathbf{a} = (10, 15, 20, 25, 30)$ and $\mathbf{b} = (17, 15, 12, 10, 7)$, respectively.

(a) Prove that the model is arbitrage-free.

(b) Find the no-arbitrage bid-ask interval for a European-style Call option on a with strike price 20.

(c) Find the no-arbitrage bid-ask interval for a European-style Put option on b with strike price 13.

8.5 Further Reading

- Richard F. Bass, *Real Analysis for Graduate Students*, Version 2.1. Published by the author (2014).

- Freddy Delbaen and Walter Schachermayer. "A General Version of the Fundamental Theorem of Asset Pricing." *Mathematische Annalen*, 300:1 (1994), pp.463–520.

- Julius (Gyula) Farkas, "Theorie der Einfachen Ungleichungen." *Journal für die Reine und Angewandte Mathematik*, 124 (1902), pp.1–27.

- Ruth J. Williams. *Introduction to the Mathematics of Finance*. Graduate Texts in Mathematics, Volume 72. American Mathematical Society, Providence (2006).

9

Project Suggestions

These larger-scale assignments supplement the chapter exercises. They combine theorems, algorithms, and experiments to test understanding while also demanding a degree of patience and fortitude. They are presented as alternatives to timed final examinations, allowing students to access and deploy computers without jeopardizing academic integrity.

For these projects, let XYZ denote the common stock of a publicly traded company that offers a dividend. The choice may be left to the students, or else made by the instructor for individuals or small groups. One of the stocks listed among the Standard and Poor's 100 (S&P100) is a good choice, as these are heavily traded with widely published data.

- Choose a time to expiry T that includes at least two dividend payments for XYZ expected to equal the most recent. Use the CRR model for American-style options with dividend to price Call and Put options with expiry T at various near-the-money strike prices. What volatility should be used? How do the modeled prices compare with the market prices for these options? How would the modeled prices change if the second dividend was 20% higher than the first?

- Find the implied volatility $\sigma(K, T)$ of XYZ using options at various near-the-money strike prices K and near-future expiry dates T. Justify your choice of options. Use both CRR and Black-Scholes pricing models and compare the results. Also compare the results to published implied volatilities and comment on any differences. Plot the implied volatility surface $\sigma(K, T)$ for your results.

- Choose a time to expiry T that contains no expected dividends. Construct an implied binomial tree from the spot price and at least five near-the-money Call option premiums on XYZ with expiry T. Use it to price five near-the-money Put options on XYZ with the same expiry T. Compare your results with the market prices of those Puts and also with the values given by the Call-Put parity formula.

A

Answers

A.1 . . . to Chapter 1 Exercises

1. Suppose that a risky asset S has spot price $S(0) = 100$ and that the riskless return to $T = 1$ year is $R = 1.0112$. Assuming there are no arbitrages, compute the following:

   (a) current zero-coupon bond discount $Z(0, T)$,

   (b) Forward price for one share of S at expiry T,

   (c) riskless annual interest rate given continuous compounding.

   Solution: (a) From the formula on p.3, $Z(0, T) = \frac{1}{R} = 0.9889$.

   (b) By Eq.1.15, the fair price is $K = RS(0) = 101.12$.

   (c) By Eq.1.5, $r = (\log R)/T = 0.0111$, or 1.11%.   □

2. With $S(0)$, R, and T as in Exercise 1, suppose that $S(T)$ is modeled by the following table:

$S(T)$	90	95	98	100	102	105	110
$\Pr(S(T))$	0.01	0.04	0.15	0.30	0.30	0.18	0.02

   (a) Use this finite probability space model to estimate premiums $C(0)$ and $P(0)$ for European-style Call and Put options, respectively, with strike price $K = 101$ and expiry T.

   (b) Does Call-Put Parity hold in this model? What might cause it to be inaccurate?

   Solution: (a) First enter the data with these Octave commands:

```
R=1.0112; S0=100; ST=[90 95 98 100 102 105 110];
K=101; Pr=[0.01 0.04 0.15 0.30 0.30 0.18 0.02];
```

   By Eq.1.8, the Call payoff at expiry is $C(T) = [S(T) - K]^+$. By Fair Price Theorem 1.4, the no-arbitrage Call premium is the present value of the expected value of this payoff:

$$C(0) = \mathrm{E}(C(T))/R = \frac{1}{R}\sum [S(T) - K]^+ \Pr(S(T)),$$

which may be computed with Octave/MATLAB commands

```
Cpayoff=max(0,ST-K); C0=(Cpayoff*Pr')/R   % C0 = 1.1867
```

Likewise, by Eq.1.9 the Put payoff is $P(T) = [K - S(T)]^+$ and the no-arbitrage Put premium is therefore:

$$P(0) = \mathrm{E}(P(T))/R = \frac{1}{R}\sum [K - S(T)]^+ \; \mathrm{Pr}(S(T)),$$

which may be computed with the commands

```
Ppayoff=max(0,K-ST); P0=(Ppayoff*Pr')/R   % P0=1.0878
```

(b) Observe that Call-Put parity does not quite hold:

```
C0-P0    % ans =  0.098892
S0-K/R   % ans =  0.11867
```

The difference is approximately 0.02 instead of exacly 0 as in Eq.1.16. This is the difference between $E(S(T)) = 101.10$, which is the model's estimate for the Forward price of S, and $RS(0) = 101.12$, which is another Forward estimate using riskless return.   □

3. Use the no-arbitrage Axiom 1 to prove that Eq.1.7 holds.

   Solution: If the asset A costs more, then short-sell A for $A(0)$, buy the sequence of zero-coupon bonds, pocket the surplus, and use the proceeds from the bonds to pay the cash flow to the A buyer over time, settling all liabilities.

   Otherwise, if the asset A costs less, then short-sell the bonds, use the money to buy A for $A(0)$, pocket the surplus, and then pay the bond buyers back over time to settle all obligations.

   In either case there is an arbitrage, contradicting Axiom 1.   □

4. Prove Corollary 1.3. (Hint: review the proof of Theorem 1.2.)

   Solution: Recall the statement of Corollary 1.3: If at some time $T > 0$, $A(T, \omega) > B(T, \omega)$ in all states ω, then $A(t) > B(t)$ for all times $0 \leq t < T$.

   Proof: Suppose not. Then there is a time t with $0 \leq t < T$ such that $A(t) \leq B(t)$, so assemble the portfolio $-A + B$ by selling B short and buying A. Observe that $-A(t) + B(t) \leq 0$ costs nothing and may even leave a surplus.

   At time T, sell A for $A(T, \omega)$ and buy B for $B(T, \omega)$ to cover the short. By hypothesis this pays off $A(T, \omega) - B(T, \omega) > 0$ in all states ω. The assembled portfolio would thus be an arbitrage opportunity, contradicting Axiom 1. Conclude that $A(t) > B(t)$ for all times $0 \leq t \leq T$.   □

5. Suppose that $C(0) - P(0) < S(0) - K/R$, in contradiction with Eq.1.16. Construct an arbitrage.

   Solution: At $t = 0$, starting with no money, do the following:

   - Short-sell P for $P(0)$ cash.
   - Short-sell S for $S(0)$ cash.
   - Buy C for $C(0)$ cash.
   - Deposit K/R cash into the bank at riskless return R.

   That leaves $S(0) - K/R - C(0) + P(0) > 0$ cash. Then at $t = T$, clear all debts as follows:

   - If $S(T) > K$ then exercise C for a profit of $S(T) - K$. Otherwise C expires worthless.
   - If $K > S(T)$, then cash-settle the short-sold P for $K - S(T) > 0$. Otherwise P expires worthless.
   - Withdraw $RK/R = K$ cash from the bank, including interest.
   - Cover the short-sale of S for $S(T)$ cash.

   That leaves $C(T) - P(T) + K - S(T) = [S(T) - K]^+ - [K - S(T)]^+ + K - S(T) = 0$ and no debts in all cases. The positive amount obtained at $t = 0$ is therefore an arbitrage profit forbidden by Axiom 1.   □

6. Prove Eq.1.20, the Call-Put Parity Formula for foreign exchange options:

$$C(0) - P(0) = \frac{X(0)}{R_f} - \frac{K}{R_d}.$$

   Use the no-arbitrage axiom.

   Solution: To prove equality, show that inequality in either direction results in an arbitrage opportunity.

   Case 1: $C(0) - P(0) > X(0)/R_f - K/R_d$.

   At $t = 0$, starting with no money, perform the following trades:

   - Short-sell C for $C(0)$ DOM cash.
   - Buy P for $P(0)$ DOM cash.
   - Borrow K/R_d DOM cash at R_d from the domestic bank.
   - Convert $X(0)/R_f$ DOM to $1/R_f$ FRN at spot exchange rate $X(0)$.
   - Deposit $1/R_f$ FRN cash into the foreign bank at R_f.

   That leaves $C(0) - P(0) - X(0)/R_f + K/R_d > 0$ DOM in cash, a net profit to keep.

   At $t = T$, clear all debts as follows:

   - Withdraw 1 FRN cash from the foreign bank, including interest.

- If $K \geq X(T)$, then
  - Short-sold C expires worthless, as $C(T) = [X(T) - K]^{+} = 0$ imposes no liability.
  - Exercise P to convert 1 FRN into K DOM.
- Else $K < X(T)$, so
  - Convert 1 FRN to $X(T)$ DOM at the market rate $X(T)$.
  - Cash settle C for $C(T) = [X(T) - K]^{+} = X(T) - K$.
  - Do not exercise P, as $P(T) = [K - X(t)]^{+} = 0$ is worthless.
- Repay the domestic bank loan with interest for K DOM cash.

That leaves $-[X(T) - K]^{+} + [K - X(T)]^{+} - K + X(T) = 0$ and no unfunded liabilities. The positive amount obtained at $t = 0$ is therefore an arbitrage profit prohibited by Axiom 1.

Case 2: $C(0) - P(0) < X(0)/R_f - K/R_d$.

At $t = 0$, starting with no money, perform the following trades:

- Short-sell P for $P(0)$ cash.
- Buy C for $C(0)$ cash.
- Borrow $1/R_f$ FRN from the foreign bank at R_f.
- Convert $1/R_f$ FRN to $X(0)/R_f$ DOM at spot exchange rate $X(0)$.
- Deposit K/R_d DOM cash into the domestic bank.

That leaves $X(0)/R_f - K/R_d - C(0) + P(0) > 0$ DOM cash, a net profit to keep.

At $t = T$, clear all debts as follows:

- Withdraw K DOM cash from the domestic bank, including interest.
- If $K \geq X(T)$, then
  - Do not exercise the worthless C, as $C(T) = [X(T) - K]^{+} = 0$.
  - Cash settle P for its value $P(T) = [K - X(T)]^{+} = K - X(T)$.
  - Convert $X(T)$ DOM to 1 FRN cash at the market rate $X(T)$.
- Else $K < X(T)$, so
  - Short-sold P expires worthless, as $P(T) = [K - X(T)]^{+} = 0$.
  - Exercise C to convert K DOM to 1 FRN cash at rate K.
- Repay the foreign bank loan with interest for 1 FRN cash.

That leaves $[K - X(T)]^{+} - [X(T) - K]^{+} + K - X(T) = 0$ and no unfunded liabilities. The positive amount obtained at $t = 0$ is therefore an arbitrage profit prohibited by Axiom 1.

Conclude that Eq.1.20 holds. $\square$

7. (a) Prove that the plus-part function satisfies Eq.1.17:

$$[X]^+ - [-X]^+ = X,$$

for any number X.

(b) Apply the identity in part (a) to the payoff values of European-style Call and Put options for S at strike price K and expiry T to show Eq.1.18:

$$C(T) - P(T) = S(T) - K.$$

Solution: (a) Let X be any number. Then either $X < 0$, $X = 0$, or $X > 0$. Check all three possibilities:

$$[X]^+ - [-X]^+ = \begin{cases} 0 - (-X) = X, & X < 0, \\ 0 + 0 = X, & X = 0, \\ X - 0 = X, & X > 0, \end{cases} = X,$$

in all cases.

(b) Let $X = S(T) - K$. Then $C(T) = [S(T) - K]^+ = [X]^+$ while $P(T) = [K - S(T)]^+ = [-X]^+$. Apply part (a) to conclude that

$$C(T) - P(T) = [X]^+ - [-X]^+ = X = S(T) - K,$$

as claimed.   □

8. Plot the payoff and profit graphs for the following colorfully named option portfolios as a function of the price $S(T)$ at expiry time T:

(a) *Long straddle:* buy one Call and one Put on S with the same expiry T and at-the-money strike price $K \approx S(0)$. For what values of $S(T)$ will this be profitable?

(b) *Long strangle:* buy one Call at K_c and one Put at K_p with the same expiry T but with out-of-the-money strike prices $K_p < S(0) < K_c$. How does its profitability compare with that of a long straddle?

Solution: (a) See the payoff and profit graph for a Long straddle in Figure A.1.

The straddle will be profitable if $S(T)$ is sufficiently far from $S(0)$ in either direction, namely more than the sum of the two premiums.

(b) See the Long strangle payoff and profit graph in Figure A.2.

Since the two strangle options are out-of-the-money, their premiums are lower than the at-the-money options in a straddle, making the strangle portfolio cheaper. However, for the strangle to be profitable, the stock price may have to move farther.   □

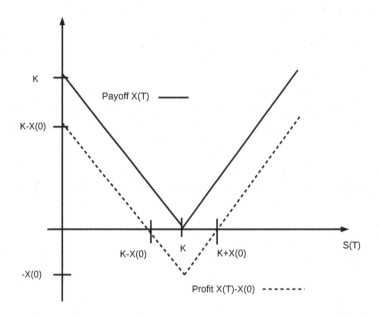

FIGURE A.1
Payoff and profit for a Long straddle portfolio $X = C + P$.

9. A *butterfly spread* is a portfolio of European-style Call options purchased at time $t = 0$ with the same expiry $t = T$ but at three strike prices $L < M < H$, where $M = \frac{1}{2}(L + H)$:

   - buy one Call C_L at strike price L for $C_L(0)$,
   - buy one Call C_H at strike price H for $C_H(0)$,
   - sell two Calls C_M short at strike price M for $2C_M(0)$.

   (a) Plot the payoff graph for the butterfly spread at expiry when its price is $C_L(T) + C_H(T) - 2C_M(T)$. Mark the three strike prices on the $S(T)$ axis.

   (b) Conclude that $C_M(0) < \frac{1}{2}[C_L(0) + C_H(0)]$. (Hint: apply the no-arbitrage Axiom 1 to the graph plotted in part (a).)

   Solution: (a) See the payoff graph for a butterfly spread and its component Calls in Figure A.3.

   (b) The net cost to assemble a butterfly spread at $t = 0$ is $C_L(0) + C_H(0) - 2C_M(0)$. If $C_M(0) \geq \frac{1}{2}[C_L(0) + C_H(0)]$, then this net cost is less than or equal to zero. But its value at $t = T$ is nonnegative in all states and positive in some states, namely if $L < S(T) < H$, so this is

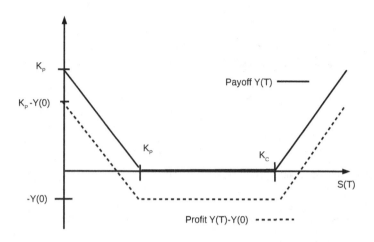

FIGURE A.2
Payoff and profit for a Long strangle portfolio $Y = C_{K_C} + P_{K_P}$.

an arbitrage opportunity. Conclude by the no-arbitrage Axiom 1 that
$C_M(0) < \frac{1}{2}[C_L(0) + C_H(0)]$. □

10. An *iron condor* is a portfolio $C_1 - C_2 - P_3 + P_4$ of four European-style options. To construct it, simultaneously buy one Call at K_1, sell one Call at K_2, sell one Put at K_3, and buy one Put at K_4, all with the same expiry T but with $K_1 < K_2 < K_3 < K_4$.

(a) Describe the payoff graph for an iron condor portfolio at expiry.

(b) Assuming no arbitrage, prove that the portfolio must have a positive net premium.

(c) Assuming no arbitrage, find inequalities bounding the maximum profit and the maximum loss of an iron condor portfolio at expiry.

Solution: (a) Let $icp(S(T))$ be the payoff at expiry T of the iron condor portfolio, as a function of the underlying asset's price $S(T)$. By Eqs.1.8

Individual European Call Option Payoffs

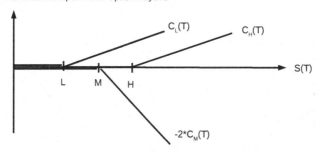

Payoff of Butterfly Spread C(T)=C$_L$(T)+C$_H$(T)-2C$_M$(T)

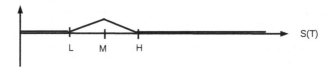

FIGURE A.3
Payoffs for the butterfly spread $C = C_L + C_H - 2C_M$ and its components.

and 1.9,

$$
\begin{aligned}
icp(s) &= C_1(T) - C_2(T) - P_3(T) + P_4(T) \\
&= [s - K_1]^+ - [s - K_2]^+ - [K_3 - s]^+ + [K_4 - s]^+ \\
&= \begin{cases}
(K_4 - K_3), & s \le K_1, \\
(s - K_1) + (K_4 - K_3), & K_1 < s < K_2, \\
(K_2 - K_1) + (K_4 - K_3), & K_2 \le s \le K_3, \\
(K_2 - K_1) + (K_4 - s), & K_3 < s < K_4, \\
(K_2 - K_1), & s \ge K_4.
\end{cases}
\end{aligned}
$$

(b) From part (a), the payoff is strictly positive. By Corollary 1.3, the net premium must therefore be positive.

(c) Let $y = C_1(0) + P_4(0) - C_2(0) - P_3(0)$ be the net premium for the iron condor portfolio. By part (b) it must be positive, and it must be subtracted from the payoff to give the profit at expiry. Hence the profit must be less than $(K_2 - K_1) + (K_4 - K_3)$.

The minimum net value at expiry, which is the worst possible loss, will be $\min\{(K_2 - K_1), (K_4 - K_3)\} - y$. But y is bounded above by the no-arbitrage assumption since if $y > (K_2 - K_1) + (K_4 - K_3)$, then the profit graph will lie entirely below the x-axis. Such an asset is guaranteed to lose, so selling it short would be an arbitrage opportunity. Hence $y \leq (K_2 - K_1) + (K_4 - K_3)$, so the loss is bounded below by

$$\min\{(K_2-K_1), (K_4-K_3)\} - y \geq -\max\{(K_2-K_1), (K_4-K_3)\}.$$

Equality holds for $S(T) \leq K_1$ if K_4-K_3 is bigger, or for $S(T) \geq K_4$ if K_2-K_1 is bigger. $\qquad\square$

A.2   ...to Chapter 2 Exercises

1. Let

$$\Phi(x) = \frac{1}{\sqrt{2\pi}} \int_{-\infty}^{x} e^{-t^2/2}\, dt$$

be the cumulative distribution function of the standard normal random variable. Prove that

$$1 - \Phi(x) = \Phi(-x)$$

for every x.

Solution: This result holds for every cumulative distribution function with a density function that is symmetric about $t = 0$. In particular, that is true for $\frac{1}{\sqrt{2\pi}} e^{-t^2/2}$:

$$
\begin{aligned}
\Phi(x) &= \frac{1}{\sqrt{2\pi}} \int_{-\infty}^{x} e^{-t^2/2}\, dt \\
&= \frac{1}{\sqrt{2\pi}} \int_{-\infty}^{\infty} e^{-t^2/2}\, dt - \frac{1}{\sqrt{2\pi}} \int_{x}^{\infty} e^{-t^2/2}\, dt \\
&= 1 - \frac{1}{\sqrt{2\pi}} \int_{x}^{\infty} e^{-t^2/2}\, dt \\
&= 1 - \frac{1}{\sqrt{2\pi}} \int_{-\infty}^{-x} e^{-t^2/2}\, dt = 1 - \Phi(-x),
\end{aligned}
$$

where the last step employs the change of variable $t \leftarrow -t$. $\qquad\square$

2. Rewrite the recursive definition of the random walk X in Section 2.2 with these normalizations:

   - Multiply by $1/\sqrt{n/4} = 2/\sqrt{n}$ to have variance 1.
   - Change the time step to $1/n$ so that the time interval is $[0, 1]$.

Solution: Renormalize $X(t,\omega)$ to get the continuous piecewise linear function $\hat{X}$ on $0 \le t \le 1$ defined recursively by:

$$\hat{X}(0,\omega) \;=\; 0, \qquad \text{all } \omega \in \Omega;$$

$$\hat{X}\left(\frac{i+1}{n},\omega\right) \;=\; \begin{cases} \hat{X}\left(\frac{i}{n},\omega\right) - \frac{2}{\sqrt{n}}, & \omega_i = 0, \\ \hat{X}\left(\frac{i}{n},\omega\right) + \frac{2}{\sqrt{n}}, & \omega_i = 1, \end{cases} \qquad i = 0,1,\ldots,n-1,$$

with values at intermediate points $\frac{i}{n} < t < \frac{i+1}{n}$ defined by linear interpolation:

$$\hat{X}(t,\omega) = \left[t - \frac{i}{n}\right]\hat{X}\left(\frac{i+1}{n},\omega\right) + \left[\frac{i+1}{n} - t\right]\hat{X}\left(\frac{i}{n},\omega\right).$$

By the same analysis as for X, renormalized $\hat{X}(1,\omega)$ takes values in this set of $n+1$ reachable points:

$$\left\{\hat{X}_k \overset{\text{def}}{=} -2\sqrt{n} + 2k\frac{2}{\sqrt{n}} : k = 0,1,\ldots,n\right\}.$$

It has the same binomial probability mass function as X on its reachable points:

$$\Pr\left(\hat{X}(1,\omega) = \hat{X}_k\right) = \frac{1}{2^n}\binom{n}{k} = \binom{n}{k}\left(\frac{1}{2}\right)^k\left(\frac{1}{2}\right)^{n-k}.$$

By symmetry, conclude that $\mathrm{E}(\hat{X}(1,\cdot)) = 0 = \mathrm{E}(X(n,\cdot))$. Then compute the variance by observing that

$$\hat{X}(1,\omega) = -2\sqrt{n} + 2k\frac{2}{\sqrt{n}} = \frac{2}{\sqrt{n}}(-n+2k) = \frac{2}{\sqrt{n}}X(n,\omega),$$

so by Eqs.2.6 and 2.7, $\mathrm{Var}(\hat{X}(1,\cdot)) = \left(\frac{2}{\sqrt{n}}\right)^2 \mathrm{Var}(X(n,\cdot)) = 1$. $\square$

3. Derive the Black-Scholes formula for European-style Put options, Eq.2.26:
$$P(0) = e^{-rT}K\Phi(-d_2) - S_0\Phi(-d_1).$$

(Hint: follow the steps in Section 2.3.1, but use the Put payoff $[K - S(T)]^+$ at expiry.)

Solution: Under the assumptions of no expected arbitrage and constant riskless rate r, the premium for a European Put option with strike price K and expiry time T is

$$P(0) = e^{-rT}\,\mathrm{E}(P(T)), \qquad \text{where} \quad P(T) = [K - S(T)]^+. \qquad \text{(A.1)}$$

Compute this expectation using the lognormal p.d.f. q assumed by the Black-Scholes model of the future. So, integrate $[K - s]^+ q(s)$ over all positive prices s that S may take:

$$
\begin{aligned}
P(0) &= e^{-rT} \int_0^\infty [K - s]^+ q(s)\, ds \\
&= e^{-rT} \int_0^K (K - s) q(s)\, ds \\
&= e^{-rT} \left[K \int_0^K q(s)\, ds - \int_0^K s q(s)\, ds \right].
\end{aligned}
$$

Apply Eq.2.16 to compute the left-hand integral:

$$
\int_0^K q(s)\, ds = Q(K) = \Phi \left(\frac{\log K - \mu}{\sigma} \right).
$$

The other integral may be evaluated by substituting $s \leftarrow e^t$ in Eq.2.13, then completing squares and factoring out constant terms:

$$
\begin{aligned}
\int_0^K s q(s)\, ds &= \frac{1}{\sigma \sqrt{2\pi}} \int_0^K \exp \left(-\frac{[\log s - \mu]^2}{2\sigma^2} \right) ds \\
&= \frac{1}{\sigma \sqrt{2\pi}} \int_{-\infty}^{\log K} \exp \left(-\frac{[t - \mu]^2}{2\sigma^2} \right) e^t\, dt \\
&= \exp \left(\mu + \frac{\sigma^2}{2} \right) \frac{1}{\sigma \sqrt{2\pi}} \int_{-\infty}^{\log K} \exp \left(-\frac{[t - (\mu + \sigma^2)]^2}{2\sigma^2} \right) dt \\
&= \exp \left(\mu + \frac{\sigma^2}{2} \right) \Phi \left(\frac{\log K - (\mu + \sigma^2)}{\sigma} \right),
\end{aligned}
$$

recognizing the c.d.f. of $\mathcal{N}((\mu + \sigma^2), \sigma^2)$. Combining the two parts gives

$$
\begin{aligned}
P(0) = e^{-rT} \Bigg[&K \Phi \left(\frac{\log K - \mu}{\sigma} \right) \\
&- \exp \left(\mu + \frac{\sigma^2}{2} \right) \Phi \left(\frac{\log K - (\mu + \sigma^2)}{\sigma} \right) \Bigg].
\end{aligned}
$$

Using Eq.2.11 with $t = T$ and the identities in Eqs.2.22 gives

$$
\begin{aligned}
\log K - \mu &= -\log \frac{S_0}{K} - \left(r - \frac{v^2}{2} \right) t, \\
\log K - (\mu + \sigma^2) &= -\log \frac{S_0}{K} - \left(r + \frac{v^2}{2} \right) T, \\
\exp \left(\mu + \frac{\sigma^2}{2} \right) &= S_0 \exp(rT),
\end{aligned}
$$

so the Put premium should be

$$P(0) = e^{-rT} K\Phi \left(\frac{-\log \frac{S_0}{K} - \left(r - \frac{v^2}{2}\right)T}{v\sqrt{T}} \right)$$

$$-S_0\Phi \left(\frac{-\log \frac{S_0}{K} - \left(r + \frac{v^2}{2}\right)T}{v\sqrt{T}} \right).$$

This may be written using d_1 and d_2 from Eq.2.24 as

$$P(0) = e^{-rT} K\Phi(-d_2) - S_0\Phi(-d_1),$$

which is the Black-Scholes formula for European-style Puts. $\square$

4. Let d_1, d_2 be defined as in Eq.2.24, and let ϕ be the standard normal p.d.f. defined in Eq.2.18. Show that

$$S_0\phi(d_1) - Ke^{-rT}\phi(d_2) = 0.$$

(Hint: Notice that $d_1 - d_2 = v\sqrt{T}$ and

$$d_1 + d_2 = 2(\log \frac{S_0}{K} + rT)/(v\sqrt{T}),$$

and thus $(d_1^2 - d_2^2)/2 = (d_1 - d_2)(d_1 + d_2)/2 = \log(S_0/K) + rT.$)

Solution: Expand $\sqrt{2\pi}\left[S_0\phi(d_1) - Ke^{-rT}\phi(d_2)\right]$ to get

$$S_0 e^{-d_1^2/2} - Ke^{-rT}e^{-d_2^2/2} = e^{-d_2^2/2}\left(S_0 e^{(d_2^2 - d_1^2)/2} - Ke^{-rT}\right).$$

Now $(d_2^2 - d_1^2)/2 = -\log(S_0/K) - rT$ from the hint, so

$$S_0\exp(-\log \frac{S_0}{K} - rT) - Ke^{-rT} = S_0\frac{K}{S_0}e^{-rT} - Ke^{-rT} = 0,$$

proving the result. $\square$

5. Derive Δ_C and Δ_P in Section 2.3.3 by differentiating the Black-Scholes formulas.

Solution: First, differentiate the Black-Scholes $C(0)$ formula, Eq.2.25, with respect to S_0:

$$\Delta_C = \frac{\partial C(0)}{\partial S_0} = \Phi(d_1) + S\Phi'(d_1)\frac{\partial d_1}{\partial S_0} - Ke^{-rt}\Phi'(d_2)\frac{\partial d_2}{\partial S_0},$$

by the product rule and the chain rule. Also, from Eq.2.24,

$$\frac{\partial d_1}{\partial S_0} = \frac{\partial d_2}{\partial S_0} = \frac{1}{S_0 v\sqrt{T}}.$$

Next, by the Fundamental Theorem of Calculus,

$$\Phi'(d_1) = \phi(d_1) = \frac{1}{\sqrt{2\pi}}e^{-d_1^2/2}, \quad \Phi'(d_2) = \phi(d_2) = \frac{1}{\sqrt{2\pi}}e^{-d_2^2/2},$$

where $\phi(x) = \exp(-x^2/2)/\sqrt{2\pi}$ is the standard normal density. These may be combined to give

$$\Delta_C = \Phi(d_1) + \frac{S_0\phi(d_1) - Ke^{-rT}\phi(d_2)}{S_0 v\sqrt{T}}.$$

But $S_0\phi(d_1) - Ke^{-rT}\phi(d_2) = 0$ by Exercise 4, so the second term vanishes. Conclude that $\Delta_C = \Phi(d_1)$, as claimed.

Finally, use the Call-Put Parity formula to find Δ_P:

$$C(0) - P(0) = S_0 - Ke^{-rT} \quad \Longrightarrow \quad P(0) = C(0) - S_0 + Ke^{-rT}.$$

The partial derivative with respect to S_0 yields $\Delta_P = \Delta_C - 1 = \Phi(d_1) - 1$, also as claimed. $\qquad\square$

6. Derive Γ_C and Γ_P in Section 2.3.3 from Black-Scholes.

Solution: Compute Gammas by differentiating the Deltas with respect to S_0. Using Eq.2.28 for Δ_C,

$$\Gamma_C = \frac{\partial \Delta_C}{\partial S_0} = \frac{\partial \Phi(d_1)}{\partial S_0} = \phi(d_1)\frac{\partial d_1}{\partial S_0} = \frac{\phi(d_1)}{S_0 v\sqrt{T}}.$$

Likewise, using Eq.2.29 for Δ_P,

$$\Gamma_P = \frac{\partial \Delta_P}{\partial S_0} = \frac{\partial}{\partial S_0}[\Phi(d_1) - 1] = \phi(d_1)\frac{\partial d_1}{\partial S_0} = \frac{\phi(d_1)}{S_0 v\sqrt{T}}.$$

These both use $\frac{\partial d_1}{\partial S_0} = \frac{1}{S_0 v\sqrt{T}}$, found while computing Deltas. $\qquad\square$

7. Derive Θ_C and Θ_P in Section 2.3.3 from Black-Scholes.

Solution: Begin by differentiating Eq.2.25 with respect to T:

$$\Theta_C = -\frac{\partial C}{\partial T} = -S_0\Phi'(d_1)\frac{\partial d_1}{\partial T} - Kre^{-rT}\Phi(d_2) + Ke^{-rT}\Phi'(d_2)\frac{\partial d_2}{\partial T}.$$

Then differentiate Eqs.2.24 with respect to T:

$$\frac{\partial d_1}{\partial T} = \frac{r + \frac{v^2}{2} - \frac{1}{T}\log\frac{S_0}{K}}{2v\sqrt{T}} \quad \text{and} \quad \frac{\partial d_2}{\partial T} = \frac{r - \frac{v^2}{2} - \frac{1}{T}\log\frac{S_0}{K}}{2v\sqrt{T}}.$$

Substitute and write $v^2/2 = v^2 - v^2/2$ to get

$$\Theta_C = -S_0\phi(d_1)\frac{r + \frac{v^2}{2} - \frac{1}{T}\log\frac{S_0}{K}}{2v\sqrt{T}} - Kre^{-rT}\Phi(d_2)$$

$$+ Ke^{-rT}\phi(d_2)\frac{r - \frac{v^2}{2} - \frac{1}{T}\log\frac{S_0}{K}}{2v\sqrt{T}}$$

$$= \frac{-S_0\phi(d_1)v^2}{2v\sqrt{T}} - Kre^{-rT}\Phi(d_2)$$

$$- \frac{[S_0\phi(d_1) - Ke^{-rT}\phi(d_2)](r - \frac{v^2}{2} - \frac{1}{T}\log\frac{S_0}{K})}{2v\sqrt{T}},$$

after rearrangement. But $S_0\phi(d_1) - Ke^{-rT}\phi(d_2) = 0$ by Exercise 4, so the last term vanishes, leaving

$$\Theta_C = \frac{-S_0\phi(d_1)v}{2\sqrt{T}} - Kre^{-rT}\Phi(d_2),$$

as claimed.

To find Θ_P, rearrange the Call-Put Parity formula for European-style options as follows:

$$P(0) = C(0) - S_0 + Ke^{-rT}.$$

Then applying $-\partial/\partial T$ to both sides yields

$$\Theta_P = \Theta_C + Kre^{-rT} = \frac{-S_0\phi(d_1)v}{2\sqrt{T}} - Kre^{-rT}[\Phi(d_2) - 1].$$

But $\Phi(x) - 1 = -\Phi(-x)$ for any x by Eq.2.19, so

$$\Theta_P = \frac{-S_0\phi(d_1)v}{2\sqrt{T}} + Kre^{-r*T}\Phi(-d_2),$$

as claimed.   □

8. Derive κ_C and κ_P in Section 2.3.3 from Black-Scholes.

Solution: For these "Vegas," differentiate Eqs.2.25 and 2.26 with respect to volatility v:

$$\kappa_C = \frac{\partial C}{\partial v} = S_0\Phi'(d_1)\frac{\partial d_1}{\partial v} - Ke^{-rT}\Phi'(d_2)\frac{\partial d_2}{\partial v}.$$

Then, either by hand or using Macsyma, compute

$$\frac{\partial d_1}{\partial v} = -\frac{d_1}{v} + \sqrt{T} \qquad \text{and} \qquad \frac{\partial d_2}{\partial v} = -\frac{d_2}{v} - \sqrt{T},$$

so
$$\kappa_C = S_0\phi(d_1)[-\frac{d_1}{v} + \sqrt{T}] - Ke^{-rT}\phi(d_2)[-\frac{d_2}{v} - \sqrt{T}],$$

and after rearrangement,

$$= -[S_0\phi(d_1) - Ke^{-rT}\phi(d_2)][-\frac{d_2}{v} - \sqrt{T}] + S_0\phi(d_1)[\frac{(d_2 - d_1)}{v} + 2\sqrt{T}].$$

The first term vanishes since $S_0\phi(d_1) - Ke^{-rT}\phi(d_2) = 0$, while the second term simplifies using $d_2 - d_1 = -v\sqrt{T}$, leaving

$$\kappa_C = S_0\phi(d_1)\sqrt{T}.$$

Differentiating the Call-Put Parity formula $C(0) - P(0) = S_0 - Ke^{-rT}$ then gives
$$\kappa_P = \kappa_C = S_0\phi(d_1)\sqrt{T},$$

since the difference does not depend on v.   □

9. Derive ρ_C and ρ_P in Section 2.3.3 from Black-Scholes.

   Solution: Differentiate Eq.2.25 with respect to interest rate r to get

$$\rho_C = \frac{\partial C}{\partial r} = S_0\Phi'(d_1)\frac{\partial d_1}{\partial r} - K(-T)e^{-rT}\Phi(d_2) - Ke^{-rT}\Phi'(d_2)\frac{\partial d_2}{\partial r}.$$

By hand or using Macsyma, compute

$$\frac{\partial d_1}{\partial r} = \frac{\partial d_2}{\partial r} = \frac{\sqrt{T}}{v},$$

so
$$\rho_C = S_0\phi(d_1)\frac{\sqrt{T}}{v} + KTe^{-rT}\Phi(d_2) - Ke^{-rT}\phi(d_2)\frac{\sqrt{T}}{v},$$

which, after simplification and rearrangement, gives

$$= [S_0\phi(d_1) - Ke^{-rT}\phi(d_2)]\frac{\sqrt{T}}{v} + KTe^{-rT}\Phi(d_2).$$

The first term vanishes because $S_0\phi(d_1) - Ke^{-rT}\phi(d_2) = 0$ by Exercise 4, leaving
$$\rho_C = KTe^{-rT}\Phi(d_2).$$

Using the Call-Put Parity Formula as $P(0) = C(0) - S_0 + Ke^{-rT}$ and differentiating both sides with respect to r gives

$$\rho_P = \rho_C - 0 - KTe^{-rT} = KTe^{-rT}[\Phi(d_2) - 1] = -KTe^{-rT}\Phi(-d_2)$$

as claimed, using Eq.2.19 at the last step.   □

10. Implement the computation of all the Black-Scholes Greeks in Octave or MATLAB and add this functionality to the program BS() in Section 2.4. Apply it to compute the premiums and all Greeks for European-style Call and Put options on a risky asset with the following parameters: spot price \$90, strike price \$95, expiry in 1 year, annual riskless rate 2%, volatility 15%.

Solution: Start with the Macsyma program on p.34 and translate it to augment the Octave program BS() into the following:

```
1   function [Op,De,Ga,Th,Ve,Rh]=BSG(T,S0,K,r,v)
2   % Octave/MATLAB function to evaluate
3   % the partial derivatives, or "Greeks"
4   % for European-style Call and Put options
5   % using the Black-Scholes formulas.
6   % INPUTS:                             (Example)
7   %    T  = time to expiry              (1 year)
8   %    S0 = asset spot price            ($90)
9   %    K  = strike price                ($95)
10  %    r  = riskless APR                (0.02)
11  %    v  = volatility                  (0.15)
12  % OUTPUTS:
13  %    Op = [C0; P0]          Option premiums
14  %    De = [DeltaC; DeltaP]  Deltas
15  %    Ga = [GammaC; GammaP]  Gammas
16  %    Th = [ThetaC; ThetaP]  Thetas
17  %    Ve = [VegaC; VegaP]    Vegas=kappas
18  %    Rh = [RhoC; RhoP]      rhos
19  % EXAMPLE:
20  %    [Op,De,Ga,Th,Ve,Rh]=BSG(1,90,95,0.02,0.15)
21  %
22      normcdf = @(x) 0.5*(1.0+erf(x/sqrt(2.0)));
23      normpdf = @(x) exp(-x^2/2)/sqrt(2*pi);
24      d1 = (log(S0/K)+T*(r+v^2/2)) / (v*sqrt(T));
25      d2 = (log(S0/K)+T*(r-v^2/2)) / (v*sqrt(T));
26      C0 = S0*normcdf(d1)-K*exp(-r*T)*normcdf(d2);
27      P0 = K*exp(-r*T)*normcdf(-d2)-S0*normcdf(-d1);
28      DeltaC = normcdf(d1); DeltaP = normcdf(d1)-1;
29      GammaP = GammaC = normpdf(d1)/(v*S0*sqrt(T));
30      boTh = -v*S0*normpdf(d1)/(2*sqrt(T)); % common
31      ThetaC = boTh - r*exp(-r*T)*K*normcdf(d2);
32      ThetaP = boTh + r*exp(-r*T)*K*normcdf(-d2);
33      VegaP = VegaC = S0*normpdf(d1)*sqrt(T);
34      RhoC =  T*exp(-r*T)*K*normcdf(d2);
35      RhoP = -T*exp(-r*T)*K*normcdf(-d2);
36      Op = [C0; P0];   De = [DeltaC; DeltaP];
37      Ga = [GammaC; GammaP]; Th = [ThetaC; ThetaP];
38      Ve = [VegaC; VegaP];   Rh = [RhoC; RhoP];
39      return
40  end
```

Use this augmented function BSG() to compute the option premiums and all the Greeks for the parameters given in this exercise:

```
T=1; S0=90; K=95; r=0.02; v=0.15;
[Op,De,Ga,Th,Ve,Rh] = BSG(T,S0,K,r,v)
```

The output is an array of six column vectors as in this table:

	Option	Δ=De	Γ=Ga	Θ=Th	κ=Ve	ρ=Rh
Call	4.0548	0.43955	0.029211	−3.3720	35.492	35.505
Put	7.1736	−0.56045	0.029211	−1.5096	35.492	−57.614

As a check for typographical errors, repeat the calculation in Macsyma after defining the Greeks as on p.34:

```
float(ev(BSCall,K=95,S0=90,r=0.02,T=1,v=0.15)); /* 4.0548 */
float(ev(DeltaC,K=95,S0=90,r=0.02,T=1,v=0.15)); /* 0.4395 */
float(ev(GammaC,K=95,S0=90,r=0.02,T=1,v=0.15)); /* 0.0292 */
float(ev(ThetaC,K=95,S0=90,r=0.02,T=1,v=0.15)); /* -3.372 */
float(ev(VegaC,K=95,S0=90,r=0.02,T=1,v=0.15));  /* 35.492 */
float(ev(RhoC,K=95,S0=90,r=0.02,T=1,v=0.15));   /* 35.505 */
```

The returned values are in agreement with those computed by Octave for the Call option premium and Greeks. □

11. (a) Verify Eq.2.34 relating Θ_C, Δ_C, and Γ_C.

(b) Find the equivalent relation for Puts and verify it.

Solution: (a) Substitute Eqs.2.28, 2.30, and 2.32 into $\Theta_C + rS_0\Delta_C + \frac{1}{2}v^2S_0^2\Gamma_C - rC(0)$ to get:

$$\frac{-S_0\phi(d_1)v}{2\sqrt{T}} - Kre^{-rT}\Phi(d_2) + rS_0\Phi(d_1) + \frac{1}{2}v^2S_0^2\frac{\phi(d_1)}{S_0v\sqrt{T}} - rC(0).$$

But this equals

$$-Kre^{-rT}\Phi(d_2) + rS_0\Phi(d_1) - rC(0) \quad = \quad 0,$$

since $C(0) = S_0\Phi(d_1) - Ke^{-rT}\Phi(d_2)$ by Eq.2.25, the Black-Scholes Call pricing formula.

(b) The equivalent relation for Puts simply replaces $C(0)$ with $P(0)$ and uses the Put Greeks:

$$\Theta_P + rS_0\Delta_P + \frac{1}{2}v^2S_0^2\Gamma_P - rP(0) = 0.$$

Check it by substituting Eqs.2.29, 2.31, and 2.33 into the left hand side to get:

$$-\frac{vS_0\phi(d_1)}{2\sqrt{T}} + re^{-rT}K\Phi(-d_2)$$

$$+rS_0[\Phi(d_1) - 1] + \frac{1}{2}v^2S_0^2\frac{\phi(d_1)}{vS_0\sqrt{T}} - rP(0)$$

$$= re^{-rT}K\Phi(-d_2) + rS_0[\Phi(d_1) - 1] - rP(0)$$

$$= rKe^{-rT}\Phi(-d_2) - rS_0\Phi(-d_1) - rP(0) \qquad = 0,$$

as claimed, since $Ke^{-rT}\Phi(-d_2) - S_0\Phi(d_1) = P(0)$ by Eq.2.26. The next-to-last step uses Eq.2.19 to rewrite $\Phi(d_1) - 1 = -\Phi(-d_1)$. □

12. (a) Find the coefficients p_1, p_2, p_3 that give the least-squares best fit

$$f(x) = p_1 + p_2e^x + p_3e^{-x}$$

to the data $\{(x, y)\} = \{(-2, 4), (-1, 1), (0, 0), (1, 1), (2, 4)\}$.

(b) Plot f at 81 equispaced points on a graph showing the data.

Solution: (a) Use the functions $f_1(x) = 1$, $f_2(x) = e^x$, and $f_3(x) = e^{-x}$ in Eq.2.48 to find the least-squares best fit to the data. Do this by modifying the example code for general least squares regression:

```
x=[-2;-1;0;1;2]; y=[4;1;0;1;4]; % data
f1=@(v)ones(size(v)); f2=@exp; f3=@(v)exp(-v); % functions
F=@(v) [f1(v),f2(v),f3(v)]; % regression matrix function
pf=(F(x)'*F(x))\(F(x)'*y) % least squares best fit coeffs
```

If the optim package is available, the last step may be done with

```
pf=LinearRegression(F(x),y) % least squares best fit coeffs
```

Both methods give $p_1 = -1.25403$, $p_2 = 0.70066$, and $p_3 = 0.70066$.

(b) Modify the example plotting code to use 81 z values equispaced between -2 and 2:

```
z=linspace(-2,2,81)'; % must be a column vector to use with F
yf=F(z)*pf; % column vector of the best-fit function at z
plot(z,yf,"b--", x,y,"k*"); % graph of best fit versus data
title("Best fit to (x,y) by 1,exp(x),exp(-x)")
legend("best fit","(x,y)");
```

The results may be seen in Figure A.4. □

13. At one instant, two days before expiry, near-the-money Call options for ABC common stock had the following prices:

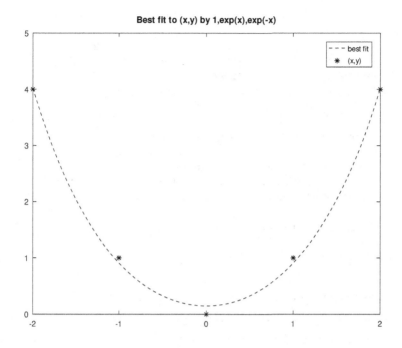

FIGURE A.4
Graph from Exercise 12(b).

Strike price	Premium	Open interest
44.00	3.80	3,260
45.00	2.77	4,499
46.00	1.77	3,862
47.00	0.78	6,271
48.00	0.18	10,156
49.00	0.03	10,619
50.00	0.01	14,219

The spot price for ABC is \$47.58. Estimate the premium for the at-the-money Call option in the following ways:

(a) Unweighted quadratic regression.

(b) Weighted quadratic regression.

(c) Polynomial interpolation.

(d) Spline interpolation.

Solution: Use the strike prices for x, the premiums for y, and the open interest for the weights w.

(a) First, enter the data and define the quadratic functions:

```
x=[44.00; 45.00; 46.00; 47.00; 48.00; 49.00; 50.00];
y=[ 3.80;  2.77;  1.77;  0.78;  0.18;  0.03;  0.01];
f1=@(v)ones(size(v)); f2=@(v)v; f3=@(v)v.*v; % 1,v,v^2
F=@(v) [f1(v),f2(v),f3(v)]; % regression matrix function
S0=47.58; % Spot price, to be used as the strike price
```

There are three equivalent quadratic regression methods:

```
pu=(F(x)'*F(x))\(F(x)'*y); F(S0)*pu
pua=LinearRegression(F(x),y); F(S0)*pua
polyval(polyfit(x,y,2),S0)
```

All three give the same estimate $0.51.

(b) With the data entered in part (a), add open-interest weights as follows:

```
w=[ 3260;  4499;  3862;  6271; 10156; 10619; 14219];
W=diag(w); % weight matrix
p=(F(x)'*W*F(x))\(F(x)'*W*y);   F(S0)*p
```

Alternatively,

```
pa=LinearRegression(F(x),y,sqrt(w)); F(S0)*pa
```

Both ways give an at-the-money premium estimate $0.47.

(c) Reusing the data entered in part (a), this is returned by

```
polyval(polyfit(x,y,length(x)-1),S0)
```

This estimate is $0.37.

(d) With the data entered in part (a), this is returned by:

```
ppval(spline(x,y),S0)
```

A feature of Octave's `spline()` function is that it evaluates the computed spline if given a third argument:

```
spline(x,y,S0)
```

Both commands give the same estimate $0.37.

NOTE: the market premium at this instant was $0.38 for the Call option at strike price $47.50.   □

A.3 ... to Chapter 3 Exercises

1. Suppose that $S(t, \omega)$, $0 \leq t \leq T$ is the price of a risky asset S, and that the riskless return over time $[0, T]$ is R. Model the future at time $t = T$ using $\Omega = \{\uparrow, \downarrow\}$ and assume that $S(T, \downarrow) < S(T, \uparrow)$.

(a) Use the no-arbitrage Axiom 1 to conclude that

$$S(T, \downarrow) < RS(0) < S(T, \uparrow).$$

(b) Use the Fair Price Theorem 1.4 to prove the same inequalities.

Solution: (a) First exclude the case $RS(0) \leq S(T, \downarrow) < S(T, \uparrow)$, as it offers the following arbitrage opportunity:

- At time $t = 0$, borrow $S(0)$ from the bank and buy one share of S. Initial cost is 0.
- At time $t = T$, sell S for $S(t, \omega)$ and repay the loan with interest for $RS(0)$. The net result is $S(T, \omega) - RS(0)$.

The net at expiry is either $S(T, \downarrow) - RS(0) \geq 0$ or $S(T, \uparrow) - RS(0) > 0$, an arbitrage opportunity as claimed, forbidden by Axiom 1.

Second, exclude the possibility $S(T, \downarrow) < S(T, \uparrow) \leq RS(0)$ as it also provides an arbitrage opportunity:

- At time $t = 0$, sell S short for $S(0)$ and deposit the money in the bank. Initial cost is again 0.
- At time $t = T$, withdraw the principal and interest $RS(0)$ from the bank and buy S to cover the short for $S(t, \omega)$. The net result is $RS(0) - S(T, \omega)$.

The net at expiry in this case is either $RS(0) - S(T, \uparrow) \geq 0$ or $RS(0) - S(T, \downarrow) > 0$ giving an arbitrage opportunity as claimed, forbidden by Axiom 1.

Conclude that $S(T, \downarrow) < RS(0) < S(T, \uparrow)$.

(b) From Theorem 1.4 and the definition of expectation,

$$RS(0) = \Pr(\downarrow)S(T, \downarrow) + \Pr(\uparrow)S(T, \uparrow)$$

But $0 < \Pr(\downarrow) < 1$ since otherwise the future is certain, and thus also $0 < \Pr(\uparrow) < 1$ since $\Pr(\uparrow) = 1 - \Pr(\downarrow)$. Conclude that $RS(0)$ lies strictly inside the interval $[S(T, \downarrow), S(T, \uparrow)]$, as claimed. □

2. In Exercise 1 above, model the future at time $t = T$ using the N-step binomial model $\Omega = \{\omega_0, \omega_1, \ldots, \omega_N\}$ and assume that

$$S(T, \omega_k) = S(0)u^k d^{N-k},$$

where $S(0) > 0$ is the spot price and $0 < d < u$ are the up factor and down factor, respectively, over one time step T/N.

(a) Use the no-arbitrage Axiom 1 to conclude that

$$d < R^{1/N} < u.$$

(b) Use the Fair Price Theorem 1.4 to prove the same inequalities.

Solution: First note that $0 < d < u$ and $S(0) > 0$ together imply that

$$0 < S(T, \omega_0) = S(0)d^N < S(T, \omega_1) < \cdots < S(T, \omega_N) = S(0)u^N$$

(a) If $R^{1/N} \le d < u$, then $R \le d^N < u^N$, which implies

$$RS(0) \le S(T, \omega_0); \qquad RS(0) < S(T, \omega_k), \quad k = 1, \ldots, N,$$

so that borrowing $S(0)$ to buy S at $t = 0$ will result in an arbitrage profit at $t = T$. Similarly, if $d < u \le R^{1/N}$, then $d^N < u^N \le R$, which implies

$$RS(0) \ge S(T, \omega_N); \qquad RS(0) > S(T, \omega_k), \quad 0 \le k < N,$$

so that selling S short for $S(0)$ and investing the money risklessly at $t = 0$ will result in an arbitrage profit at $t = T$.

Both cases are forbidden by Axiom 1, so $d < R^{1/N} < u$.

(b) From Theorem 1.4 and the definition of expectation,

$$RS(0) = \mathrm{E}(S(T)) = \sum_{k=0}^{N} \Pr(\omega_k) S(T, \omega_k), = S(0) \sum_{k=0}^{N} \Pr(\omega_k) u^k d^{N-k}.$$

Dividing by $S(0) > 0$ simplifies this to

$$R = \sum_{k=0}^{N} \Pr(\omega_k) u^k d^{N-k}.$$

Now $0 < d < u$ implies that $d^N < ud^{N-1} < \cdots < u^{N-1}d < u^N$, and the probabilities lie in $[0, 1]$ and sum to 1, so the right-hand side is a convex combination of points in the interval $[d^N, u^N]$. Since the asset is risky, at least two of the states have positive probabilities. The convex combination must therefore lie strictly inside the interval, so

$$d^N < R < u^N,$$

from which it follows that $d < R^{1/N} < u$ as claimed.   □

3. Suppose that a portfolio X contains risky stock S and riskless bond B in amounts h_0, h_1:

$$X(t, \omega) = h_0 B(t, \omega) + h_1 S(t, \omega).$$

Model the future at time $t = T$ using $\Omega = \{\uparrow, \downarrow\}$, assuming only that $S(T, \uparrow) \neq S(T, \downarrow)$ and that $B(T, \uparrow) = B(T, \downarrow) = R$. Compute h_0 and h_1 in terms of all the other quantities. (Hint: use Macsyma to derive Eq.3.1.)

Solution: Set up the system of equations at $t = T$:

$$
\begin{aligned}
X(T, \uparrow) &= h_0 B(T, \uparrow) + h_1 S(T, \uparrow) = h_0 R + h_1 S(T, \uparrow) \\
X(T, \downarrow) &= h_0 B(T, \downarrow) + h_1 S(T, \downarrow) = h_0 R + h_1 S(T, \downarrow)
\end{aligned}
$$

Use these Macsyma commands to solve the system:

```
eq1: xTu=h0*R+h1*sTu;   /* Up state equation */
eq2: xTd=h0*R+h1*sTd;   /* Down state equation */
hOh1: solve([eq1,eq2],[h0,h1]);   /* Solve for h0,h1 */
```

That results in this output:

```
[[h0=-(sTd*xTu-sTu*xTd)/((sTu-sTd)*R),
   h1=(xTu-xTd)/(sTu-sTd)]]
```

Writing the Macsyma solution in the orginal notation gives

$$h_0 = \frac{S(T, \uparrow)X(T, \downarrow) - S(T, \downarrow)X(T, \uparrow)}{(S(T, \uparrow) - S(T, \downarrow))R}; \quad h_1 = \frac{X(T, \uparrow) - X(T, \downarrow)}{S(T, \uparrow) - S(T, \downarrow)}.$$

□

4. In Exercise 3 above, suppose that X is a European-style Call option for S with expiry T and strike price K. Use the payoff formula $X(T) = [S(T) - K]^+$ in the equation for h_1 to prove that

$$0 \leq h_1 \leq 1.$$

Conclude that, in this model of the future, a European-style Call option for S is equivalent to a portfolio containing part of a share of S plus or minus some cash.

Solution: Substitute the payoff formula into the equation for h_1, then add and subtract K in the denominator to get:

$$h_1 = \frac{X(T, \uparrow) - X(T, \downarrow)}{S(T, \uparrow) - S(T, \downarrow)} = \frac{[S(T, \uparrow) - K]^+ - [S(T, \downarrow) - K]^+}{(S(T, \uparrow) - K) - (S(T, \downarrow) - K)}.$$

It may be assumed that $S(T, \uparrow) > S(T, \downarrow)$, since the states can be switched without changing the value of h_1. Then there are three cases to consider:

Case 1: If $S(T,\uparrow) > S(T,\downarrow) > K$, then both plus-parts are positive, so

$$h_1 = \frac{(S(T,\uparrow) - K) - (S(T,\downarrow) - K)}{(S(T,\uparrow) - K) - (S(T,\downarrow) - K)} = 1.$$

Case 2: If $K \geq S(T,\uparrow) > S(T,\downarrow)$, then both plus-parts are zero, so

$$h_1 = \frac{0 - 0}{(S(T,\uparrow) - K) - (S(T,\downarrow) - K)} = 0.$$

Case 3: If $S(T,\uparrow) > K \geq S(T,\downarrow)$, then the first plus-part is positive but the second is zero, so

$$h_1 = \frac{(S(T,\uparrow) - K) - 0}{(S(T,\uparrow) - K) - (S(T,\downarrow) - K)}.$$

But the denominator is $(S(T,\downarrow) - K) \leq 0$, which implies

$$(S(T,\uparrow) - K) - (S(T,\downarrow) - K) \geq (S(T,\uparrow) - K) > 0,$$

so the positive denominator is no smaller than the positive numerator, so $0 < h_1 \leq 1$.

Conclude that $0 \leq h_1 \leq 1$ in all cases. $\square$

5. In Exercise 3 above, suppose that X is a European-style Put option for S with expiry T and strike price K. Use the payoff formula $X(T) = [K - S(T)]^+$ in the equation for h_1 to prove that

$$-1 \leq h_1 \leq 0.$$

Conclude that, in this model of the future, a European-style Put option for S is equivalent to a portfolio containing part of a share of S sold short plus or minus some cash.

Solution: Substitute the payoff formula into the equation for h_1, multiply numerator and denominator by -1, and then add and subtract K in the denominator to get:

$$h_1 = \frac{X(T,\uparrow) - X(T,\downarrow)}{S(T,\uparrow) - S(T,\downarrow)} = -\frac{[K - S(T,\uparrow)]^+ - [K - S(T,\downarrow)]^+}{(K - S(T,\uparrow)) - (K - S(T,\downarrow))}.$$

It may be assumed that $S(T,\uparrow) > S(T,\downarrow)$, since the states can be switched without changing the value of h_1. Then there are three cases to consider:

Case 1: If $K > S(T,\uparrow) > S(T,\downarrow)$, then both plus-parts are positive, so

$$h_1 = -\frac{(K - S(T,\uparrow)) - (K - S(T,\downarrow))}{(K - S(T,\uparrow)) - (K - S(T,\downarrow))} = -1.$$

Case 2: If $S(T,\uparrow) > S(T,\downarrow) \geq K$, then both plus-parts are zero, so

$$h_1 = -\frac{0 - 0}{(K - S(T,\uparrow)) - (K - S(T,\downarrow))} = 0.$$

Case 3: If $S(T,\uparrow) \geq K > S(T,\downarrow)$, then the first plus-part is zero but the second is positive, so

$$\begin{aligned}
h_1 &= -\frac{0 - (K - S(T,\downarrow))}{(K - S(T,\uparrow)) - (K - S(T,\downarrow))} \\
&= \frac{(K - S(T,\downarrow))}{(K - S(T,\uparrow)) - (K - S(T,\downarrow))}.
\end{aligned}$$

But the denominator is $(K - S(T,\uparrow)) \leq 0$, which implies

$$(K - S(T,\uparrow)) - (K - S(T,\downarrow)) \leq -(K - S(T,\downarrow)) < 0,$$

so the negative denominator has no smaller absolute value than the positive numerator, so $-1 \leq h_1 < 0$.

Conclude that $-1 \leq h_1 \leq 0$ in all cases.

Remark. An alternative proof uses the identity $y = [y]^+ - [-y]^+$ which is true for any y. Then

$$(S(T,\omega) - K) = [S(T,\omega) - K]^+ - [K - S(T,\omega)]^+,$$

so

$$[K - S(T,\omega)]^+ = [S(T,\omega) - K]^+ - (S(T,\omega) - K),$$

and thus

$$h_1 = \frac{[S(T,\uparrow) - K]^+ - [S(T,\downarrow) - K]^+}{(S(T,\uparrow) - K) - (S(T,\downarrow) - K)} - 1.$$

The result now follows from the Call h_1 inequalities. $\qquad\square$

6. Suppose that $C(0)$ and $P(0)$ are the premiums for European-style Call and Put options, respectively, on an asset S with the following parameters: expiry at $T = 1$ year, spot price $S(0) = 90$, strike price $K = 95$. Assume that the riskless annual percentage rate is $r = 0.02$, and the volatility for S is $\sigma = 0.15$, and that these will remain constant from now until expiry.

(a) Use a LibreOffice Calc spreadsheet to implement the Cox-Ross-Rubinstein (CRR) model to compute $C(0)$ and $P(0)$ with $N = 10$ time steps, using the backward pricing formula in Eq.3.18. (Hint: compare output with CRReurAD() to check for bugs.)

(b) Use the Octave function CRReurAD() with $N = 10$, $N = 100$, and $N = 1000$ time steps to compute $C(0)$ and $P(0)$.

(c) Repeat part (b) with the Octave function CRReur() on p.88, again using $N = 10$, $N = 100$, and $N = 1000$ time steps to compute $C(0)$ and $P(0)$. Profile the time required to compute them, and compare the time and the output with that of CRReurAD().

(d) Compare the prices from parts (b) and (c). Is it justified to use $N = 1000$? Is $N = 10$ sufficiently accurate?

Solution: (a) See the spreadsheet CRR.ods in the programs archive. With $N = 10$ time steps, rounding to five significant digits, it computes $C(0) = \$4.1733$ and $P(0) = \$7.2922$.

(b) Implement the program CRReurAD() on p.76, input the parameters, and compute the CRR approximations at $N = 10$, $N = 100$, and $N = 1000$ with the commands

```
T=1; S0=90; K=95; r=0.02; v=0.15;
[C0,P0]=CRReurAD(T,S0,K,r,v,10) % 4.1733, 7.2922
[C0,P0]=CRReurAD(T,S0,K,r,v,100) % 4.0572, 7.1761
[C0,P0]=CRReurAD(T,S0,K,r,v,1000) % 4.0555, 7.1744
```

At $N = 10$ time steps it returns $C(0) = \$4.1733$ and $P(0) = \$7.2922$, in agreement with the spreadsheet.

After $N = 100$ time steps, $C(0) = \$4.0572$ and $P(0) = \$7.1761$.

After $N = 1000$ time steps, $C(0) = \$4.0555$ and $P(0) = \$7.1744$.

In all cases the computations are almost instantaneous, requiring no noticeable time.

(c) The Octave program CRReur()m on p.88 uses the backward induction formula to compute Call and Put premiums. It therefore fills two recombining binomial trees of depth N at a cost of $O(N^2)$ compared with the $O(N)$ cost of the Arrow-Debreu expansion method.

Input the parameters and compute the CRR approximations at $N = 10$, $N = 100$, and $N = 1000$ with the commands

```
T=1; S0=90; K=95; r=0.02; v=0.15;
[C,P]=CRReur(T,S0,K,r,v,10); C(1,1),P(1,1) % 4.1733, 7.2922
profile on; CRReur(T,S0,K,r,v,10); profshow % ~0.013 seconds
[C,P]=CRReur(T,S0,K,r,v,100); C(1,1),P(1,1) % 4.0572, 7.1761
profile on; CRReur(T,S0,K,r,v,100); profshow % ~1.05 seconds
[C,P]=CRReur(T,S0,K,r,v,1000); C(1,1),P(1,1) % 4.0555, 7.1744
profile on; CRReur(T,S0,K,r,v,1000); profshow % ~103 seconds
```

Note that the outputs are pairs of matrices, so that to get just the premiums it is necessary to extract just the $(1,1)$ element.

At $N = 10$ time steps it almost instantaneously returns $C(0) = \$4.1733$ and $P(0) = \$7.2922$, in agreement with the spreadsheet. Profiled time was 0.013 seconds.

With $N = 100$ time steps it very quickly computes $C(0) = \$4.0572$ and $P(0) = \$7.1761$. Profiled time was 1.05 seconds.

With $N = 1000$ time steps it takes a considerably longer time to compute $C(0) = \$4.0555$ and $P(0) = \$7.1744$. Profiled time was 102 seconds.

Note that the ratios of profiled times agree with the $O(N^2)$ complexity estimate.

(d) $N = 100$ seems justified since the prices are quite different from the $N = 10$ values. It seems unjustified to use $N = 1000$, which costs much more time and space (using the backward induction algorithm) but gives almost the same result as $N = 100$.   □

7. Compare the prices from parts (a) and (b) of previous Exercise 6 with the Black-Scholes prices computed using Eqs.2.25 and 2.26. Plot the logarithm of the differences against $\log N$ to estimate the rate of convergence. (Hint: Use the programs in Chapter 2, Section 2.4.)

Solution: Use the parameters from Exercise 6 in the Octave program BS() on p.34, as follows:

```
T=1; S0=90; K=95; r=0.02; v=0.15; [C0,P0]=BS(T,S0,K,r,v)
```

This returns $C_{BS} = \text{C0} = 4.0548$ and $P_{BS} = \text{P0} = 7.1736$. Now compute the logarithms of the differences as a function of $\log N$:

```
Ns=[10,100,1000]; log(Ns)
C0=4.0548; CCRR=[4.1733,4.0572,4.0555]; log(abs(CCRR-C0))
P0=7.1736; PCRR=[7.2922,7.1761,7.1744]; log(abs(PCRR-P0))
```

The output is tabulated below:

N	$\log N$	$\log|C_{CRR} - C_{BS}|$	$\log|P_{CRR} - P_{BS}|$
10	2.3026	-2.1328	-2.1320
100	4.6052	-6.0323	-5.9915
1000	6.9078	-7.2644	-7.1309

Finally, generate the log-log plots:

```
plot(log(Ns),log(abs(CCRR-C0))); title("Call Difference");
xlabel("log N"); ylabel("log|CRR(N)-BS|");    figure;
plot(log(Ns),log(abs(PCRR-P0))); title("Put Difference");
xlabel("log N"); ylabel("log|CRR(N)-BS|");
```

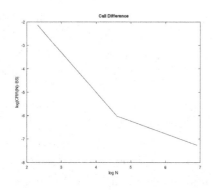 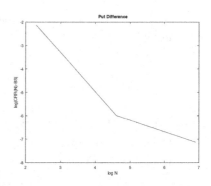

FIGURE A.5
(From Exercise 7) Log-log plots showing the differences between Black-Scholes prices and their N-step CRR approximations, for certain European-style Call and Put options, as a function of N.

The results may be seen in Figure A.5. For both Call and Put differences, the graphs are close to lines of slope -1, suggesting that the difference between Black-Scholes and its N-step CRR approximation is $O(N^{-1})$. This may be quantified by regression using `polyfit(x,y,1)`:

```
polyfit(log(Ns),log(abs(CCRR-C0)),1) % -1.114310  -0.011598
polyfit(log(Ns),log(abs(PCRR-P0)),1) % -1.085497  -0.085887
```

The first output number is the slope of the least-squares line fitting the data, in both cases close to -1. The second is the intercept; it is an estimate for the logarithm of the constant in the $O(N^{-1})$ rate.

Remark. Using only 5 significant digits introduces substantial round-off error at large N, where the differences are small. This is unavoidable since the parameters are only specified to 2 or 3 significant digits. □

8. Derive Eq.3.32 on p.79:

$$q = \frac{1}{2} + \frac{r + \frac{\sigma^2}{2}}{2\sigma}\sqrt{\frac{T}{N}} + O\left(\frac{T}{N}\right).$$

Solution: Recall that $q = (u - 1/R)/(u - 1/u)$. Using Taylor's approximation in the numerator gives

$$[1 + \sigma\sqrt{\frac{T}{N}} + \frac{\sigma^2}{2}\frac{T}{N} + O(\sqrt{\frac{T}{N}}^3)] - [1 - \frac{rT}{N} + O([\frac{T}{N}]^2)],$$

while in the denominator it gives

$$[1 + \sigma\sqrt{\frac{T}{N}} + \frac{\sigma^2}{2}\frac{T}{N} + O(\sqrt{\frac{T}{N}}^3)] - [1 - \sigma\sqrt{\frac{T}{N}} + \frac{\sigma^2}{2}\frac{T}{N} + O(\sqrt{\frac{T}{N}}^3)].$$

Canceling terms and simplifying the ratio gives

$$
q = \frac{\sigma\sqrt{\frac{T}{N}} + \left(r + \frac{\sigma^2}{2}\right)\frac{T}{N} + O\left(\sqrt{\frac{T}{N}}^{3}\right) + O\left(\left[\frac{T}{N}\right]^2\right)}{2\sigma\sqrt{\frac{T}{N}} + O\left(\sqrt{\frac{T}{N}}^{3}\right)}
$$

$$
= \frac{1}{2} + \frac{r + \frac{\sigma^2}{2}}{2\sigma}\sqrt{\frac{T}{N}} + O\left(\frac{T}{N}\right),
$$

as claimed.   □

9. Use the CRR approximation with $N = 4$ to compute the European-style Call option premiums at several hundred equally spaced spot prices $75 \leq S_0 \leq 115$, with expiry $T = 1$, strike $K = 95$, $r = 0.02$, and $\sigma = 0.15$.

(a) Plot the values against S_0.

(b) At what values of S_0 in that range does the graph appear to be nonsmooth?

(c) Compute the points of nondifferentiability for S_0 in $[75, 115]$.

Solution: (a) Use `CRReurAD()` in the following Octave code:

```
T=1; K=95; r=0.02; v=0.15; N=4; m=401;
S=linspace(75,115,m); CS=zeros(size(S));
for i=1:m
  S0=S(i); [CO,PO]=CRReurAD(T,S(i),K,r,v,N);   CS(i)=CO;
end
plot(S,CS); title("CRR with N=4");xlabel("S0");ylabel("CO");
```

See the result in Figure A.6.

(b) The graph appears piecewise linear with joints $\hat{S}_0 \in \{82, 95, 110\}$ where the Call premium is not differentiable with respect to S_0.

(c) Compute the joints, or points of nondifferentiability $\hat{S}_0$ nearest K, using Eq.3.39 and $j \in \{N/2 - 1, N/2, N/2 + 1\} = \{1, 2, 3\}$:

$$
\hat{S}_0 \in \left\{\frac{K}{u^{2(j+1)-N}}, \frac{K}{u^{2j-N}}, \frac{K}{u^{2(j-1)-N}}\right\} = \{d^2 K, K, u^2 K\},
$$

where $1/d = u = \exp(\sigma\sqrt{T/N})$. With the given parameters,

$$
u = \exp(0.15\sqrt{1/4}) = 1.0779, \implies \hat{S}_0 \in \{81.767,\ 95,\ 110.37\},
$$

in good agreement with the visual estimate.   □

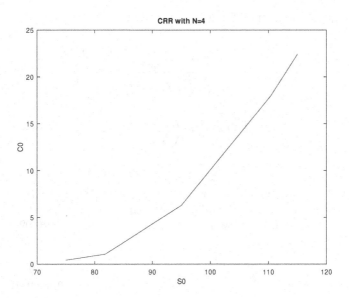

FIGURE A.6

(From Exercise 9) CRR approximation with $N = 4$ to the European-style Call option premium $C(0)$, as a function of spot price $S(0)$.

10. Compute the CRR option premiums and Greeks for European-style Call and Put options on a risky asset with the following parameters: spot price \$90, strike price \$95, expiry in 1 year, annual riskless rate 2%, and volatility 15%. Use $N = 100$ steps. Justify the method used.

    Solution: First compute the option premiums with CRReurAD():

    ```
    T=1; S0=90; K=95; r=0.02; v=0.15; N=100;
    [C0,P0]=CRReurAD(T,S0,K,r,v,N)
    ```

    For Δ and Γ, use the interpolation method on p.83:

    ```
    h0=2*S0*v*sqrt(T/N);  % critical h
    u2=exp(2*v*sqrt(T/N)); % squared up factor
    x=[S0/u2, S0, S0*u2]-S0; % shifted abscissas
    [C0,P0]=CRReurAD(T,S0,K,r,v,N);
    [C0u,P0u]=CRReurAD(T,S0*u2,K,r,v,N); % ...at S0*u^2
    [C0d,P0d]=CRReurAD(T,S0/u2,K,r,v,N); % ...at S0/u^2
    yC=[C0d, C0, C0u]; % Call ordinates
    p=polyfit(x,yC,2); DeltaC=p(2), GammaC=2*p(1)
    yP=[P0d, P0, P0u]; % Put ordinates
    p=polyfit(x,yP,2); DeltaP=p(2), GammaP=2*p(1)
    ```

This is necessary because the approximation is not differentiable with respect to S_0.

For the other Greeks, use the centered difference approximation to the derivative with h set to 10% of the abscissa value:

```
h=0.10*T; % for ThetaC, ThetaP
[COu,POu]=CRReurAD(T+h,S0,K,r,v,N);
[COd,POd]=CRReurAD(T-h,S0,K,r,v,N);
ThetaC=-(COu-COd)/(2*h), ThetaP=-(POu-POd)/(2*h)
h=0.10*v; % for VegaC, VegaP
[COu,POu]=CRReurAD(T,S0,K,r,v+h,N);
[COd,POd]=CRReurAD(T,S0,K,r,v-h,N);
VegaC=(COu-COd)/(2*h), VegaP=(POu-POd)/(2*h)
h=0.10*r; % for RhoC, RhoP
[COu,POu]=CRReurAD(T,S0,K,r+h,v,N);
[COd,POd]=CRReurAD(T,S0,K,r-h,v,N);
RhoC=(COu-COd)/(2*h), RhoP=(POu-POd)/(2*h)
```

Since the centered difference approximation has $O(h^2) \approx 1\%$ relative error, expect roughly two significant digits of accuracy. Much smaller values of h are not justified since the option premiums are only $O(1/N) \approx 1\%$ accurate as shown in Exercise 7.

The results are tabulated below:

	Option	Delta	Gamma	Theta	Vega	Rho
Call	4.0572	0.43987	0.029063	−3.4348	36.298	35.457
Put	7.1761	−0.56013	0.029063	−1.5724	36.298	−57.661

Comparison with the Black-Scholes premiums and Greeks computed in Chapter 2, Exercise 10 shows good agreement.   □

A.4   ...to Chapter 4 Exercises

1. Implement a compound Put option, the option to purchase a European-style Put option, with expiry T and strike price K, for price L at time T_1 satisfying $0 < T_1 < T$. Use your code to price such an option with parameters $(T, T_1, S_0, K, L, r, v, N) = (1, 0.5, 90, 95, 4.50, 0.02, 0.15, 20)$. (Hint: modify CRRcc.m.)

Solution: The Octave function CRRcp() differs from CRRcc() at line 25 and in a few updated comments.

250   *Answers*

```
1   function [W,P] = CRRcp(T,T1,S0,K,L,r,v,N)
2   %   Octave/MATLAB function to price a compound Put
3   %   in the Cox−Ross−Rubinstein (CRR) binomial model.
4   % INPUTS:                                          (Example)
5   %   T  = expiration time in years                  (1)
6   %   T1 = choice time; must have 0<T1<T             (0.5)
7   %   S0 = spot stock price                          (90)
8   %   K  = stock strike price at expiry T            (95)
9   %   L  = option strike price at time T1            (4.50)
10  %   r  = risk−free yield                           (0.02)
11  %   v  = volatility; must be >0                    (0.15)
12  %   N  = height of the binomial tree               (20)
13  % OUTPUT:
14  %   W  =  price of the compound Put option
15  %   P  =  price of the vanilla European Put option
16  % EXAMPLE:
17  %   [W,P] = CRRcp(1,0.5,90,95,4.50,0.02,0.15,20);
18  %   W(1,1),P(1,1)  % to get just W(0) and P(0)
19  %
20     [pu,up,R]=CRRparams(T,r,v,N); % Use CRR values
21     [C,P]=CRReur(T,S0,K,r,v,N); % vanilla options at T
22     M=round(T1*N/T); % number of time steps to time T1
23     W=zeros(M+1,M+1);   % smaller output matrix
24     for j=0:M % Set terminal values at expiry T1
25       W(M+1,j+1) = max(L−P(M+1,j+1),0); % plus part
26     end
27     for n=M−1:−1:0   % Backward induction
28       for j=0:n % Binomial pricing model value
29       W(n+1,j+1)=(pu*W(n+2,j+2)+(1−pu)*W(n+2,j+1))/R;
30       end
31     end
32     return  % Prices in matrices W and P are defined.
33  end
```

Compute the compound Put premium with the requested parameters as follows:

```
[W,P] = CRRcp(1,0.5,90,95,4.50,0.02,0.15,20);
W(1,1),P(1,1)  % 0.96243  7.1857
```

Conclude that the premium is about 0.96.   □

2. Let $\mathcal{M}(n)$ be the number of paths in a recombining binomial tree to depth n, as defined in Eq.4.6. Prove that

$$\mathcal{M}(n+1) = [2\mathcal{M}(n)] \cup [2\mathcal{M}(n)+1],$$

where $aX + b \stackrel{\text{def}}{=} \{ax + b : x \in X\}$ for sets X of numbers.

Solution: Apply the definitions:

$$\begin{aligned}
\mathcal{M}(n) &= \{2^n, 2^n + 1, \ldots, 2^n + (2^n - 1)\}, \\
2\mathcal{M}(n) &= \{2^{n+1}, 2^{n+1} + 2, \ldots, 2^{n+1} + (2^{n+1} - 2)\}, \\
2\mathcal{M}(n) + 1 &= \{2^{n+1} + 1, 2^{n+1} + 3, \ldots, 2^{n+1} + (2^{n+1} - 1)\},
\end{aligned}$$

so the union of the last two sets is evidently

$$\mathcal{M}(n+1) = \{2^{n+1}, 2^{n+1} + 1, \ldots, 2^{n+1} + (2^{n+1} - 1)\},$$

as claimed. $\qquad\qquad\qquad\qquad\qquad\qquad\qquad\qquad\qquad\qquad\qquad\qquad$ $\square$

3. Implement floating strike option pricing in the CRR model using geometric means instead of arithmetic means, as in `CRRgro` versus `CRRaro`. Compare the results on the suggested example inputs.

Solution: The Octave function `CRRflg()` differs from `CRRflt()` at lines 21 and 24, which are directly copied from lines 19 and 22 of `CRRgro()` where `lU` is computed and used. It also differs in the use of `Geom` instead of `Avg` and in a few updated comments.

```
1   function [C, P] = CRRflg(T, S0, r, v, N)
2   % Octave/MATLAB function to price floating strike
3   % Call and Put options with geometric means and the
4   % Cox-Ross-Rubinstein (CRR) binomial pricing model.
5   % INPUTS:                                  (Example)
6   %    T  =  expiration time in years        (1)
7   %    S0 =  spot stock price                (90)
8   %    r  =  riskless yield per year         (0.02)
9   %    v  =  volatility; must be >0          (0.20)
10  %    N  =  height of the tree              (4)
11  % OUTPUTS:
12  %    C0 = Call option premium at t=0.
13  %    P0 = Put option premium at t=0.
14  % EXAMPLE:
15  %    [C, P] = CRRflg(1, 90, 0.02, 0.20, 4);
16  % To get just the premiums at t=0, use
17  %    C(1), P(1)
18  %
19      [pu,up,R] = CRRparams(T,r,v,N);
20      Sbar=NRTCRR(S0,up,1/up,N); % expanded S tree
21      lU=NRTpsums(log(Sbar),N); % partial sums of logs
22      C=zeros(size(Sbar));  P=zeros(size(Sbar));
23      for m=2^N:(2*2^N -1) % m indices at expiry
24          Geom = exp(lU(m)/(N+1));  % geometric mean
25          C(m) = max(0,Sbar(m)-Geom); % Call payoff
26          P(m) = max(0,Geom-Sbar(m)); % Put payoff
27      end  % ...prices set at expiry.
28      for m=(2^N-1):-1:1 % recursive previous indices
```

```
29       C(m) = (pu*C(2*m+1) + (1-pu)*C(2*m))/R;
30       P(m) = (pu*P(2*m+1) + (1-pu)*P(2*m))/R;
31     end % ...all prices set by backward induction.
32     return
33   end
```

Compare (and test for bugs) by running the geometric mean and arithmetic mean functions on the same suggested example inputs:

```
[Cg,Pg]=CRRflg(1,90,0.02,0.20,4);Cg(1),Pg(1) % 4.6711, 3.4209
[Ca,Pa]=CRRflt(1,90,0.02,0.20,4);Ca(1),Pa(1) % 4.4853, 3.5920
```

The Call and Put premiums are similar, though not identical. □

4. Use CRRaro to compute the average-rate Call and Put premiums in the CRR model for $S_0 = K = 100$, $T = 1$, $r = 0.05$, $v = 0.15$, and different values of N. Compare $\overline{C}(0) - \overline{P}(0)$ with the limit value in Eq.4.20.

   Solution: The Octave commands below perform the computations with $N = 3, 5, 7, 9, 11$:

```
r=0.05; T=1; K=100; S0=100; v=0.15;
limit=S0*(1-exp(-r*T))/(r*T)-K/exp(r*T)    % limit =  2.4182
[C,P]=CRRaro(T,S0,K,r,v,3);  N3=C(1)-P(1)   % N3   =  2.4250
[C,P]=CRRaro(T,S0,K,r,v,5);  N5=C(1)-P(1)   % N5   =  2.4223
[C,P]=CRRaro(T,S0,K,r,v,7);  N7=C(1)-P(1)   % N7   =  2.4211
[C,P]=CRRaro(T,S0,K,r,v,9);  N9=C(1)-P(1)   % N9   =  2.4205
[C,P]=CRRaro(T,S0,K,r,v,11); N11=C(1)-P(1) % N11  =  2.4201
```

   The outputs suggest that every $N \geq 5$ achieves an approximation within one cent. □

5. Implement floating strike option pricing in the CRR model using geometric means instead of arithmetic means, as in CRRgro versus CRRaro. Compare the results on the suggested example inputs.

   Solution: The Octave function CRRflg() differs from CRRflt() at lines 21 and 24, which are directly copied from lines 19 and 22 of CRRgro() where 1U is computed and used. It also differs in the use of Geom instead of Avg and in a few updated comments.

```
1  function [C, P] = CRRflg(T, S0, r, v, N)
2  % Octave/MATLAB function to price floating strike
3  % Call and Put options with geometric means and the
4  % Cox-Ross-Rubinstein (CRR) binomial pricing model.
5  % INPUTS:                            (Example)
6  %    T  =  expiration time in years    (1)
7  %    S0 =  spot stock price            (90)
8  %    r  =  riskless yield per year     (0.02)
9  %    v  =  volatility; must be >0      (0.20)
```

```
10  %   N  =  height  of  the  tree              (4)
11  % OUTPUTS:
12  %     C0 = Call option premium at t=0.
13  %     P0 = Put option premium at t=0.
14  % EXAMPLE:
15  %     [C, P] = CRRflg(1, 90, 0.02, 0.20, 4);
16  % To get just the premiums at t=0, use
17  %     C(1), P(1)
18  %
19    [pu,up,R] = CRRparams(T,r,v,N);
20    Sbar=NRTCRR(S0,up,1/up,N); % expanded S tree
21    lU=NRTpsums(log(Sbar),N); % partial sums of logs
22    C=zeros(size(Sbar));  P=zeros(size(Sbar));
23    for m=2^N:(2*2^N −1) % m indices at expiry
24      Geom = exp(lU(m)/(N+1)); % geometric mean
25      C(m) = max(0,Sbar(m)−Geom); % Call payoff
26      P(m) = max(0,Geom−Sbar(m)); % Put payoff
27    end % ...prices set at expiry.
28    for m=(2^N−1):−1:1 % recursive previous indices
29      C(m) = (pu*C(2*m+1) + (1−pu)*C(2*m))/R;
30      P(m) = (pu*P(2*m+1) + (1−pu)*P(2*m))/R;
31    end % ...all prices set by backward induction.
32    return
33  end
```

Compare (and test for bugs) by running both the geometric mean and the arithmetic mean functions on the same suggested example inputs:

```
[Cg,Pg]=CRRflg(1,90,0.02,0.20,4);Cg(1),Pg(1) % 4.6711, 3.4209
[Ca,Pa]=CRRflt(1,90,0.02,0.20,4);Ca(1),Pa(1) % 4.4853, 3.5920
```

The Call and Put premiums are similar, though not identical. □

6. Implement floating strike option pricing in the CRR model using path-dependent Arrow-Debreu securities. Check that the results agree with CRRflt.

   Solution: The Octave function CRRfltAD() differs from CRRflt() in three ways, just as CRRaroAD() differs from CRRaro():

```
1  function [C0, P0] = CRRfltAD(T, S0, r, v, N)
2  % Octave/MATLAB function to price floating strike
3  % Call and Put options using path−dependent
4  % Arrow−Debreu expansions with the Cox−Ross−
5  % Rubinstein (CRR) binomial pricing model.
6  % INPUTS:                              (Example)
7  %    T  =  expiration time in years      (1)
8  %    S0 =  spot stock price              (90)
9  %    r  =  riskless yield per year       (0.02)
10 %    v  =  volatility; must be >0        (0.20)
```

```
11  %   N  =  height of the tree                (4)
12  % OUTPUTS:
13  %     C0 = Call option premium at t=0.
14  %     P0 = Put option premium at t=0.
15  % EXAMPLE:
16  %     [C0,P0]=CRRfltAD(1,90,0.02,0.20,4);
17  %
18     [pu,up,R]=CRRparams(T,r,v,N);
19     Sbar=NRTCRR(S0,up,1/up,N); % expanded S tree
20     Ubar=NRTpsums(Sbar,N); % all partial sums
21     Lbar=PathAD(pu*ones(N,N),R*ones(N,N),N);
22     mN=2^N:(2*2^N-1); % all m-indices at expiry
23     AvgN=Ubar(mN)/(N+1); % S averages at expiry
24     C0=max(0,Sbar(mN)-AvgN)*Lbar(mN) '; % Call
25     P0=max(0,AvgN-Sbar(mN))*Lbar(mN) '; % Put
26     return % ..inner producs give A–D expansions
27  end
```

Compare (and test for bugs) by running both functions on the same suggested example inputs:

```
[C0,P0]=CRRfltAD(1,90,0.02,0.20,4)
[C,P]=CRRflt(1,90,0.02,0.20,4); C(1),P(1)
```

The outputs agree: `C0=4.4853=C(1)` and `P0=3.5920=P(1)`.   □

7. Write an Octave program to compute the maximums along all paths in a non-recombining binary tree of depth N. (Hint: modify NRTmin().)

Solution: Let $\bar{S}_{\max}(m)$ be the maximum value of $S(n, j)$ along the path indexed by m. It may be computed from the non-recombining tree $\bar{S}$ by the following recursion:

$$
\begin{aligned}
\bar{S}_{\max}(1) &\stackrel{\text{def}}{=} \bar{S}(1), \\
\bar{S}_{\max}(2m) &= \max\{\bar{S}_{\max}(m), \bar{S}(2m)\}, \qquad (A.2) \\
\bar{S}_{\max}(2m+1) &= \max\{\bar{S}_{\max}(m), \bar{S}(2m+1)\}.
\end{aligned}
$$

Eq.A.2 is easily implemented in Octave. Following the hint, start with NRTmin() and modify lines 15 and 16, replacing **min** with **max**.

```
1  function Maxb = NRTmax(Sbar, N)
2  % Octave/MATLAB function to compute maximums
3  % along paths in a non–recombining tree (NRT).
4  % INPUTS:                              (Example)
5  %     Sbar = NRT array of length 2*2^N-1   (1:15)
6  %     N = tree depth, must be >=0          (3)
7  % OUTPUT:
8  %     Maxb = NRT of maximums, same size as Sbar
9  % EXAMPLE:
```

```
10  %   Maxb = NRTmax(1:15,3)
11  %
12     Maxb=zeros(size(Sbar)); % allocate the output
13     Maxb(1)=Sbar(1); % max along the trivial path
14     for m=1:2^N-1   % all future times up to N-1
15       Maxb(2*m)=max(Maxb(m),Sbar(2*m));    % down
16       Maxb(2*m+1)=max(Maxb(m),Sbar(2*m+1)); % up
17     end
18     return
19  end
```

□

8. Implement lookback option pricing in the CRR model using path-dependent Arrow-Debreu securities. Check that the results agree with CRRlb.

   Solution: The Octave function CRRlbAD() differs from CRRlb() in three ways, just as CRRaroAD() differs from CRRaro():

```
1  function [C0,P0] = CRRlbAD(T,S0,r,v,N)
2  % Octave/MATLAB function to price Lookback Call
3  % and Put options using path-dependent Arrow-Debreu
4  % expansions with the Cox-Ross-Rubinstein (CRR)
5  % binomial pricing model.
6  % INPUTS:                             (Example)
7  %   T  =  expiration time in years     (1)
8  %   S0 =  spot stock price             (100)
9  %   r  =  riskless yield per year      (0.05)
10 %   v  =  volatility; must be >0       (0.20)
11 %   N  =  height of the tree           (4)
12 % OUTPUTS:
13 %    C0 = Call option premium at t=0
14 %    P0 = Put option premium at t=0
15 % EXAMPLE:
16 %    [C0,P0]=CRRlb(1,100,0.05,0.20,4)
17 %
18    [pu,up,R]=CRRparams(T,r,v,N);
19    Sbar=NRTCRR(S0,up,1/up,N); % expanded S tree
20    MinS=NRTmin(Sbar,N); MaxS=NRTmax(Sbar,N);
21    Lbar=PathAD(pu*ones(N,N),R*ones(N,N),N);
22    mN=2^N:(2*2^N-1); % all m-indices at expiry
23    C0=(Sbar(mN)-MinS(mN))*Lbar(mN)'; % Call
24    P0=(MaxS(mN)-Sbar(mN))*Lbar(mN)'; % Put
25    return % ..inner producs give A-D expansions
26 end
```

Compare (and test for bugs) by running both functions on the same suggested example inputs:

```
[CO,PO]=CRR1bAD(1,100,0.05,0.20,4)
[C,P]=CRR1b(1,100,0.05,0.20,4); C(1),P(1)
```

The outputs agree: CO=13.758=C(1) and PO=9.6589=P(1).   □

9. Implement ladder Put option pricing in the CRR model by modifying
   CRRladC appropriately. Check that the premium is at least as great as
   that for the vanilla European-style Put with the same parameters.

   Solution: For the ladder Put option, there is a specified strike price K
   and a decreasing ladder $\{L_i\}$ of in-the-money prices below K:

   $$S(0) \geq \min\{S(0), K\} > L_1 > \cdots > L_k.$$

   Let $M(T) = \min\{S(t) : 0 \leq t \leq T\}$ be the minimum price of S up to
   expiry. Then the payoff is

   $$P(T) \stackrel{\text{def}}{=} \max\{[K - S(T)]^+, K - L_i\}, \qquad (A.3)$$

   where L_i is the lowest rung reached by $S(t)$ as recorded by $M(T)$:

   $$L_{i+1} < M(T) \leq L_i.$$

   In the discrete CRR implementation with N time steps to expiry, both S
   and M are stored as linear arrays with m indexing. The paths of length
   N are indexed by $m \in \mathcal{M}(N) = \{2^N, \ldots, 2^N + (2^N - 1)\}$, so those are
   the indices in the non-recombining tree P where terminal values are set
   using Eq.A.3. Then the option premium is found by backward induction
   as usual:

   $$P(m) = \frac{pP(2m+1) + (1-p)P(2m)}{R}, \quad m = 2^N - 1, 2^N - 2, \ldots, 2, 1,$$

   where p is the risk neutral up probability from the CRR model, and R
   is the riskless return per time step.

```
1   function LadP = CRRladP(T,S0,K,L,r,v,N)
2   % Octave/MATLAB function to price a ladder Put
3   % option using a Cox-Ross-Rubinstein (CRR) model.
4   % INPUTS:                              (Example)
5   %    T  = expiration time in years     (1)
6   %    S0 = spot price                   (50)
7   %    K  = strike price                 (55)
8   %    L  = decreasing prices < S0,K     ([45,40,35])
9   %    r  = riskless yield per year      (0.05)
10  %    v  = volatility; must be >0       (0.20)
11  %    N  = height of the tree           (4)
12  % OUTPUTS:
13  %    LadP = Ladder Put option price array.
14  % EXAMPLE:
```

```
15  %     LadP=CRRladP(1,50,55,[45,40,35],0.05,0.20,4)
16  %
17      [pu,up,R] = CRRparams(T,r,v,N);
18      Sbar=NRTCRR(S0,up,1/up,N); % S as an NRT
19      MinS=NRTmin(Sbar,N); % non-recombining min tree
20      LadP=zeros(size(Sbar)); % Put prices NRT
21      % Initialize with the payoffs at expiry:
22      k=length(L);           % number of ladder levels, >1
23      for m=2^N:2*2^N -1 % state indexes at expiry
24        if(MinS(m)>L(1)) % above level L(1) is special
25          LadP(m)=max(K-Sbar(m),0);
26        else % MinS =< L(1)
27          if(MinS(m)>L(k)) % ... use levels L(2)>...>L(k)
28            for l=2:k       % loop to find the level
29              if(MinS(m)>L(l))
30                LadP(m)=max(max(K-Sbar(m),K-L(l-1)),0);
31                break; % found l: L(l)<MinS=<L(l-1),
32              end
33            end             % ... so exit the "l" loop
34          else % MinS=<L(k), another special case
35            LadP(m)=max(max(K-Sbar(m),K-L(k)),0);
36          end
37        end
38      end
39      for m=(2^N-1):-1:1 % backward recursion
40        LadP(m)=(pu*LadP(2*m+1)+(1-pu)*LadP(2*m))/R;
41      end
42      return % LadP(1) is the option premium
43  end
```

Note that, as with the Call,

$$P(T) \geq [K - S(T)]^{+} \geq 0,$$

so this exotic option cannot cost less than a vanilla European Put with the same parameters. A few experiments will show how big the differences can be, for example,

```
LadP=CRRladP(1,50,55,[45,40,35],0.05,0.20,4);LadP(1) % 5.8784
[eC,eP]=CRReur(1,50,55,0.05,0.20,4);eP(1,1)          % 5.5372
```

The returned values round to \$5.88 for the ladder Put versus \$5.54 for the vanilla European Put, ordered as expected.   □

A.5   ... to Chapter 5 Exercises

1. Suppose that S and R are modeled with a recombining binomial tree of $N \geq 1$ levels. Prove that

$$G(N-1,j) \;=\; F(N-1,j) \;=\; S(N-1,j)R(N-1,j)$$

for all states j.

Solution: From Eq.5.4 it follows that

$$F(N-1,j) = S(N-1,j)R(N-1,j).$$

It remains to compute $G(N-1,j)$ with the backward induction formula. Since $G(N,j) = S(N,j)$ for all j,

$$
\begin{aligned}
G(N-1,j) &= \pi G(N,j+1) + (1-\pi)G(N,j) \\
&= \pi S(N,j+1) + (1-\pi)S(N,j) \\
&= \frac{\pi S(N,j+1) + (1-\pi)S(N,j)}{R(N-1,j)} R(N-1,j) \\
&= S(N-1,j)R(N-1,j),
\end{aligned}
$$

where $\pi \stackrel{\text{def}}{=} p(N-1,j)$ is the risk neutral up probability at $n = N-1$ in state j. Conclude that

$$F(N-1,j) = S(N-1,j)R(N-1,j) = G(N-1,j).$$

Remark. Since $F(N,j) = G(N,j) = S(N,j)$ for all states j, both the ultimate $(n = N)$ and the penultimate $(n = N-1)$ no-arbitrage prices of Forwards and Futures must be equal in all states of a recombining binomial model.   □

2. Suppose that S and R are modeled with a recombining binomial tree of $N = 2$ levels. Prove that if $(R(1,1) - R(1,0))(S(1,1) - S(1,0)) > 0$, then $G(0,0) > F(0,0)$.

Solution: Use the results from Exercise 1 above:

$$
\begin{aligned}
F(N,j) &= G(N,j) &= S(N,j) \\
F(N-1,j) &= G(N-1,j) &= S(N-1,j)R(N-1,j).
\end{aligned}
$$

Now fix $N = 2$ and suppose that

$$(R(1,1) - R(1,0))(S(1,1) - S(1,0)) > 0.$$

It may be assumed that $R(1,1) > R(1,0) > 0$ and $S(1,1) > S(1,0) > 0$, since these quantities must be positive, relabeling the up and down states if necessary. But then with the above equations,

$$G(1,1) = R(1,1)S(1,1) > R(1,0)S(1,0) = G(1,0)$$

Write $\pi \stackrel{\text{def}}{=} p(0,0)$, the risk neutral up probability, to compute

$$G(0,0) = \pi G(1,1) + (1-\pi)G(1,0),$$

using Eq.5.8. Likewise, compute the Forward price by backward induction for S and D with that π:

$$
\begin{aligned}
F(0,0,2) &= S(0,0)/D(0,0) \\
&= \frac{[\pi S(1,1) + (1-\pi)S(1,0)]/R(0,0)}{[\pi D(1,1) + (1-\pi)D(1,0)]/R(0,0)} \\
&= \frac{\pi S(1,1) + (1-\pi)S(1,0)}{\pi D(1,1) + (1-\pi)D(1,0)} \\
&= \frac{\pi S(1,1) + (1-\pi)S(1,0)}{\pi/R(1,1) + (1-\pi)/R(1,0)} \\
&= \frac{\pi G(1,1)/R(1,1) + (1-\pi)G(1,0)/R(1,0)}{\pi/R(1,1) + (1-\pi)/R(1,0)} \\
&= \theta G(1,1) + (1-\theta)G(1,0),
\end{aligned}
$$

where

$$
\begin{aligned}
\theta &\stackrel{\text{def}}{=} [\pi/R(1,1)]/[\pi/R(1,1) + (1-\pi)/R(1,0)] \\
&= \pi R(1,0)/[\pi R(1,0) + (1-\pi)R(1,1)] \\
&< \pi,
\end{aligned}
$$

since the denominator is between $R(1,0)$ and $R(1,1) > R(1,0)$. Thus also

$$1 - \theta > 1 - \pi.$$

Note that $\{\theta, 1 - \theta\}$ is a probability function with a lesser weight $\theta < \pi$ for $G(1,1,2)$ and greater weight $1 - \theta > 1 - \pi$ for $G(1,0,2)$. But since $G(1,1,2) > G(1,0,2) > 0$, it follows that

$$
\begin{aligned}
F(0,0,2) &= \theta G(1,1,2) + (1-\theta)G(1,0,2) \\
&< \pi G(1,1,2) + (1-\pi)G(1,0,2) \\
&= G(0,0,2),
\end{aligned}
$$

so $G(0,0,2) > F(0,0,2)$, as claimed. $\qquad\square$

3. Suppose that S and R are modeled with a recombining binomial tree of $N > 2$ levels. Prove that if

$$(R(N-1,j+1) - R(N-1,j))(S(N-1,j+1) - S(N-1,j)) > 0$$

for all $j = 0, 1, ..., N - 2$, then $G(0,0) > F(0,0)$.

Solution: For each $0 \leq j \leq N - 2$, there is a two-level recombining binomial tree below $(N - 2, j)$ with this portion of the S and R model:

$$S(N-2, j)$$
$$S(N-1, j) \quad S(N-1, j+1) \qquad\qquad R(N-2, j)$$
$$S(N, j) \quad\quad S(N, j+1) \quad\quad S(N, j+2) \qquad R(N-1, j) \quad R(N-1, j+1)$$

Adjusting the time to $N \leftarrow 2$ and renumbering the state to $j \leftarrow 0$ results in the two-level model

$$S(0,0)$$
$$S(1,0) \quad S(1,1) \qquad\qquad R(0,0)$$
$$S(2,0) \quad S(2,1) \quad S(2,2) \qquad R(1,0) \quad R(1,1)$$

Likewise, the hypothesis thus renumbered goes from

$$(R(N - 1, j + 1) - R(N - 1, j))(S(N - 1, j + 1) - S(N - 1, j)) > 0$$

to

$$(R(1, 1) - R(1, 0))(S(1, 1) - S(1, 0)) > 0$$

Then apply the $N = 2$ result from Exercise 2 above to conclude that

$$G(N - 2, j) > F(N - 2, j)$$

for all $0 \leq j \leq N - 2$.

Now suppose that $G(0,0) \leq F(0,0)$. Then there is an arbitrage:

- At time 0, sell one Short Forward at strike price $F(0,0)$ and buy one Long Future at strike price $G(0,0)$. Total expense is \$0.
- At time $N - 2$, in whatever state j, buy one Long Forward at $F(N-2, j)$ and sell one Short Future at $G(N-2, j)$. Total expense is again \$0.
- At time N,
  - borrow $G(0,0)$ cash from the bank,
  - buy a share of S for $G(0,0)$ to settle the Long Future,
  - sell the share of S to the Short Forward counterparty for $F(0,0)$,
  - buy a share of S from the Long Forward counterparty for $F(N - 2, j)$,
  - sell the share of S to the Short Future counterparty for $G(N - 2, j)$.
  - Return $G(0,0)$ to the bank.

Net proceeds will be

$$F(0,0) - F(N-2,j) + G(N-2,j) - G(0,0) > 0,$$

with no remaining liabilities.

Conclude by the no-arbitrage axiom that $F(0,0) < G(0,0)$. □

4. Fix $N = 4$ in the function FwdFut() defined on p.135.

(a) Find inputs S and R such that $F(0) > G(0)$, namely F(1,1)>G(1,1) in the output of [F,G]=FwdFut(S,R,N).

(b) Find S and R such that $F(0) < G(0)$, namely F(1,1)<G(1,1) in the output of [F,G]=FwdFut(S,R,N).

Note: To be valid, solutions S and R in parts (a) and (b) must be positive and must satisfy the no-arbitrage condition

$$S(n+1,j) < R(n,j)S(n,j) < S(n+1,j+1)$$

for all $0 \le n < N$ and all $0 \le j \le n$.

Solution: First note that the function pu=RiskNeut(S,R,N) will return risk neutral up probabilities satisfying

$$0 < \text{pu}(n+1,j+1) = \frac{R(n,j)S(n,j) - S(n+1,j)}{S(n+1,j+1) - S(n+1,j)} < 1,$$

iff the no-arbitrage conditions hold for all $0 \le n < N$ and $0 \le j \le n$. Thus it may be used to check candidate solutions. Roughly speaking, the conditions will hold if $R(n,j) \approx 1$ and $S(n,j)$ has sufficient volatility so that the up and down factors are far from 1.

Note also that solutions are not unique. For example, if S solves part (a) for some particular R, then so does cS for any $c > 0$ since both F and G are linear functions of S.

(a) Adding a row to the example on p.135 gives

$S =$	100					$R =$	1.025			
	85	115					1.03	1.02		
	75	100	130				1.035	1.025	1.015	
	65	85	115	150			1.04	1.03	1.02	1.01
	55	80	100	130	170					

Enter this into Octave as follows:

```
Sa=[100,0,0,0,0; 85,115,0,0,0; 75,100,130,0,0;
    65,85,115,150,0; 55,80,100,130,170];
Ra=[1.025,0,0,0; 1.03,1.02,0,0; 1.035,1.025,1.015,0;
    1.04,1.03,1.02,1.01];
```

Check the no-arbitrage conditions by running `RiskNeut(Sa,Ra,4)`, which returns valid probabilities:

```
0.58333
0.50200   0.57667
0.63125   0.58333   0.48429
0.50400   0.37750   0.57667   0.53750
```

Finally, compute $F(0) = $ `Fa(1,1)` $= 109.92 > 109.52 = $ `Ga(1,1)` $= G(0)$:

```
[Fa,Ga]=FwdFut(Sa,Ra,4); Fa(1,1),Ga(1,1)
```

(b) It is suggested by the previous exercises that if S, R solves part (a), then reversing the numbering of the R states will solve part (b). So, put

```
Sb=Sa; Rb=[1.025, 0,     0,     0;
           1.02,  1.03,  0,     0;
           1.015, 1.025, 1.035, 0;
           1.01,  1.02,  1.03,  1.04];
```

Confirm the no-arbitrage conditions with `RiskNeut(Sb,Rb,4)`, and then compute:

```
[Fb,Gb]=FwdFut(Sb,Rb,4); Fb(1,1),Gb(1,1)
```

which returns $F(0) = $ `Fb(1,1)` $= 110.81 < 111.32 = $ `Gb(1,1)` $= G(0)$. □

5. Suppose that a commodities exchange wishes to broker Futures contracts on an asset S, expiring in 0.5 years while riskless annual interest rates are expected to remain constant at 0.02%. Under consideration are margin requirements of 20, 30, 50, 80, and 150% of S_0 for each contract.

(a) Compute the probability of a margin call, which will occur if and only if the margin balance falls below 0, for both Long and Short Futures contracts, with these five margin requirements. Use $N = 6$ time steps and volatilities $\sigma \in \{0.10, 0.15, 0.20, 0.25, 0.30, 0.40, 0.50, 0.70\}$. Tabulate the results and compare Long and Short margin requirements.

(b) Profile the computation with $N = 6$ and again with $N = 13$ to compare the run times. Include in the count all times above 1% of the total. Compare the run times to test the $O(2^N)$ order of complexity.

Solution: (a) The following Octave commands use the programs in the textbook to fill tables `TabL` and `TabS` with the Long and Short margin call probabilities, respectively:

```
T=0.5; r=0.02; N=6; S0=100; % S0>0 is arbitrary
as=[0.20,0.30,0.50,0.80,1.50]; % alphas
vs=[0.10,0.15,0.20,0.25,0.30,0.40,0.50,0.70]; % volatilities
TabL=zeros(length(vs),length(as)); % output table - Longs
```

```
TabS=zeros(length(vs),length(as)); % output table - Shorts
for i=1:length(vs)
  v=vs(i);
  for j=1:length(as)
    a=as(j);
    [ML,MS,Pr] = CRRmargin(T,S0,a*S0,a*S0,r,v,N);
    TabL(i,j) = PrNeg(ML, Pr, 1, N);
    TabS(i,j) = PrNeg(MS, Pr, 1, N);
  end
end
```

The results, tabulated by volatility σ and margin deposit multiple, are in the table below:

| σ (%) | α (%) TabL | | | | α (%) TabS | | | | |
	20	30	50	80	20	30	50	80	150
10	0	0	0	0	0	0	0	0	0
15	0.029	0	0	0	0.034	0	0	0	0
20	0.125	0.016	0	0	0.125	0.031	0	0	0
25	0.227	0.033	0	0	0.210	0.119	0.014	0	0
30	0.235	0.121	0	0	0.203	0.113	0.028	0	0
40	0.493	0.182	0.020	0	0.414	0.191	0.105	0.012	0
50	0.508	0.261	0.042	0	0.399	0.399	0.181	0.023	0
70	0.538	0.538	0.177	0	0.622	0.371	0.163	0.163	0.02

Long Futures (TabL) and Short Futures (TabS) margin call probabilities ($N = 6$).

The probability of a margin call mostly increases with volatility and decreases with increasing margin deposit. This experiment suggests that, to avoid margin calls at the same volatiliy, the Short margin account needs a larger initial deposit than the Long.

(b) To profile the run times, the Octave commands in part (a) may be enclosed as follows:

```
profile off; profile clear; profile on
T=0.5; r=0.02; N=13; S0=100; % S0>0 is arbitrary
% (insert code from above)
profile off; profshow  % N=13 results
profile clear; profile on
T=0.5; r=0.02; N=6;  S0=100; % S0>0 is arbitrary
% (insert code from above)
profile off; profshow  % N=6 results
```

The results will be machine dependent. One example gave $5.4s$ for $N = 6$ and $590s$ for $N = 13$. The ratio is approximately 110, close to the expected ratio $2^{13}/2^6 = 2^7 = 128$ for $O(2^N)$ complexity. $\square$

A.6   ... to Chapter 6 Exercises

1. Suppose that S is a dividend paying stock with a declared dividend amount D at ex-dividend date t^e. Let $S(t^e-)$ denote the closing price of S on the day before the ex-dividend date, and $S(t^e+)$ denote the opening price of S on the ex-dividend date. Show that

$$S(t^e+) = S(t^e-) - D$$

by constructing an arbitrage otherwise.

Solution: If $S(t^e+) > S(t^e-) - D$, then

- buy S at time t^e-: $-S(t^e-)$
- collect D at time t^p: $+D$
- sell S at t^e+: $+S(t^e+)$

Net: $S(t^e+) + D - S(t^e-) > 0$.
Else $S(t^e+) < S(t^e-) - D$, so

- sell S short at time t^e-: $+S(t^e-)$
- pay out D at time t^p: $-D$
- buy S to cover at t^e+: $-S(t^e+)$

Net: $S(t^e-) - D - S(t^e+) > 0$.

In either case there is profit with no risk. Here it is assumed that any borrowing is cost-free since t^e+, t^e-, and t^p are nearly equal. Note that the borrower of shares sold short is responsible for paying any dividend owed to the lender. □

2. Prove Eq.6.9 from Eq.6.8.

Solution: Substitute $n \leftarrow n + 1$ into Eq.6.8 to get

$$I(n+1) = \sum_{k=1}^{\infty} \frac{d_{n+1+k}}{R_{n+1} \cdots R_{n+1+k-1}} = \sum_{k=2}^{\infty} \frac{d_{n+k}}{R_{n+1} \cdots R_{n+k-1}},$$

where the sums are actually finite because $d_{N+2} = d_{N+3} = \cdots = 0$. It follows that

$$\frac{I(n+1)}{R_n} = \sum_{k=2}^{\infty} \frac{d_{n+k}}{R_n R_{n+1} \cdots R_{n+k-1}} = I(n) - \frac{d_{n+1}}{R_n}.$$

Adding d_{n+1}/R_n to both sides gives Eq.6.9. □

3. Use the data in Table 6.1 (p.146) and Table 6.2 (p.154) to compute the present value sequence of BAC dividends from December 2014 through March 2021, namely time indices 0–25. Disregard any dividends outside of that time period.

(a) Find the values quarterly.

(b) Find the values monthly.

(c) Plot the two dividend present value sequences, and the dividends themselves, on the same graph.

Solution: Reuse the Octave commands that produced `BACdivQ`, `ri`, and so on.

(a) First, create the array `BACdivQ(1:26)`, the sequence of 26 quarterly dividends from Table 6.1, using the code on p.157.

Next, create the array `ri(1:25)`, the annualized riskless rates starting in December 2014 and sampled quarterly thereafter, using the code on p.158. These are exponentiated to give quarterly returns.

Finally, apply the recursion formula from Lemma 6.4 to find the quarterly present value sequence `IpvQ`:

```
% Part (a):
Na=24; IpvQ=zeros(1,Na+1); % allocate the output array
RQ=exp(ri/4); % annualized rates ==> quarterly returns
% Backward recursion for IpvQ
IpvQ(Na+1)=BACdivQ(Na+2)/RQ(Na+1); % ignore later dividends
for n = (Na-1):(-1):0        % textbook times {Na-1,...,1,0}
   IpvQ(n+1)=(IpvQ(n+2)+BACdivQ(n+2))/RQ(n+1); % from Lemma
end
```

Remark. These computations are all done in the function `CRRmD()` which may be reused as follows:

```
T=6; Na=24; S0=17.90; sigma=0.1779;
[S,Sx,IpvQ]=CRRmD(T,S0,BACdivQ,ri,sigma,Na);
```

The values S_0 and σ are not used when computing `IpvQ`. Other nominal values such as $S_0 = \sigma = 1$ will give the same result.

(b) Create the array `BACdivM(1:26)`, the sequence of 72 monthly dividends, using the code on p.158. Create the array `r3i`, the annualized riskless rates replicated monthly, using the code on p.159. These are exponentiated to give monthly returns. Then apply the recursion formula from Lemma 6.4 to find `IpvM`:

```
% Part (b):
Nb=72; IpvM=zeros(1,Nb+1); % allocate the output array
RM=exp(r3i/12); % annualized rates ==> monthly returns
```

```
% Backward recursion for IpvM
IpvM(Nb+1) = BACdivM(Nb+2)/RM(Nb+1); % ignore later dividends
for n = (Nb-1):(-1):0 % textbook times {Nb-1,...,1,0}
   IpvM(n+1) = (IpvM(n+2)+BACdivM(n+2))/RM(n+1);
end
```

CRRmD() may be used instead, as follows, with the nominal values $S_0 = K = \sigma = 1$:

```
T=6; Nb=72; S0=1; sigma=1;
[S,Sx,IpvM]=CRRmD(T,S0,BACdivM,r3i,sigma,Nb);
```

(c) Use the plotting function as in the Octave code below.

```
% Part (c):
tM=0:72; tQ=0:3:72; tD=0:3:75; % time indices
plot(tM,IpvM,"b--", tQ,IpvQ,"r-", tD,BACdivQ,"k*");
xlabel("Months from start of 2015");
ylabel("Dollars"); legend("IpvM","IpvQ","BACdivQ");
title("BAC dividends as cash flow present values");
```

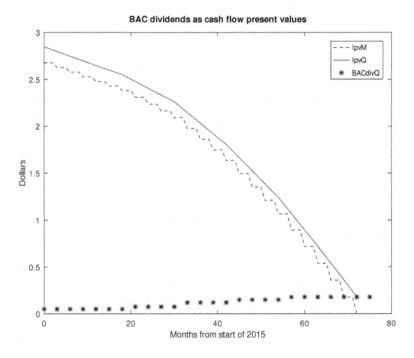

FIGURE A.7
Graph from Exercise 3.

The results may be seen in Figure A.7. Note that the quarterly present value sequence lies strictly above the monthly, since it includes the March 2021 dividend that the monthly sequence does not.   □

4. For the experiments below, use CRRDaeC() with the parameters $T = 1$, $S_0 = 100$, $K = 101$, $r = 0.02$, $v = 0.20$, and $N = 12$.

(a) Find a nonzero dividend sequence for which early Call exercise is sometimes optimal.

(b) Find another nonzero dividend sequence for which early Call exercise is never optimal.

Solution: (a) Sufficiently large dividends will make early exercise optimal. For the given parameters $r = 0.02$, $T = 1$, $K = 101$, and $N = 12$ of this experiment, quarterly \$1 dividends are sufficient:

```
T=1; S0=100; K=101; r=0.02; v=0.20, N=12;
Diq1=[0 1 0 0 1 0 0 1 0 0 1 0 0 1]; % Quarterly 1.00
[Ca,Ce,EE]=CRRDaeC(T,S0,K,Diq1,r,v,N);
EE     % shows where early exercise is optimal
Ca>Ce  % shows how early exercise adds value
```

From EE it may be seen that early exercise is optimal at states $(n, j) \in \{(6, 5), (6, 6), (9, 7), (9, 8), (9, 9)\}$ before expiry.

(b) By the Remark on p.151 concerning Lemma 6.2, very small positive dividends will not make early exercise optimal for American-style Calls. The threshold is roughly

$$d < (R - 1)K = (\exp(rT/N) - 1)K \approx 0.17,$$

So even with quarterly dividends of 0.16, early exercise will never be optimal in this 12-step model. Check this with the previously entered parameters:

```
Dtiny = (exp(r*T/N)-1)*K   % about 0.16847
[Ca,Ce,EE]=CRRDaeC(T,S0,K,Dtiny*Diq1,r,v,N); % Qrtrly Dtiny
EE     % shows that early exercise is never optimal
Ca>Ce  % shows how early exercise adds no value
```

Of course, these solutions are not unique. □

5. Implement CRR pricing for American and European Put options on an asset with dividends, using decomposition into risky ex-dividend and riskless cash flow portions. (Hint: Make a few changes to CRRDaeC().)

For the experiments below, set $T = 1$, $S_0 = 100$, $K = 101$, $r = 0.02$, $v = 0.20$, and $N = 12$.

(a) Find a nonzero dividend sequence for which early Put exercise is sometimes optimal.

(b) Find a nonzero dividend sequence for which early Put exercise is never optimal.

Solution: Following the hint, change the payoffs from $[S - K]^+$ to $[K - S]^+$ in CRRDaeC() to get the CRR pricing function for Puts:

```
1   function [Pa,Pe,EE] = CRRDaeP(T,S0,K,Di,r,v,N)
2   % Octave/MATLAB function to price American and
3   % European Put options on an asset decomposed
4   % into S=Sx+Dpv by CRRD(), using the CRR model.
5   % INPUTS:                                    (Example)
6   %    T =  expiration time                    (1 year)
7   %    S0 = stock price                        ($100)
8   %    K =  strike price                       ($101)
9   %    Di = dividend sequence              (see below)
10  %    r =  risk−free yield                    (0.02)
11  %    v =  volatility; must be >0             (0.15)
12  %    N =  height of the binomial tree          (12)
13  % OUTPUT:
14  %    Pa =  price of American Put at all (n,j).
15  %    Pe =  price of European Put at all (n,j).
16  %    EE =  Is early exercise optimal at (n,j)?
17  % EXAMPLE:
18  %    Di = [0 1 0 0 1 0 0 2 0 0 2 0 0 2]; % Di(k)=D_k
19  %    [Pa,Pe,EE]=CRRDaeP(1,100,101,Di,0.02,0.15,12);
20  %
21      [pu,up,R]=CRRparams(T,r,v,N); % Use CRR values
22      [S,Sx,Dpv]=CRRD(T,S0,Di,r,v,N); % decompose
23      Pa=zeros(N+1,N+1); Pe=zeros(N+1,N+1); % outputs
24      EE=zeros(N,N); % early exercise T/F
25      for j = 0:N % to set terminal values at (N,j)
26        xP=K−S(N+1,j+1); % Put payoff at expiry
27        Pa(N+1,j+1) = Pe(N+1,j+1) = max(xP,0);
28      end
29      % Use backward induction pricing:
30      for n = (N−1):(−1):0 % times n={N−1,...,1,0}
31        for j = 0:n % states j={0,1,...,n} at time n
32          % Backward pricing for A and E:
33          bPe=(pu*Pe(n+2,j+2)+(1−pu)*Pe(n+2,j+1))/R;
34          bPa=(pu*Pa(n+2,j+2)+(1−pu)*Pa(n+2,j+1))/R;
35          xP=K−S(n+1,j+1); % Put exercise value
36          % Set prices at node (n,j):
37          Pe(n+1,j+1)=bPe; % always binomial price
38          Pa(n+1,j+1)=max(bPa,xP); % highest price
39          % Is early exercise optimal?
40          EE(n+1,j+1)=(xP>bPa); % Yes, if xP>bPa
41        end
42      end
43      return; % Pa, Pe, and EE are defined.
44  end
```

(It is a good idea to update the comments, as is done here.)

(a) Quarterly \$1 dividends provide an example where early exercise is sometimes optimal:

```
T=1; SO=100; K=101; r=0.02; v=0.20, N=12;
Diq1 = [0 1 0 0 1 0 0 1 0 0 1 0 0 1]; % Quarterly 1.00
[Pa,Pe,EE]=CRRDaeP(T,SO,K,Diq1,r,v,N);
EE      % shows where early exercise is optimal
Pa>Pe   % shows how early exercise adds value
```

From EE it may be seen that early exercise is optimal at states $(n, j) \in$ $\{(10,0), \ldots, (10,4); (11,0), \ldots, (11,5)\}$ before expiry. This is also clear from the premiums: Pa(1,1)=9.55 versus Pe(1,1)=9.41, so the early exercise option is worth a little more.

(b) Since there is no similar result as Lemma 6.2, it is not clear what makes early exercise nonoptimal for Puts. However, with a little experimentation one might discover that a \$1 monthly dividend suffices:

```
Dim1=[0 1 1 1 1 1 1 1 1 1 1 1 1]; % Monthly 1.00
[Pa,Pe,EE]=CRRDaeP(T,SO,K,Dim1,r,v,N);
EE      % all zeros, so early exercise is never optimal
Pa>Pe   % shows how early exercise adds no value
```

Remark.   Increasing the monthly dividend, say to \$2, preserves the never-optimal property:

```
[Pa,Pe,EE]=CRRDaeP(T,SO,K,2*Dim1,r,v,N); EE % still all zeros
```

Reducing the monthly dividend to the Call threshold Dtiny, on the other hand, has an effect:

```
Dtiny = (exp(r*T/N)-1)*K   % about 0.16847
[Pa,Pe,EE]=CRRDaeP(T,SO,K,Dtiny*Dim1,r,v,N); EE % some ones!
```

The presence of nonzeros in EE shows that early exercise is optimal at states $(n, j) \in \{(10,2), (10,3), (10,4), (11,2), (11,5)\}$ before expiry. Reducing the monthly dividend still further, say to Dtiny/2, adds even more early exercise states.   □

6. Suppose that government bonds are available with maturities of 1, 2, 3, and 4 years, with annual coupons for 0.5, 0.6, 0.8, and 1.1% of face value, respectively. Their spot prices are respectively 0.9994, 0.9992, 0.9989, and 0.9985 times face value.

(a) Compute the zero-coupon bond discounts $Z(0,1)$, $Z(0,2)$, $Z(0,3)$, and $Z(0,4)$ implied by these inputs.

(b) Plot the yield curve implied by $\{Z(0,T) : T = 1, 2, 3, 4\}$.

Solution: For this case, the linear system relating coupon bond discounts to zero coupon bond discounts is nonsingular.

(a) Write an Octave/MATLAB program to perform the calculation using forward substitution as in Eq.6.17:

```
c=[0.5,0.6,0.8,1.1]/100;  % coupon rates
B=[0.9994,0.9992,0.9989,0.9985]; % spot prices
Z=B; Z(1)=Z(1)/(1+c(1)); % allocate Z, set Z(1)
for i=2:4
 Z(i) = (Z(i)-c(i)*sum(Z(1:i-1)))/(1+c(i));
end
```

This returns $Z(i) = Z(0,i)$ values corresponding to

T	1	2	3	4
$Z(0,T)$	0.99443	0.98731	0.97524	0.95546

(b) Plot the yield curve with Octave/MATLAB after converting discounts to yields:

```
rtT=@(Z,t,T)log(Z)./(t-T); % convert discounts to yields
plot(1:4,100*rtT(Z,0,1:4)); % use percentages
title("Yield Curve"); xlabel("T (years)"); ylabel("APR (%)");
```

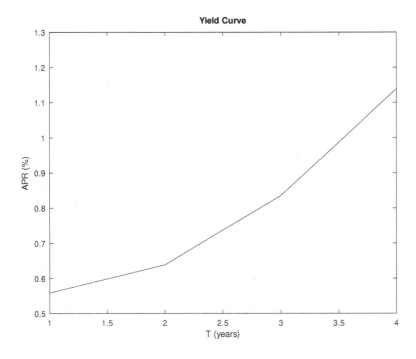

FIGURE A.8
Graph from Exercise 6.

The results may be seen in Figure A.8. □

7. Suppose that a company offers coupon bonds with maturities of 1, 2, 3, and 4 years, with semiannual coupons at 0.3, 0.4, 0.5, and 0.6% of face value, respectively. Their spot prices are respectively 0.9994, 0.9992, 0.9989, and 0.9985 times face value.

(a) Compute the zero-coupon bond discounts $\hat{Z}(0,1)$, $\hat{Z}(0,2)$, $\hat{Z}(0,3)$, and $\hat{Z}(0,4)$ implied by these inputs.

(b) Plot the yield curve implied by $\{\hat{Z}(0,T) : 0 < T \leq 4\}$.

Solution: For this case, the linear system relating coupon bond discounts to zero-coupon bond discounts is underdetermined, so use polynomial interpolation as in Eq.6.18.

(a) Write an Octave/MATLAB program to perform the calculation using built-in commands:

```
B=[0.9994,0.9992,0.9989,0.9985]; % spot discounts B(i)
c=[0.3,0.4,0.5,0.6]/100;  % coupon rates c(i)
T=[1,2,3,4]; % bond maturities T(i), in years
Del=0.5; N=T/Del; % number N(i) of semiannual coupons
bhat=(B-1-N.*c)'; % right-hand side column vector
M=4; Ahat=zeros(M,M); % allocate matrix for M=4 bonds
for i=1:M
  nDel=(1:N(i))*Del; % [Del, 2*Del, ..., N(i)*Del]
  for j=1:M
    Ahat(i,j)= (N(i)*Del)^j+c(i)*sum(nDel.^j);
  end
end
ahat=Ahat\bhat; % column of coefficients of Z polynomial
Zhat=@(t) 1+ahat'*[t; t.^2; t.^3; t.^4]; % use row vector t
Zhat(1:4)  % returns Z^(0,1),Z^(0,2),Z^(0,3),Z^(0,4)
```

This returns `Zhat(i)` $= \hat{Z}(0,i)$ values corresponding to

T	1	2	3	4
$\hat{Z}(0,T)$	0.99343	0.98335	0.96936	0.95157

(b) Plot the yield curve with Octave/MATLAB at many values of $T \in (0,4]$ after converting discounts to yields:

```
rtT=@(Z,t,T)log(Z)./(t-T); % convert discounts to yields
t=linspace(0.5,4,36); % 36 equispaced points in [0.5, 4]
plot(t,100*rtT(Zhat(t),0,t)); % use percentages
title("Company Yield"); xlabel("T (years)"); ylabel("APR (%)");
```

The results may be seen in Figure A.9.

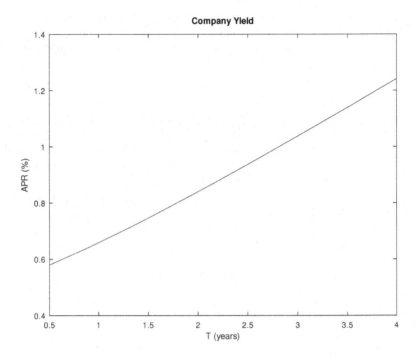

FIGURE A.9
Graph from Exercise 7.

□

8. Compute the price at issuance of a US Treasury Note with the following parameters:

   - Maturity in 7 years.
   - Semiannual coupon at 1.500% annual interest.
   - Face value $1000.
   - Yield to maturity 1.414%.

   Use both e^x and its approximations $1 + x$ and $1 - x$ in the present value calculation and compare the results.

   Solution: Implement Eq.6.16 using Octave commands.

```
T=7; N=2; I=1.5/100; F=1000; r=1.414/100;
P1=F*exp(-r*T)+(F*I/N)*sum(exp(-r*(1:(N*T))/N)); % 1005.38
P2=F/(1+r)^T+(F*I/N)*sum((1+r/N).^(-(1:(N*T)))); % 1006.03
P3=F*(1-r)^T+(F*I/N)*sum((1-r/N).^(1:(N*T)));    % 1004.72
```

Here P1 uses the exponential formula, while P2 and P3 use the two Taylor approximations $e^x \approx 1 + x$ and $e^{-x} \approx 1 - x$. All three formulas indicate a premium rather than a discount, but P2 overweights the future payments while P3 underweights them.   □

9. Compute the expected monthly riskless rates and returns over 6 months for two currencies using tabulated benchmarks:

(a) Use the data in Table 6.5 for 11 March 2022 to compute the Australian dollar values.

(b) Use the LIBOR data in Table 6.7 for 11 March 2022 to compute the US dollar values. Justify your interpolation method.

Solution: (a) Get monthly expected riskless rates from the 11 March 2022 row (11/03/2022), then exponentiate it to get riskless returns with these Octave commands:

```
bbsw=[0.0150,0.0661,0.1450,0.2550,0.3859,0.5100]; % 11/3/2022
rba=12*diff([0,bbsw]) % 0.18 0.6132 0.94680 1.32 1.5708 1.4892
Rba=exp(rba/1200) % 1.0002 1.0005 1.0008 1.0011 1.0013 1.0012
```

Note that BBSW is given as annualized percentages, so that it is necessary to divide by 100 as well as by 12 before exponentiation to get the monthly riskless returns $\bar{R}_a$.

(b) There are three given values in the tenor range 1–6 months, so use quadratic interpolation with the 1, 3, and 6 month tenors to find the 2, 4, and 5 tenors:

```
libor=[0.39657,0.82600,1.13057]; % 11 Mar 2022, 1,3,6 mos.
tenor=[1,3,6]; months=[1,2,3,4,5,6]; % evaluation months
p=polyfit(tenor,libor,2); bench=polyval(p,months)
% bench = 0.39657 0.63392 0.82600 0.97280 1.07432 1.13057
rbb=12*diff([0,bench])
% rbb= 4.75884  2.84824  2.30492  1.76160  1.21828  0.67496
Rbb=exp(rbb/1200)
% Rbb= 1.0040   1.0024   1.0019   1.0015   1.0010   1.0006
```

As in part (a), divide the annualized percentages by 1200 and exponentiate to get the monthly riskless returns $\bar{R}_b$.   □

A.7 ...to Chapter 7 Exercises

1. Suppose that C and P are European-style Call and Put options, respectively, at strike price K and expiry T, for a risky underlying asset S

with spot price S_0. Show that $0 < C(0) \le S_0$ and $0 < P(0) \le K$. (Hint: construct an arbitrage otherwise.)

Solution: First note that $C(0) > 0$ and $P(0) > 0$ are consequences of Theorem 1.1.

Next, suppose that $C(0) > S_0$. Sell one C and buy one S at time $t = 0$, keeping the surplus $C(0) - S_0 > 0$. At expiry T, if the buyer exercises C, collect $K > 0$ in exchange for S. Otherwise, sell S for $S(T) > 0$.

Finally, suppose that $P(0) > K$. Sell one P for $P(0)$ at $t = 0$ and keep the proceeds. At expiry $t = T$, if the buyer exercises P, receive that S at strike price K, leaving a surplus $P(0) - K > 0$ and sell it for $S(T) > 0$, netting additional profit. Otherwise, keep the original $P(0)$ with no further obligations.

In all cases there is a positive payoff with no initial investment. Conclude by the no-arbitrage axiom that $0 < C(0) \le S_0$ and $0 < P(0) \le K$. □

2. Show that the Eq.7.11 and Eq.7.12 probabilities produce the Arrow-Debreu spot prices $\lambda(n, j)$ in Eq.7.10 using Jamshidian's forward induction, Eq.3.21 on p.69.

Solution: This may be proved by induction on n. For $n = 0$, the only Arrow-Debreu spot price is $\lambda(0,0) = 1$, so it agrees with the value $Q(0,0)/R^0 = 1$ from Jackwerth's construction.

Now suppose that the values $\{\lambda(n-1, j) : j = 0, 1, \ldots, n-1\}$ produced by Jamshidian's induction agree with Jackwerth's values and that p and $1 - p$ are given by Eq.7.11 and Eq.7.12, respectively. (As usual, take $\lambda(n, j) = 0$ if $j < 0$ or $j > n$.) Compute $\lambda(n, j)$ for $j = 0, 1, \ldots, n$ by substituting the expressions from Jackwerth's construction into Eq.3.21:

$$
\begin{aligned}
\lambda(n, j) &= \frac{1 - p(n-1, j)}{R}\lambda(n-1, j) + \frac{p(n-1, j-1)}{R}\lambda(n-1, j-1) \\
&= \frac{1}{R}\left(1 - w\left(\frac{j}{n}\right)\right)\left[\frac{Q(n, j)}{Q(n-1, j)}\right]\frac{Q(n-1, j)}{R^{n-1}} \\
&\quad + \frac{1}{R}w\left(\frac{j}{n}\right)\left[\frac{Q(n, j)}{Q(n-1, j-1)}\right]\frac{Q(n-1, j-1)}{R^{n-1}} \\
&= \left[\left(1 - w\left(\frac{j}{n}\right)\right) + w\left(\frac{j}{n}\right)\right]\frac{Q(n, j)}{R^n} = \frac{Q(n, j)}{R^n}.
\end{aligned}
$$

Thus Jamshidian's forward induction with Jackwerth's probabilities produces Jackwerth's Arrow-Debreu spot prices. □

3. Compute implied volatility for the data in Table 7.1 using both Black-Scholes and CRR with $N = 20$. Tabulate and compare the results.

Solution: Use the Octave code that produced Figure 7.1, with the following modifications:

```
Ks=40:46; Ts=[3,9,21,35,63]; % strikes and days to expiry
C=[4.14 4.35 4.85 4.60 5.15; ...
   3.20 3.42 3.50 4.00 4.35; ...
   2.24 2.72 2.89 3.15 3.65; ...
   1.30 1.51 1.96 2.44 2.87; ...
   0.63 0.86 1.34 1.71 2.39; ...
   0.23 0.45 0.84 1.24 1.76; ...
   0.07 0.22 0.52 0.88 1.39]; % Call premiums on 2021-12-14
S0=44.13; r=0.05; % spot price and riskless APR on 2021-12-14
VimpBS=zeros(length(Ks),length(Ts)); % BS implied volatilities
VimpCRR=zeros(length(Ks),length(Ts)); % CRR implied volatilities
minv=0.01; maxv=0.99; tol=0.00001; % bisection parameters
for col=1:length(Ts)
  T=Ts(col)/365; % time to expiry in years
  for row=1:length(Ks)
    f=@(v) BS(T,S0,Ks(row),r/100,v); % Black-Scholes Call
    VimpBS(row,col)=bisection(f,C(row,col),minv,maxv,tol);
    g=@(v) CRReur(T,S0,Ks(row),r/100,v,20)(1,1); % CRR Call
    VimpCRR(row,col)=bisection(g,C(row,col),minv,maxv,tol);
  end
end
```

The output is in Table A.1:

TABLE A.1
Implied volatilities by CRR ($N = 20$) and Black-Scholes methods, from December 14, 2021 closing prices for BAC American-style Call options at $r = 0.05\%$ and $S_0 = \$44.13$.

K	Dec17 CRR	BS	Dec23 CRR	BS	Jan7 CRR	BS	Jan21 CRR	BS	Feb18 CRR	BS
40	.50	.50	.54	.54	.54	.55	.36	.35	.37	.37
41	.56	.56	.49	.49	.35	.35	.39	.39	.35	.35
42	.47	.47	.52	.52	.39	.39	.36	.36	.34	.34
43	.36	.36	.30	.30	.31	.32	.33	.34	.31	.31
44	.35	.35	.29	.29	.30	.30	.30	.30	.32	.32
45	.34	.35	.29	.29	.28	.29	.29	.30	.29	.29
46	.36	.36	.30	.30	.28	.29	.30	.30	.29	.29

Remark. As expected, the two methods produce nearly identical values. To further check the results and the code, compare the Black-Scholes (BS) values with the σ columns of Table 7.1. □

4. The table below gives part of the options chain for American-style Calls on Bank of America common stock (BAC) as of closing on March 17, 2022, when the spot price was $43.03:

Strike	T= 1 d	8 d	15 d	21 d	27 d

price K	(3/18)	(3/25)	(4/01)	(4/08)	(4/14)
42.00	1.10	1.44	1.76	1.96	2.18
43.00	0.34	0.86	1.10	1.33	1.58
44.00	0.06	0.44	0.69	0.88	1.06
45.00	0.02	0.18	0.35	0.53	0.71
46.00	0.01	0.07	0.16	0.31	0.44
47.00	0.01	0.02	0.09	0.17	0.27

Also, the US T-bill rates for various maturities were

Date	4 wk	8 wk	13 wk	26 wk	52 wk
03/14/2022	0.22	0.30	0.45	0.84	1.20
03/15/2022	0.21	0.29	0.46	0.84	1.19
03/16/2022	0.23	0.28	0.43	0.84	1.26
03/17/2022	0.20	0.30	0.40	0.79	1.20

Use this data to compute and plot the volatility surface for BAC.

Solution: Use the commands that produced Figure 7.1, but with the data for this problem. Set the riskless rate for all calculations to be the 4-week rate averaged over the 4 days sampled, which is 0.215% APR.

```
Ks=42:47; Ts=[1,8,15,21,27]; % strikes and days to expiry
C=[1.10 1.44 1.76 1.96 2.18 ;...
   0.34 0.86 1.10 1.33 1.58 ;...
   0.06 0.44 0.69 0.88 1.06 ;...
   0.02 0.18 0.35 0.53 0.71 ;...
   0.01 0.07 0.16 0.31 0.44 ;...
   0.01 0.02 0.09 0.17 0.27 ]; % Call premiums on 2022-03-17
S0=43.03; r=0.215; % spot price and riskless APR
Vimp=zeros(length(Ks),length(Ts)); % implied volatilities
minv=0.01; maxv=0.99; tol=0.00001; % bisection parameters
for col=1:length(Ts)
 T=Ts(col)/365; % time to expiry in years
 for row=1:length(Ks)
   f=@(v) BS(T,S0,Ks(row),r/100,v); % Black-Scholes Call
   Vimp(row,col)=bisection(f,C(row,col),minv,maxv,tol);
 end
end
mesh(Ts,Ks,Vimp); % note the transposed order (T,K,V(K,T))
title("Implied Volatility Surface");
xlabel("T (days)"); ylabel("K");
```

The results may be seen in Figure A.10.   □

5. Suppose that a share of XYZ has a spot price of $47.12, that riskless interest rates for the next month are expected to be a constant 0.66% APR, and that the premiums for European-style Call options expiring in 4 weeks ($T = 4/52$) are as follows:

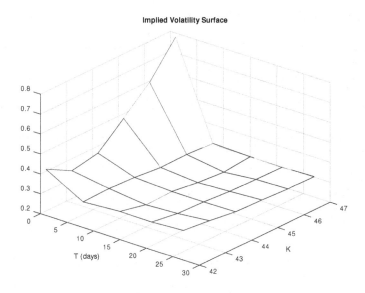

FIGURE A.10
Graph from Exercise 4.

```
Strike price:  45.00  46.00  47.00  48.00  49.00
Call premium:   3.52   2.78   2.10   1.44   1.37
```

(a) Construct an implied binomial tree for these inputs using Rubinstein's 1-2-3 algorithm. Display it along with the implied risk neutral up probabilities.

(b) Plot the three weight functions $w_1(x) = \sqrt{x}$, $w_2(x) = x^2$, and $w_3(x) = (1 - \cos(x/\pi))/2$, for $0 \le x \le 1$, on the same graph.

(c) Apply Rubinstein's 1-2-3 algorithm with Jackwerth's generalization to the data, using weights w_1, w_2, w_3 from part (b). Compare S, p, and Q for the three weights.

Solution:

(a) Apply Rubinstein's original 1-2-3 algorithm using IBT123J.m with the following Octave commands:

```
Ks=[45,46,47,48,49]; Cs=[3.52 2.78 2.10 1.44 1.37];
S0=47.12; r=0.66/100; rho=exp(r); w=@(x)x;
[S,Q,up,down,pu,N1]=IBT123J(S0,Ks,Cs,rho,w); S,pu
```

That produces the following output, edited for space savings:

```
S = % implied binomial tree    pu = % risk neutral up probs.
```

```
47.12                            0.5411
43.48 50.36                      0.3710 0.6853
41.14 47.63 51.73                0.0639 0.8916 0.5905
40.88 46.11 47.90 54.53          0.0559 0.1818 0.9779 0.3218
40.64 46.00 47.00 48.00 68.57
```

(b) The three weight functions all differ from $w(x) = x$: $w_1(x) = \sqrt{(x)}$ is concave, $w_2(x) = x^2$ is convex, and $w_3(x) = (1 + \cos(\pi x))/2$ has a unique inflection point. They may be plotted as in Figure A.11 using the following Octave/MATLAB commands:

```
w1=@(x)sqrt(x); w2=@(x)x.^2; w3=@(x)(1-cos(pi*x))/2;
t=0:0.01:1; plot(t,w1(t),"r--",t,w2(t),"b..",t,w3(t),"k-");
legend("w1","w2","w3","location","southeast");
title("Weight Functions for Jackwerth's Generalization");
```

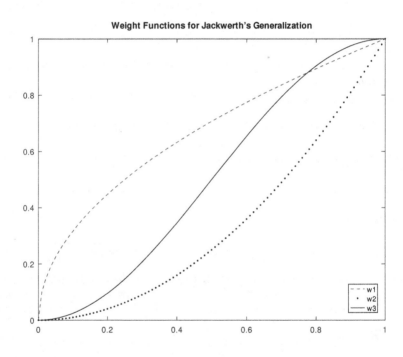

FIGURE A.11
Weight functions for Jackwerth's generalization in Rubinstein's 1-2-3 algorithm: $w_1(x) = \sqrt{(x)}$, $w_2(x) = x^2$, and $w_3(x) = (1 + \cos(\pi x))/2$.

(c) Run three experiments as follows, reusing previously assigned variables from parts (a) and (b):

```
[S1,Q1,up,down,pu1,N1]=IBT123J(S0,Ks,Cs,rho,w1); S1,Q1,pu1
```

```
[S2,Q2,up,down,pu2,N1]=IBT123J(S0,Ks,Cs,rho,w2); S2,Q2,pu2
[S3,Q3,up,down,pu3,N1]=IBT123J(S0,Ks,Cs,rho,w3); S3,Q3,pu3
```

The results, side-by-side, are:

```
S1 =                         S2 =                         S3 =
47.12                        47.12                        47.12
44.57 51.57                  41.74 49.74                  43.82 50.53
41.50 47.76 53.72            40.73 47.31 50.58            40.98 47.63 52.87
41.14 46.24 47.91 57.57      40.65 46.01 47.88 52.30      40.76 46.09 47.90 57.11
% ...each last S row is 40.64,  46.00,  47.00,  48.00,  68.57
Q1 =                         Q2 =                         Q3 =
1                            1                            1
.6246 .3754                  .3182 .6818                  .4963 .5037
.3110 .4436 .2455            .2657 .2100 .5243            .2793 .4340 .2867
.2853 .0444 .5202 .1500      .2589 .0617 .3492 .3303      .2639 .0616 .5170 .1574
% ...each last Q row is  0.2551,  0.0604,  0.0201,  0.5939,  0.0705
pu1 =                        pu2 =                        pu3 =
.3754                        .6818                        .5037
.5021 .6539                  .1650 .7690                  .4372 .5692
.0825 .9577 .6111            .0258 .7390 .6300            .0551 .8935 .5492
.1058 .3204 .9887 .4697      .0146 .0816 .9568 .2133      .0335 .1634 .9805 .4476
```

Note that there is one less row for the up probabilities. □

A.8 . . . to Chapter 8 Exercises

1. Prove that any subspace $V \subset \mathbf{R}^n$ is a closed convex cone.

   Solution: Check the needed properties:

   Cone: $\mathbf{v} \in V \implies \lambda\mathbf{v} \in V$ for any $\lambda > 0$, since any multiple of a vector is still in the subspace.

   Convex: For any $\mathbf{x}, \mathbf{y} \in V$ and any $\lambda \in [0,1]$, both $\lambda\mathbf{x} \in V$ and $(1-\lambda)\mathbf{y} \in V$ just as above. Then $\lambda\mathbf{x} + (1-\lambda)\mathbf{y} \in V$ because sums of vectors in V remain in V.

   Closed: If $V = \mathbf{R}^n$, then its complement $\mathbf{R}^n \setminus V = \emptyset$ is open, so V is closed. Otherwise, let $\mathbf{x} \in \mathbf{R}^n \setminus V$ be any point. By Lemma 8.12, there is some $\epsilon > 0$ such that $\|\mathbf{x} - \mathbf{y}\| \geq \epsilon$ for all $\mathbf{y} \in V$. Hence $B_\epsilon(\mathbf{x}) \subset \mathbf{R}^n \setminus V$, so the complement of V is open by definition, so V is closed. □

2. Prove that the closed orthant $K \in \mathbf{R}^n$ of vectors with nonnegative coordinates is a closed convex cone.

   Solution: Check the needed properties:

   Cone: $\mathbf{k} = (k_1, \ldots, k_n) \in K \iff (\forall i)k_i \geq 0$, so for any $\lambda > 0$, $\lambda\mathbf{k} = (\lambda k_1, \ldots, \lambda k_n) \in K$ because $(\forall i)\lambda k_i \geq 0$.

Convex: For any $\mathbf{x}, \mathbf{y} \in K$ and any $\lambda \in [0,1]$, the coordinates of $\lambda\mathbf{x} + (1-\lambda)\mathbf{y}$ will be

$$\lambda x_i + (1-\lambda)y_i \geq 0, \qquad i = 1, \ldots, n,$$

since x_i, y_i, λ, and $1-\lambda$ are all nonnegative. Hence $\lambda\mathbf{x} + (1-\lambda)\mathbf{y} \in K$.

Closed: It suffices to prove that the complement of K is open. But $\mathbf{x} = (x_1, \ldots, x_n) \in \mathbf{R}^n \setminus K \iff (\exists i)x_i < 0$. Suppose without loss of generality that $x_1 < 0$. Let $\epsilon = |x_1|/2$. Then for every $\mathbf{y} = (y_1, \ldots, y_n) \in B_\epsilon(\mathbf{x})$, it must be that $|y_1 - x_1| < \epsilon = |x_1|/2$, so $y_1 < 0$ as well. Conclude that $B_\epsilon(\mathbf{x}) \subset \mathbf{R}^n \setminus K$, so $\mathbf{R}^n \setminus K$ is an open set, so K is closed. $\quad\square$

3. Prove that the pointless orthant $K \setminus \mathbf{0}$ is a convex cone but is neither open nor closed.

Solution: Check the needed properties:

Cone: $\mathbf{k} = (k_1, \ldots, k_n)$ belongs to $K \setminus \mathbf{0}$ if and only if all coordinates k_i are nonnegative and at least one of them is positive. This property is preserved by multiplication by $\lambda > 0$.

Convex: For any $\mathbf{x}, \mathbf{y} \in K \setminus \mathbf{0}$ and any $\lambda \in [0,1]$, the coordinates of $\lambda\mathbf{x} + (1-\lambda)\mathbf{y}$ will be

$$\lambda x_i + (1-\lambda)y_i \geq 0, \qquad i = 1, \ldots, n,$$

since x_i, y_i, λ, and $1-\lambda$ are all nonnegative. Hence $\lambda\mathbf{x} + (1-\lambda)\mathbf{y} \in K$. It remains to show that some coordinate is positive, which can be done by checking two cases for $\lambda \in [0,1]$. By hypothesis, there is some i such that $x_i > 0$ and some j such that $y_j > 0$.

- If $\lambda > 0$, then $\lambda x_i + (1-\lambda)y_i > 0$.
- Else $\lambda = 0$, so $1 - \lambda = 1 > 0$, so $\lambda x_j + (1-\lambda)y_j = y_j > 0$.

Conclude that $\lambda\mathbf{x} + (1-\lambda)\mathbf{y} \in K \setminus \mathbf{0}$.

Not open: Every open ball centered at the point $(1, 0, \ldots, 0) \in K \setminus \mathbf{0}$ contains points with some negative coordinates which are therefore not in $K \setminus \mathbf{0}$. Hence $K \setminus \mathbf{0}$ is not open.

Not closed: Every open ball $B_\epsilon(\mathbf{0})$ centered at the point $\mathbf{0}$ in the complement of $K \setminus \mathbf{0}$ contains points with all positive coordinates which are therefore in $K \setminus \mathbf{0}$. Hence the complement of $K \setminus \mathbf{0}$ is not open. Hence $K \setminus \mathbf{0}$ is not closed. $\quad\square$

4. Prove that K^o is an open convex cone.

Solution: Check the needed properties:

Cone: $\mathbf{k} = (k_1, \ldots, k_n)$ belongs to K^o if and only if all coordinates k_i are positive. This property is preserved by multiplication by $\lambda > 0$.

Convex: For any $\mathbf{x}, \mathbf{y} \in K^o$ and any $\lambda \in [0,1]$, the coordinates of $\lambda\mathbf{x} + (1-\lambda)\mathbf{y}$ will be

$$\lambda x_i + (1-\lambda)y_i > 0, \qquad i = 1, \ldots, n,$$

since at least one of λx_i or $(1-\lambda)y_i$ must be positive for every i. Hence $\lambda\mathbf{x} + (1-\lambda)\mathbf{y} \in K^o$.

Open: Choose any $\mathbf{x} = (x_1, \ldots, x_n) \in K^o$. Let

$$\epsilon = \min\{x_i/2 : i = 1, \ldots, n\}.$$

Then $\epsilon > 0$ since $x_i > 0$ for every i. But then every point $\mathbf{y} \in B_\epsilon(\mathbf{x})$ has coordinates satisfying

$$|y_i - x_i| < \epsilon, \quad \Longrightarrow \quad y_i > x_i - \epsilon > 0.$$

Thus $\mathbf{y} \in K^o$, so $B_\epsilon(\mathbf{x}) \subset K^o$, so K^o is open. $\square$

5. Prove that the intersection of any collection of convex sets is convex.

   Solution: Suppose that $\{C_\alpha : \alpha \in I\}$ is an arbitrary collection of convex sets. Let $S = \cap_{\alpha \in I}$ be the intersection of all of them.

   If $S = \emptyset$, then S is convex as there is nothing to check.

   Otherwise, suppose $x, y \in S$, fix $t \in [0,1]$, and let $z = tx + (1-t)x$. For every $\alpha \in I$, $x, y \in C_\alpha$ implies $z \in C_\alpha$, since C_α is convex. But then $z \in S = \cap_\alpha C_\alpha$. Conclude that S is convex. $\square$

6. Prove Theorem 8.16 on p.209:

   (a) $K' = K$, that is, the nonnegative orthant is a self-dual cone.

   (b) $(K^o)' = K$ and $(K^o)^* = K \setminus \mathbf{0}$.

   (c) $(K \setminus \mathbf{0})' = K$ and $(K \setminus \mathbf{0})^* = K^o$.

   (d) $((K^o)^*)^* = K^o$, that is, the open positive orthant is its own strict double dual cone.

   Solution: (a) Choose any $\mathbf{x} = (x_1, \ldots, x_n) \in K'$. Take $\mathbf{k} = (1, 0, \ldots, 0) \in K$ to compute $x_1 = \mathbf{x}^T\mathbf{k} \geq 0$. Corresponding arguments show that $x_i \geq 0$ for every $i = 1, 2, \ldots, n$. Thus $K' \subset K$.

   Conversely, if $\mathbf{x} \in K$, then for every $\mathbf{k} \in K$ compute

$$\mathbf{x}^T\mathbf{k} = x_1 k_1 + \cdots + x_n k_n \geq 0,$$

   since all terms are nonnegative. Thus $K \subset K'$. Conclude that $K = K'$.

   (b) To find the dual, suppose $\mathbf{x} = (x_1, \ldots, x_n) \in (K^o)'$. Take $\mathbf{k} = (1, \epsilon, \ldots, \epsilon) \in K^o$ to compute $x_1 + \epsilon(x_2 + \cdots x_n) = \mathbf{x}^T\mathbf{k} \geq 0$. If $x_1 < 0$, then for sufficiently small $\epsilon > 0$ this inequality will be violated. Thus it must be that $x_1 \geq 0$. Corresponding arguments show that $x_i \geq 0$ for every $i = 1, 2, \ldots, n$. Thus $(K^o)' \subset K$.

Conversely, if $\mathbf{x} \in K$, then for every $\mathbf{k} \in K^o$ compute

$$\mathbf{x}^T \mathbf{k} = x_1 k_1 + \cdots + x_n k_n \geq 0,$$

since all factors and summands are nonnegative. Thus $K \subset (K^o)'$. Conclude that $K = (K^o)'$.

To find the strict dual, suppose $\mathbf{x} = (x_1, \ldots, x_n) \in (K^o)^*$. Take $\mathbf{k} = (1, \epsilon, \ldots, \epsilon) \in K^o$ to compute $x_1 + \epsilon(x_2 + \cdots x_n) = \mathbf{x}^T \mathbf{k} > 0$. If $x_1 < 0$, then, as before, for sufficiently small $\epsilon > 0$ this inequality will be violated. Thus $x_1 \geq 0$. Corresponding arguments show that $x_i \geq 0$ for every $i = 1, 2, \ldots, n$. But also, if $\mathbf{x} = \mathbf{0}$, then $\mathbf{x}^T \mathbf{k} = 0$ so the inequality will be violated. Thus $(K^o)^* \subset K \setminus \mathbf{0}$.

Conversely, if $\mathbf{x} \in K \setminus \mathbf{0}$, then for every $\mathbf{k} \in K^o$ compute

$$\mathbf{x}^T \mathbf{k} = x_1 k_1 + \cdots + x_n k_n > 0,$$

since all factors and summands are nonnegative and at least one of them must be positive. Thus $K \setminus \mathbf{0} \subset (K^o)^*$. Conclude that $K \setminus \mathbf{0} = (K^o)^*$.

Remark.   Parts (a) and (b) show that $A' = B'$ does not imply $A = B$.

(c) First, to find the dual, suppose $\mathbf{x} = (x_1, \ldots, x_n) \in (K \setminus \mathbf{0})'$. Take $\mathbf{k} = (1, 0, \ldots, 0) \in K \setminus \mathbf{0}$ to compute $x_1 = \mathbf{x}^T \mathbf{k} \geq 0$. Corresponding arguments show that $x_i \geq 0$ for every $i = 1, 2, \ldots, n$. Thus $(K \setminus \mathbf{0})' \subset K$.

Conversely, if $\mathbf{x} \in K$, then for every $\mathbf{k} \in K \setminus \mathbf{0}$ compute

$$\mathbf{x}^T \mathbf{k} = x_1 k_1 + \cdots + x_n k_n \geq 0,$$

since all factors and summands are nonnegative. Thus $K \subset (K \setminus \mathbf{0})'$. Conclude that $K = (K \setminus \mathbf{0})'$.

Second, to find the strict dual, suppose $\mathbf{x} = (x_1, \ldots, x_n) \in (K \setminus \mathbf{0})^*$. Take $\mathbf{k} = (1, 0, \ldots, 0) \in K \setminus \mathbf{0}$ to compute $x_1 = \mathbf{x}^T \mathbf{k} > 0$. Similarly, compute $x_i > 0$ for every $i = 1, 2, \ldots, n$. Thus $(K \setminus \mathbf{0})^* \subset K^o$.

Conversely, if $\mathbf{x} \in K^o$, then for every $\mathbf{k} \in (K \setminus \mathbf{0})^*$ compute

$$\mathbf{x}^T \mathbf{k} = x_1 k_1 + \cdots + x_n k_n > 0,$$

since all factors and summands are nonnegative and at least one of them must be positive. Thus $K \setminus \mathbf{0} \subset (K^o)^*$. Conclude that $K \setminus \mathbf{0} = (K^o)^*$.

(d) Observe that if $\mathbf{x} \in K^o$, then $\mathbf{x}^T \mathbf{k} > 0$ for every $\mathbf{k} \in (K^o)^*$. Thus $K^o \subset ((K^o)^*)^*$.

Conversely, if $\mathbf{x} \in ((K^o)^*)^*$, then choosing $\mathbf{k} = (1, 0, \ldots, 0) \in K \setminus \mathbf{0} = (K^o)^*$, as shown in part (b), gives $x_1 = \mathbf{x}^T \mathbf{k} > 0$. Similarly, $x_i > 0$ for all $i = 1, 2, \ldots, n$. Thus $\mathbf{x} \in K^o$, and since $\mathbf{x}$ was arbitrary, $((K^o)^*)^* \subset K^o$. Conclude that $((K^o)^*)^* = K^o$. $\square$

7. Prove Eq.8.6:

$$AK = \sum_{i=1}^{n} \bar{V}_i; \qquad AK^o = \sum_{i=1}^{n} V_i,$$

where $A \in \mathbf{R}^{m \times n}$, and K, K^o are the orthants of Definition 6.

Solution: Recall that $\mathbf{v}_i \in \mathbf{R}^m$ is the ith column of A, defining the rays $V_i = \{c\mathbf{v}_i : c > 0\}$ and $\bar{V}_i = \{c\mathbf{v}_i : c \geq 0\}$ for $i = 1, \ldots, n$. Then $\mathbf{x} \in AK$ iff there exists $k = (k_1, \ldots, k_n) \in K$ such that $\mathbf{x} = Ak$. But then,

$$\mathbf{x} = Ak = \sum_{i=1}^{n} k_i \mathbf{v}_i \in \sum_{i=1}^{n} \bar{V}_i,$$

since $k_i \mathbf{v}_i \in \bar{V}_i$ because $k_i \geq 0$ for all i.

Likewise, $\mathbf{x} \in AK^o$ iff there exists $\mathbf{k}^o = (k_1^o, \ldots, k_n^o) \in K^o$ such that $\mathbf{x} = Ak^o$, But then

$$\mathbf{x} = Ak^o = \sum_{i=1}^{n} k_i^o \mathbf{v}_i \in \sum_{i=1}^{n} V_i,$$

since $k_i^o \mathbf{v}_i \in V_i$ because $k_i^o > 0$ for all i. $\qquad\square$

8. Prove Corollary 8.18 on p.209: The set S of strictly profitable portfolios is a strict dual cone: $S = (AK^o)^*$

Solution: Modify the proof of Corollary 8.17 as follows:

Proof: Since $(K \setminus \mathbf{0})^* = K^o$ by Theorem 8.16(c),

$$
\begin{aligned}
\mathbf{s} \in S \quad &\Longleftrightarrow \quad \mathbf{s}^T A \in K \setminus \mathbf{0} \\
&\Longleftrightarrow \quad (\forall \mathbf{k} \in (K \setminus \mathbf{0})^*)(\mathbf{s}^T A)\mathbf{k} > 0 \\
&\Longleftrightarrow \quad (\forall \mathbf{k} \in K^o)(\mathbf{s}^T A)\mathbf{k} > 0 \\
&\Longleftrightarrow \quad (\forall \mathbf{k} \in K^o)\mathbf{s}^T (A\mathbf{k}) > 0 \\
&\Longleftrightarrow \quad (\forall \mathbf{v} \in AK^o)\mathbf{s}^T \mathbf{v} > 0 \\
&\Longleftrightarrow \quad \mathbf{s} \in (AK^o)^*,
\end{aligned}
$$

since $AK^o = \{A\mathbf{k} : \mathbf{k} \in K^o\}$, so that the next to last condition is just the definition of membership in the strict dual cone $(AK^o)^*$. $\qquad\square$

9. Suppose $S \subset \mathbf{R}^n$ is any set. Prove the following:

(a) $S^\perp$ is a subspace.

(b) $S^* \subset S'$ and thus $S^* \cap S' = S^*$.

(c) $S^\perp \subset S'$ and thus $S^\perp \cap S' = S^\perp$.

(d) $S^\perp \cap S^* = \emptyset$.

(e) $S^\perp$, S', and S^* are all convex cones.

(f) If $\mathbf{0} \in S$, then $S^* = \emptyset$. Thus if S is a subspace, then $S^* = \emptyset$.

Solution: (a) Check the definition:

- $\mathbf{0} \in S^\perp$ since $\mathbf{0}^T\mathbf{s} = 0$ for any $\mathbf{s} \in S$.
- Given $\mathbf{x}, \mathbf{y} \in S^\perp$, take any $\mathbf{s} \in S$ and compute

$$\mathbf{s}^T(\mathbf{x} + \mathbf{y}) = \mathbf{s}^T\mathbf{x} + \mathbf{s}^T\mathbf{y} = 0 + 0 = 0.$$

Conclude that $\mathbf{x} + \mathbf{y} \in S^\perp$.

- Given $\mathbf{x} \in S^\perp$ and $c \in \mathbf{R}$, take any $\mathbf{s} \in S$ and compute

$$\mathbf{s}^T(c\mathbf{x}) = c\mathbf{s}^T\mathbf{x} = c0 = c0.$$

Conclude that $c\mathbf{x} \in S^\perp$.

(b) $\mathbf{v}^T\mathbf{s} > 0 \implies \mathbf{v}^T\mathbf{s} \geq 0$, so every $\mathbf{v} \in S^*$ also belongs to S'.

(c) $\mathbf{v}^T\mathbf{s} = 0 \implies \mathbf{v}^T\mathbf{s} \geq 0$, so every $\mathbf{v} \in S^\perp$ also belongs to S'.

(d) It is impossible to have both $\mathbf{v}^T\mathbf{s} = 0$ and $\mathbf{v}^T\mathbf{s} > 0$, so there are no vectors $\mathbf{v}$ in both S^* and $S^\perp$.

(e) Check the two needed properties for $S^\perp$, S', and S^*:

Cones: Let $\mathbf{x}$ be a vector in $\mathbf{R}^n$ and let λ be a positive real number.

- $(\forall \mathbf{s} \in S)\mathbf{s}^T\mathbf{x} = 0 \implies (\forall \mathbf{s} \in S)\mathbf{s}^T(\lambda\mathbf{x}) = \lambda 0 = 0$;
- $(\forall \mathbf{s} \in S)\mathbf{s}^T\mathbf{x} \geq 0 \implies (\forall \mathbf{s} \in S)\mathbf{s}^T(\lambda\mathbf{x}) = \lambda\mathbf{s}^T\mathbf{x} \geq 0$;
- $(\forall \mathbf{s} \in S)\mathbf{s}^T\mathbf{x} > 0 \implies (\forall \mathbf{s} \in S)\mathbf{s}^T(\lambda\mathbf{x}) = \lambda\mathbf{s}^T\mathbf{x} > 0.$

Convex: Let $\mathbf{x}, \mathbf{y}$ be vectors in $\mathbf{R}^n$ and let $\lambda \in [0, 1]$ be a real number. Let $\mathbf{s} \in S$ be arbitrary.

- $\mathbf{s}^T[\lambda\mathbf{x} + (1-\lambda)\mathbf{y}] = \lambda\mathbf{s}^T\mathbf{x} + (1-\lambda)\mathbf{s}^T\mathbf{y} = 0$ if $\mathbf{x}, \mathbf{y} \in S^\perp$;
- $\mathbf{s}^T[\lambda\mathbf{x} + (1-\lambda)\mathbf{y}] = \lambda\mathbf{s}^T\mathbf{x} + (1-\lambda)\mathbf{s}^T\mathbf{y} \geq 0$ if $\mathbf{x}, \mathbf{y} \in S'$;
- $\mathbf{s}^T[\lambda\mathbf{x} + (1-\lambda)\mathbf{y}] = \lambda\mathbf{s}^T\mathbf{x} + (1-\lambda)\mathbf{s}^T\mathbf{y} > 0$ if $\mathbf{x}, \mathbf{y} \in S^*$.

(f) If $\mathbf{0} \in S$, then every $\mathbf{x} \in \mathbf{R}^n$ gives $\mathbf{x}^T\mathbf{0} = 0$, so there is no $\mathbf{x} \in \mathbf{R}^n$ such that $\mathbf{x}^T\mathbf{0} > 0$, so $S^* = \emptyset$. □

10. Suppose that $n > 2$ and market model $A, \mathbf{q}$ has

$$A = \begin{pmatrix} R & \cdots & R \\ a_1 & \cdots & a_n \end{pmatrix},$$

where $R > 1$ is the riskless return and $\mathbf{a} = (a_1, \ldots, a_n)$ is a nonconstant payoff vector for the sole risky asset.

(a) Find necessary and sufficient conditions on $\mathbf{q}$ such that $A, \mathbf{q}$ is arbitrage-free. (Hint: use the Fundamental Theorem.)

(b) Exhibit a derivative payoff $\mathbf{d}$ for which no exact hedge exists. (This shows that A is not a complete market.)

(c) Exhibit a derivative $\mathbf{d}$ for which an exact hedge does exist.

Solution: (a) Following the hint, observe that $A, \mathbf{q}$ is arbitrage-free if and only if there exists a positive vector $\mathbf{k} = [k_1, \ldots, k_n]^T$ such that $\mathbf{q} = A\mathbf{k}$. But $\mathbf{q} = [1, S_0]^T$, so $R\mathbf{k}$ is a p.d.f., so if $\mathbf{q} = A\mathbf{k}$, then

$$S_0 = \mathbf{q}(2) = A\mathbf{k}(2) = \sum_{i=1}^{n} a_i k_i = \frac{1}{R} \sum_{i=1}^{n} a_i (Rk_i),$$

which is possible if and only if $\min\{a_i\} < RS_0 < \max\{a_i\}$.

Remark. Thus RS_0 must be inside the range of the payoffs $\{a_i\}$.

(b) It may be assumed that $a_1 < a_2 \le a_3$. (Otherwise, simply renumber the states.) Submatrix

$$A_2 = \begin{pmatrix} R & R \\ a_1 & a_2 \end{pmatrix}$$

is invertible, so the numbers h_0, h_1 are uniquely determined by

$$\begin{pmatrix} h_0 \\ h_1 \end{pmatrix} = A_2^{-1} \begin{pmatrix} d_1 \\ d_2 \end{pmatrix},$$

where $\mathbf{d} = [d_1, d_2, d_3, \ldots]^T$ is the payoff vector for the derivative to be hedged. Now choose $d_1 = d_2 = 1$, determine h_0, h_1, and choose d_3 different from $h_0 R + h_1 a_3$. The derivative with this payoff cannot be exactly hedged in this market.

(c) As in part (b), choose $d_1 = d_2 = 1$, determine h_0, h_1, but choose

$$d_i = h_0 R + h_1 a_i, \qquad i = 3, \ldots, n.$$

The row vector $\mathbf{d} = [1, 1, d_3, \ldots, d_n]$ lies in the row space of A and thus is exactly hedged by $\mathbf{h} = [h_0, h_1]$. $\square$

11. Suppose that a market model has five states, a riskless asset returning $R = 1.02$, and two risky assets a, b with spot prices $a_0 = 20$ and $b_0 = 12$ and payoffs $\mathbf{a} = (10, 15, 20, 25, 30)$ and $\mathbf{b} = (17, 15, 12, 10, 7)$, respectively.

(a) Prove that the model is arbitrage-free.

(b) Find the no-arbitrage bid-ask interval for a European-style Call option on a with strike price 20.

(c) Find the no-arbitrage bid-ask interval for a European-style Put option on b with strike price 13.

Solution: The computations may be done in Octave with the `glpk` package. Begin by putting the data into a finite market model Aq:

```
a=[10,15,20,25,30]; b=[17,15,12,10,7]; a0=20; b0=12;
R=1.02; A=[R R R R R; a; b]; q=[1; a0; b0]; A,q
```

(This is in the notation of Definition 4.)

(a) To show that A, q is arbitrage-free, by Fundamental Theorem 8.4 it suffices to find $k > 0$ such that $q = Ak$. This 3×5 linear system may be placed into row echelon form, yielding

```
   rref([A R*q])
%  1.00000    0.00000    0.00000   -1.00000   -1.00000   -0.48000
%  0.00000    1.00000    0.00000    1.00000   -0.00000    0.88000
%  0.00000    0.00000    1.00000    1.00000    2.00000    0.60000
```

(Use R*q in the augmented matrix to get the risk neutral probabilities $p = Rk$, which sum to 1, instead of the discounted vector k which will sum to $\frac{1}{R}$.) The complete set of three pivot rows shows that the system is consistent but underdetermined and thus has a two-parameter family of solutions. By Corollary 8.8, it suffices to check that the unique minimal norm solution is positive. Octave computes it with

```
A\q*R % p =  0.024   0.432   0.040   0.448   0.056
A\q   % k = 0.023529   0.423529   0.039216   0.439216   0.054902
```

Conclude that A, q is arbitrage-free.

Remark. The row echelon form of [A R*q] shows how to find the complete solution set of positive vectors $Rk \overset{\text{def}}{=} p = (p_1, p_2, p_3, p_4, p_5)$. Identify the pivot variables p_1, p_2, p_3, so p_4 and p_5 are free variables. The general solution may be expressed as

$$
\begin{aligned}
p_1 &= p_4 + p_5 - 0.48 \\
p_2 &= -p_4 + 0.88 \\
p_3 &= -p_4 - 2p_5 + 0.60
\end{aligned}
$$

(Notice that $\sum_i p_i = 1$.) To find a positive solution $p > 0$ requires solving the simultaneous inequalities

$$p_4 > 0, \ p_5 > 0, \ p_4 + p_5 > 0.48, \ p_4 < 0.88, \ p_4 + 2p_5 < 0.60,$$

which reduce to the intervals $0 < p_5 < 0.12$ and $0.48 - p_5 < p_4 < 0.60 - 2p_5$. One of the infinitely many solutions is thus parametrized by the midpoints:

```
p5=0.12/2; p4=((0.48-p5)+(0.60-2*p5))/2;
p1=p4+p5-0.48; p2= -p4+0.88; p3= -p4-2*p5+0.60;
p=[p1 p2 p3 p4 p5] % p = 0.03   0.43   0.03   0.45   0.06
k=p/R % k = 0.029412  0.421569  0.029412  0.441176  0.058824
```

This gives another explicit positive solution $\mathbf{k} > \mathbf{0}$, proving that $A, \mathbf{q}$ is arbitrage-free without using Corollary 8.8.

For parts (b) and (c), reuse the Octave code from Section 8.2.3, first putting the parameters, market matrix and spot prices into GLPK format:

```
sellctype="LLLLL"; % Lower constraint type A'*x(j)>=bb(j), j=1:5
sellsense=1; % Optimization direction for q'*x: "1" ==> "min"
buyctype="UUUUU"; % Upper constraint type A'*x(j)=<bb(j), j=1:5
buysense=-1; % Optimization direction for q'*x: "-1" ==> "max"
vartype="CCC";    % ==> x(j) is Continuous, j=1:3
param.msglev=1;  % ==> use a low verbosity level
huge=1000; infty=huge*(abs(a0)+abs(b0)); % huge and huger
param.itlim=huge; % ==> huge maximum number of iterations
lb=[-infty; -huge; -huge]; % huge Lower bounds on x
ub=[ infty;  huge;  huge]; % huge Upper bounds on x
```

(b) Compute the Call payoff on asset a, then find a superreplication and a subreplication, taking the derivative seller's and buyer's perspectives, respectively:

```
Ka=20; ba=max(a-Ka,0); % C(T): payoff for "a" Call
[hs,ask]=glpk(q,A',ba,lb,ub,sellctype,vartype,sellsense,param)
% hs = [102.9412; -2.00; -5.00]; Cost-minimizing hedge portfolio
% ask = 2.9412;  minimum cost to superreplicate the Call
[hb,bid]=glpk(q,A',ba,lb,ub,buyctype,vartype,buysense,param)
% hb = [ -117.6471; 3.00; 5.00]; Cost-maximizing hedge portfolio
% bid = 2.3529;  maximum cost to subreplicate the Call
```

Since bid is strictly less than ask, by Corollary 8.9 there is no exact hedge for this derivative in this market.

(c) Compute the Put payoff on asset b, then find a superreplication and a subreplication, taking the derivative seller's and buyer's perspectives, respectively:

```
Kb=13; bb=max(Kb-b,0); % P(T): payoff for "b" Put
[hs,ask]=glpk(q,A',bb,lb,ub,sellctype,vartype,sellsense,param)
% hs = [ 61.7647; -1.20; -3.00]; Cost-minimizing hedge portfolio
% ask = 1.7647;  minimum cost to superreplicate the Call
```

```
[hb,bid]=glpk(q,A',bb,lb,ub,buyctype,vartype,buysense,param)
% hb = [ -26.47059; 0.80; 1.00]; Cost-maximizing hedge portfolio
% bid = 1.5294;  maximum cost to subreplicate the Call
```

Since bid is strictly less than ask, by Corollary 8.9 there is no exact
hedge for this derivative in this market.   □

Index

Printed in the United States
by Baker & Taylor Publisher Services